Investing in Development
A Practical Plan to Achieve the Millennium Development Goals

UN Millennium Project
Jeffrey D. Sachs, Director

Task force coordinators

Agnes Binagwaho
Nancy Birdsall
Jaap Broekmans
Mushtaque Chowdhury
Pietro Garau
Geeta Rao Gupta
Amina J. Ibrahim
Calestous Juma
Yolanda Kakabadse Navarro
Lee Yee-Cheong
Roberto Lenton
Jeff McNeely
Don J. Melnick
Patrick Messerlin
Paula Munderi
Mari Pangestu
Allan Rosenfield
Josh Ruxin
Pedro Sanchez
Elliott D. Sclar
Burton Singer
M.S. Swaminathan
Awash Teklehaimanot
Albert Wright
Ernesto Zedillo

Secretariat

John W. McArthur, Project Manager
Chandrika Bahadur
Stan Bernstein
Yassine Fall
Eric Kashambuzi
Margaret Kruk
Guido Schmidt-Traub

Routledge
Taylor & Francis Group
LONDON AND NEW YORK

earthscan
from Routledge

Millennium Project

First published 2005 by Earthscan

Published 2022 by Routledge
4 Park Square, Milton Park, Abingdon, Oxon OX14 4RN
605 Third Avenue, New York, NY 10017

Routledge is an imprint of the Taylor & Francis Group, an informa business

Publisher's Note

The publisher has gone to great lengths to ensure the quality of this reprint but poinst out that some imperfections in the original copies may be apparent.

A catalogue record for this book is available from the British Library

Library of Congress Cataloging-in-Publication Data

A catalog record has been requested

This publication should be cited as: UN Millennium Project. 2005. *Investing in Development: A Practical Plan to Achieve the Millennium Development Goals.* New York.

Photos: Front cover, top to bottom and left to right, Christopher Dowswell/UNDP, Pedro Cote/UNDP, Giacomo Pirozzi/Panos Pictures, Liba Taylor/Panos Pictures, Jørgen Schytte/UNDP, UN Photo Library, Giacomo Pirozzi/UNICEF, Curt Carnemark/World Bank, Pedro Cote/UNDP, Franck Charton/UNICEF, Paul Chesley/Getty Images, Ray Witlin/World Bank, Pete Turner/Getty Images, B. Marquez/UNDP.

This book was edited, designed, and produced by Communications Development Inc., Washington, D.C., and its UK design partner, Grundy & Northedge.

The UN Millennium Project was commissioned by the UN Secretary-General and supported by the UN Development Group, which is chaired by the Administrator of the United Nations Development Programme. The report is an independent publication. This publication does not necessarily reflect the views of the United Nations, the United Nations Development Programme, or their Member States.

ISBN 13: 978-1-84407-217-0 (pbk)
ISBN 13: 978-1-138-47193-1 (hbk)

Contents

Part 2 Country-level processes to achieve the Millennium Development Goals

Boxes

Figures

Maps

Tables

Preface

The UN Millennium Project has been a unique undertaking. Its 10 task forces, Secretariat, and broad array of participants from academia, government, UN agencies, international financial institutions, nongovernmental organizations, donor agencies, and the private sector created a worldwide network of development practitioners and experts across an enormous range of countries, disciplines, and organizations. The Project was made possible by the unique commitment, skills, and convictions of the task force coordinators, who led their groups to take on some of the most challenging development questions of our generation, and by the task force members, who gave remarkably of their time. This has been a global effort, in the service of a great global cause—the Millennium Development Goals (MDGs). Our Project has been a microcosm of a larger truth: achieving the Millennium Development Goals will require a global partnership suitable for an interconnected world. The world truly shares a common fate.

This has been a labor of love for the many participants in the task forces and Secretariat. Individuals have volunteered vast amounts of effort and expertise to the Project. Their contributions, far beyond any reasonable expectation, have immeasurably sharpened and strengthened the messages contained in the Project's many outputs, including this report, the task force final reports, the newly developed tools for needs assessment, and the advisory support for MDG-based planning in several countries.

I believe that all of the participants have contributed in such a manner because they recognize the special nature of this effort. Part of that is the honor and privilege of working on behalf of UN Secretary-General Kofi Annan, who launched the UN Millennium Project and who has played an unparalleled role in promoting the global fight against extreme poverty. Part is the delight of working for and together with UNDP Administrator Mark Malloch Brown, whose tenacity, vision, and leadership have guided the UN Development Group's efforts for several years. We have also enjoyed, admired, and richly benefited from the leaders of other UN agencies, who direct teams that save lives and ease burdens of poverty and despair throughout the world on a daily basis.

Another special aspect of the Project is the rare and powerful opportunity to help give voice to the hopes, aspirations, and vital needs of the world's poorest and most voiceless people. We have met countless heroes and heroines of development in the three years of our work—in the villages and slums of Africa, Asia, Latin America, and other parts of the developing world. We have seen people preserving their spirit, integrity, commitment, and hope for the future even when they have little else, when tragic circumstances have left them bereft of health, education, possessions, and a means of livelihood. The human spirit, we have seen on innumerable occasions, is truly indomitable.

This triumph of the human spirit gives us the hope and confidence that extreme poverty can be cut by half by the year 2015, and indeed ended altogether within the coming years. The world community has at its disposal the proven technologies, policies, financial resources, and most importantly, the human courage and compassion to make it happen.

Jeffrey D. Sachs
January 2005

Contributors

This document presents the findings and recommendations of the UN Millennium Project, an independent advisory body to UN Secretary-General Kofi Annan. We are grateful to the Secretary-General for initiating and supporting the UN Millennium Project, for his unswerving commitment to the objective of global poverty reduction, and for his remarkable and tireless leadership of the United Nations system. We also thank Mark Malloch Brown, Administrator of the United Nations Development Programme (UNDP) and Chair of the United Nations Development Group, for his sage guidance and support at every step of the project. We also wish to express our profound appreciation to members of the UN Development Group's Executive Committee for their ongoing support: Carol Bellamy, United Nations Children's Fund; Jim Morris, World Food Programme; and Thoraya Obaid, United Nations Population Fund. We are further grateful to Louise Fréchette, Deputy Secretary-General, José Antonio Ocampo, Department of Social and Economic Affairs; his predecessor Nitin Desai; and Shashi Tharoor, Department of Public Information, for their generous guidance and advice.

This report has been co-authored by the coordinators of the UN Millennium Project's 10 task forces and Secretariat, building on the contributions made by hundreds of scholars, development practitioners, scientists, political leaders, and policy leaders involved since the Project's inception in July 2002. A large number of task force associates and task force members made tremendous contributions to many parts of this report, including Gabriella Carolini, Glenn Denning, Helen de Pinho, Philip Dobie, Lisa Dreier, Lynn Freedman, Caren Grown, Ruth Levine, Kristen Lewis, Joan Paluzzi, Robin Sears, Smita Srinivas, Yesim Tozan, Ron Waldman, Haynie Wheeler, Paul Wilson, and Nalan Yuksel. In the UN Millennium Project Secretariat, Albert Hyunbae Cho, Michael Faye, Michael Krouse, Fatou Lo, Gordon McCord, Luis Javier Montero, Rohit

Wanchoo, Emily White, and Alice Wiemers worked around the clock for many months to provide invaluable research assistance. Erin Trowbridge provided extremely important comments and inputs. Prarthna Dayal, Rafael Flor, Maria Beatriz Orlando, Nora Simpson, Kelly Tobin, and Brian Torpy also made essential efforts in further supporting task force contributions.

This report also directly integrates many of the ideas developed by the UN Millennium Project's Task Force on Poverty and Economic Development, including many outlined by its interim report of February 2004. The members and associates of the Task Force on Poverty and Economic Development who contributed include Kwesi Botchwey, Haidari Amani, Ernest Aryeetey, George Cahuzac, Andrew Cassels, Jamie Drummond, Richard Freeman, Rebecca Grynspan, Pekka Haavisto, Aynul Hasan, Peter Heller, Macartan Humphreys, John Langmore, Ruth Jacoby, Carlos Jarque, Allan Jury, Eddy Lee, Zhu Ling, Thomas Merrick, Vijay Modi, John Okidi, Hafiz Pasha, Michael Platzer, Steven Radelet, Atiqur Rahman, Frederic Richard, Ana-Teresa Romero, Rabbi Royan, Ratna Sahay, Francisco Sercovich, Sudhir Shetty, David Simon, Suresh Tendulkar, Michael Usnick, Ashutosh Varshney, and Xianbin Yao. Several of them wrote crucial background papers that formed the basis for important sections of the text. These include Philip Alston (human rights), Macartan Humphreys and Ashutosh Varshney (conflict), Vijay Modi (rural infrastructure), David Simon (aid flows), and Steven Radelet (governance and official development assistance). The Economic Commission for Africa (together with UNDP Ethiopia) and the Economic and Social Commission for Asia and the Pacific (in collaboration with UNDP Thailand) hosted the task force meetings where many of the ideas in this report were developed.

UNDP offered a gracious home to the UN Millennium Project in addition to providing instrumental financial, in-kind, and intellectual contributions on behalf of the UN Development Group. For this we thank all members of the senior management team including Abdoulie Janneh, Rima Khala Hunaidi, Elena Martinez, Jan Mattson, Kalman Mizsei, Shoji Nishimoto, Hafiz Pasha, Julia Taft, and their respective bureau teams. Under the leadership of Bruce Jenks, the Bureau for Resources and Strategic Partnerships provided tireless support to the project, particularly from Turhan Saleh, Amina Tirana, and the rest of the MDGs Unit. Mark Suzman in the Office of the Administrator provided daily doses of patient and thoughtful advice.

The Millennium Trust Fund at UNDP was supported by several governments and foundations and provided the core financial support for the Project since its inception in 2002. The Project is also deeply grateful for the contributions from the Ford Foundation, the Bill and Melinda Gates Foundation, the William and Flora Hewlett Foundation, the John D. and Catherine T. MacArthur Foundation, the David and Lucile Packard Foundation, the Open Society Institute, and the Rockefeller Foundation.

The UN Millennium Project would like to give special thanks to the Earth Institute at Columbia University for its intellectual, administrative, financial, and in-kind support to the Project Secretariat and to many of the task forces. The project also thanks the Center for Global Development, the John F. Kennedy School of Government at Harvard University, Partners in Health, the International Centre for Research on Women, the Mailman School of Public Health at Columbia University, and the Yale Center for the Study of Globalization for their crucial role in supporting the activities of the Project's task forces.

The UN Millennium Project is indebted to the staff of the United Nations system and to members of governments, civil society, and the private sector, too numerous to mention by name, who have contributed directly or indirectly to the report. All errors and omissions remain the responsibility of the authors.

Task force reports

Task Force on Hunger
Halving hunger: it can be done

Task Force on Education and Gender Equality
Toward universal primary education: investments, incentives, and institutions

Task Force on Education and Gender Equality
Taking action: achieving gender equality and empowering women

Task Force on Child Health and Maternal Health
Who's got the power? Transforming health systems for women and children

Task Force on HIV/AIDS, Malaria, TB, and Access to Essential Medicines, Working Group on HIV/AIDS
Combating AIDS in the developing world

Task Force on HIV/AIDS, Malaria, TB, and Access to Essential Medicines, Working Group on Malaria
Coming to grips with malaria in the new millennium

Task Force on HIV/AIDS, Malaria, TB, and Access to Essential Medicines, Working Group on TB
Investing in strategies to reverse the global incidence of TB

Task Force on HIV/AIDS, Malaria, TB, and Access to Essential Medicines, Working Group on Access to Essential Medicines
Prescription for healthy development: increasing access to medicines

Task Force on Environmental Sustainability
Environment and human well-being: a practical strategy

Task Force on Water and Sanitation
Health, dignity, and development: what will it take?

Task Force on Improving the Lives of Slum Dwellers
A home in the city

Task Force on Trade
Trade for development

Task Force on Science, Technology, and Innovation
Innovation: applying knowledge in development

Abbreviations

APRM	African Peer Review Mechanism
ASEAN	Association of Southeast Asian Nations
CARICOM	Caribbean Community
CGIAR	Consultative Group on International Agricultural Research
CIS	Commonwealth of Independent States
COMESA	Common Market for Eastern and Southern Africa
CSO	civil society organization
DAC	Development Assistance Committee
DFID	Department for International Development (UK)
DOTS	directly observed treatment, short course
ECOWAS	Economic Community of West African States
EIA	Energy Information Administration
EITI	Extractive Industries Transparency Initiative
ERS	Economic Recovery Strategy for Wealth and Employment Creation
EU	European Union
FAO	Food and Agriculture Organization
FDI	foreign direct investment
FTA	free trade agreement
GATS	General Agreement on Trade in Services
GATT	General Agreement on Tariffs and Trade
GAVI	Global Alliance for Vaccines and Immunization
GDP	gross domestic product
GEMS	Global Environment Monitoring System
GNI	gross national income
GNP	gross national product
HIPC	heavily indebted poor country
ICT	information and communication technologies
IDA	International Development Association
IFAD	International Fund for Agricultural Development

IFF	International Financing Facility
ILO	International Labour Organization
IMF	International Monetary Fund
IMR	infant mortality rate
IPCC	Intergovernmental Panel on Climate Change
LDC	Least Developed Country
MCA	Millennium Challenge Account
MCC	Millennium Challenge Corporation
MDG	Millennium Development Goal
MFN	most-favored nation
MTEF	medium-term expenditure framework
NEPAD	New Partnership for Africa's Development
NGO	nongovernmental organization
OECD	Organisation for Economic Co-operation and Development
ODA	official development assistance
ODS	overall trade-distorting domestic support
PRS	poverty reduction strategy
PRSP	Poverty Reduction Strategy Paper
R&D	research and development
RAE	Rehabilitation of Arid Environments Charitable Trust (Kenya)
SADC	Southern African Development Community
SDI	Slum/Shack Dwellers International
SDT	special and differential treatment
SPARC	Society for Promotion of Area Resource Centers (India)
SWAps	sectorwide approaches
TRIPS	Agreement on Trade-Related Aspects of Intellectual Property Rights
UNAIDS	Joint United Nations Programme on HIV/AIDS
UNCTAD	United Nations Conference on Trade and Development
UNDAF	UN Development Assistance Framework
UNDP	United Nations Development Programme
UNDESA	United Nations Department of Economic and Social Affairs
UNECA	United Nations Economic Commission for Africa
UNEP	United Nations Environment Programme
UNESCAP	United Nations Economic and Social Commission for Asia and the Pacific
UNESCO	United Nations Educational, Scientific and Cultural Organization
UNFPA	United Nations Population Fund
UNICEF	United Nations Children's Fund
UNIFEM	United Nations Development Fund for Women
UNIDO	United Nations Industrial Development Organisation
USAID	U.S. Agency for International Development
WBCSD	World Business Council for Sustainable Development
WIPO	World Intellectual Property Organization
WTO	World Trade Organization

goals

Millennium Development Goals

Goal 1

Eradicate extreme poverty and hunger

Target 1.
Halve, between 1990 and 2015, the proportion of people whose income is less than $1 a day

Target 2.
Halve, between 1990 and 2015, the proportion of people who suffer from hunger

Goal 2

Achieve universal primary education

Target 3.
Ensure that, by 2015, children everywhere, boys and girls alike, will be able to complete a full course of primary schooling

Goal 3

Promote gender equality and empower women

Target 4.
Eliminate gender disparity in primary and secondary education, preferably by 2005, and in all levels of education no later than 2015

Goal 4

Reduce child mortality

Target 5.
Reduce by two-thirds, between 1990 and 2015, the under-five mortality rate

Goal 5

Improve maternal health

Target 6.
Reduce by three-quarters, between 1990 and 2015, the maternal mortality ratio

Goal 6

Combat HIV/AIDS, malaria, and other diseases

Target 7.
Have halted by 2015 and begun to reverse the spread of HIV/AIDS

Target 8.
Have halted by 2015 and begun to reverse the incidence of malaria and other major diseases

Goal 7

Ensure environmental sustainability

Target 9.
Integrate the principles of sustainable development into country policies and programs and reverse the loss of environmental resources

Target 10.
Halve, by 2015, the proportion of people without sustainable access to safe drinking water and basic sanitation

Target 11.
Have achieved by 2020 a significant improvement in the lives of at least 100 million slum dwellers

Goal 8

Develop a global partnership for development

Target 12.
Develop further an open, rule-based, predictable, nondiscriminatory trading and financial system (includes a commitment to good governance, development, and poverty reduction—both nationally and internationally)

Target 13.
Address the special needs of the Least Developed Countries (includes tariff- and quota-free access for Least Developed Countries' exports, enhanced program of debt relief for heavily indebted poor countries [HIPCs] and cancellation of official bilateral debt, and more generous official development assistance for countries committed to poverty reduction)

Target 14.
Address the special needs of landlocked developing countries and small island developing states (through the Program of Action for the Sustainable Development of Small Island Developing States and 22nd General Assembly provisions)

Target 15.
Deal comprehensively with the debt problems of developing countries through national and international measures in order to make debt sustainable in the long term

Some of the indicators are monitored separately for the least developed countries, Africa, landlocked developing countries, and small island developing states

Target 16.
In cooperation with developing countries, develop and implement strategies for decent and productive work for youth

Target 17.
In cooperation with pharmaceutical companies, provide access to affordable essential drugs in developing countries

Target 18.
In cooperation with the private sector, make available the benefits of new technologies, especially information and communications technologies

Ten key recommendations

Recommendation 1

Developing country governments should adopt development strategies bold enough to meet the Millennium Development Goal (MDG) targets for 2015. We term them MDG-based poverty reduction strategies. To meet the 2015 deadline, we recommend that all countries have these strategies in place by 2006. Where Poverty Reduction Strategy Papers (PRSPs) already exist, those should be aligned with the MDGs.

Recommendation 1

The MDG-based poverty reduction strategies should anchor the scaling up of public investments, capacity building, domestic resource mobilization, and official development assistance. They should also provide a framework for strengthening governance, promoting human rights, engaging civil society, and promoting the private sector. The MDG-based poverty reduction strategies should:

- Be based on an assessment of investments and policies needed to reach the Goals by 2015.
- Spell out the detailed national investments, policies, and budgets for the coming three to five years.
- Focus on rural productivity, urban productivity, health, education, gender equality, water and sanitation, environmental sustainability, and science, technology, and innovation.
- Focus on women's and girls' health (including reproductive health) and education outcomes, access to economic and political opportunities, right to control assets, and freedom from violence.
- Promote mechanisms for transparent and decentralized governance.
- Include operational strategies for scale-up, such as training and retaining skilled workers.
- Involve civil society organizations in decisionmaking and service delivery, and provide resources for monitoring and evaluation.
- Outline a private sector promotion strategy and an income generation strategy for poor people.
- Be tailored, as appropriate, to the special needs of landlocked, small island developing, least developed, and fragile states.
- Mobilize increased domestic resources by up to four percentage points of GNP by 2015.
- Calculate the need for official development assistance.
- Describe an "exit strategy" to end aid dependency, appropriate to the country's situation.

Note: Recommendations for sector-specific policies and investments are summarized in this report and described at length in the individual reports of the UN Millennium Project task forces.

ecommendation 3

Developing country governments should craft and implement the MDG-based poverty reduction strategies in transparent and inclusive processes, working closely with civil society organizations, the domestic private sector, and international partners.

- Civil society organizations should contribute actively to designing policies, delivering services, and monitoring progress.
- Private sector firms and organizations should contribute actively to policy design, transparency initiatives and, where appropriate, public-private partnerships.

ecommendation 4

International donors should identify at least a dozen MDG "fast-track" countries for a rapid scale-up of official development assistance (ODA) in 2005, recognizing that many countries are already in a position for a massive scale-up on the basis of their good governance and absorptive capacity.

ecommendation 5

Developed and developing countries should jointly launch, in 2005, a group of Quick Win actions to save and improve millions of lives and to promote economic growth. They should also launch a massive effort to build expertise at the community level.

The Quick Wins include but are not limited to:

- Free mass distribution of malaria bed-nets and effective antimalaria medicines for all children in regions of malaria transmission by the end of 2007.
- Ending user fees for primary schools and essential health services, compensated by increased donor aid as necessary, no later than the end of 2006.
- Successful completion of the 3 by 5 campaign to bring 3 million AIDS patients in developing countries onto antiretroviral treatment by the end of 2005.
- Expansion of school meals programs to cover all children in hunger hotspots using locally produced foods by no later than the end of 2006.
- A massive replenishment of soil nutrients for smallholder farmers on lands with nutrient-depleted soils, through free or subsidized distribution of chemical fertilizers and agroforestry, by no later than the end of 2006.

The massive training program of community-based workers should aim to ensure, by 2015, that each local community has:

- Expertise in health, education, agriculture, nutrition, infrastructure, water supply and sanitation, and environmental management.
- Expertise in public sector management.
- Appropriate training to promote gender equality and participation.

ecommendation 6

Developing country governments should align national strategies with such regional initiatives as the New Partnership for Africa's Development and the Caribbean Community (and Common Market), and regional groups should receive increased direct donor support for regional projects. Regional development groups should:

- Be supported to identify, plan, and implement high-priority cross-border infrastructure projects (roads, railways, watershed management).
- Receive direct donor support to implement cross-border projects.
- Be encouraged to introduce and implement peer-review mechanisms to promote best practices and good governance.

Recommendation 7

High-income countries should increase official development assistance (ODA) from 0.25 percent of donor GNP in 2003 to around 0.44 percent in 2006 and 0.54 percent in 2015 to support the Millennium Development Goals, particularly in low-income countries, with improved ODA quality (including aid that is harmonized, predictable, and largely in the form of grants-based budget support). Each donor should reach 0.7 percent no later than 2015 to support the Goals and other development assistance priorities. Debt relief should be more extensive and generous.

- ODA should be based on actual needs to meet the Millennium Development Goals and on countries' readiness to use the ODA effectively.
- Criteria for evaluating the sustainability of a country's debt burden must be consistent with the achievement of the Goals.
- Aid should be oriented to support the MDG-based poverty reduction strategy, rather than to support donor-driven projects.
- Donors should measure and report the share of their ODA that supports the actual scale-up of MDG-related investments.
- Middle-income countries should also seek opportunities to become providers of ODA and give technical support to low-income countries.

Recommendation 8

High-income countries should open their markets to developing country exports through the Doha trade round and help Least Developed Countries raise export competitiveness through investments in critical trade-related infrastructure, including electricity, roads, and ports. The Doha Development Agenda should be fulfilled and the Doha Round completed no later than 2006.

Recommendation 9

International donors should mobilize support for global scientific research and development to address special needs of the poor in areas of health, agriculture, natural resource and environmental management, energy, and climate. We estimate the total needs to rise to approximately $7 billion a year by 2015.

Recommendation 10

The UN Secretary-General and the UN Development Group should strengthen the coordination of UN agencies, funds, and programs to support the MDGs, at headquarters and country level. The UN Country Teams should be strengthened and should work closely with the international financial institutions to support the Goals.

- The UN Country Teams should be properly trained, staffed, and funded to support program countries to achieve the Goals.
- The UN Country Team and the international financial institutions (World Bank, International Monetary Fund, regional development banks) should work closely at country level to improve the quality of technical advice.

1 Why the MDGs are important, where we stand, and why we're falling short

Chapter 1

The Millennium Development Goals and why they matter

We have the opportunity in the coming decade to cut world poverty by half. Billions more people could enjoy the fruits of the global economy. Tens of millions of lives can be saved. The practical solutions exist. The political framework is established. And for the first time, the cost is utterly affordable. Whatever one's motivation for attacking the crisis of extreme poverty—human rights, religious values, security, fiscal prudence, ideology—the solutions are the same. All that is needed is action.

This report recommends the way forward. It outlines a way to attain this bold ambition. It describes how to achieve the Millennium Development Goals.

What are the Millennium Development Goals?

The Millennium Development Goals are the world's time-bound and quantified targets for addressing extreme poverty in its many dimensions—income poverty, hunger, disease, lack of adequate shelter, and exclusion—while promoting gender equality, education, and environmental sustainability. They are also basic human rights—the rights of each person on the planet to health, education, shelter, and security as pledged in the Universal Declaration of Human Rights and the UN Millennium Declaration.

How will the world look in 2015 if the Goals are achieved? More than 500 million people will be lifted out of extreme poverty. More than 300 million will no longer suffer from hunger. There will also be dramatic progress in child health. Rather than die before reaching their fifth birthdays, 30 million children will be saved. So will the lives of more than 2 million mothers.

There's more. Achieving the Goals will mean safe drinking water for another 350 million people, and the benefits of basic sanitation for 650 million, allowing them to lead healthier and more dignified lives. Hundreds of millions more women and girls will lead their lives in freedom, with more security and more opportunity. Behind these large numbers are the lives and hopes of people seeking new opportunities to end the burden of grinding poverty.

Many countries are reaping the benefits of globalization and are on track to achieve at least some of the Goals by the appointed deadline of 2015. Between 1990 and 2001, according to World Bank estimates, the proportion of people living in extreme poverty fell from 28 percent to 21 percent in the developing world. The number of people in extreme poverty dropped from 1.21 billion to 1.09 billion (Chen and Ravallion 2004). Many regions, especially large parts of East Asia and South Asia, experienced dramatic economic and social progress.

Yet broad regions are far off track. Sub-Saharan Africa, most dramatically, has been in a downward spiral of AIDS, resurgent malaria, falling food output per person, deteriorating shelter conditions, and environmental degradation, so that most countries in the region are on a trajectory to miss most or all of the Goals. Climate change could worsen the situation by increasing food insecurity, spreading vector-borne diseases, and increasing the likelihood of natural disasters; a prolonged decline in rainfall in parts of Africa has already wreaked havoc. Meanwhile, for some Goals, such as reducing maternal mortality and reversing the loss of environmental resources, most of the world is off track. The early target for gender parity in primary and secondary education—with a deadline of 2005—will be missed in many countries.

It is time to put the Goals on the fast-track they require and deserve. The Goals need to be achieved at the country—not just the global or regional—level (box 1.1). This report presents a practical plan for doing so, one based on work conducted by more than 250 of the world's leading development practitioners over the past two years in the context of the UN Millennium Project. Throughout, we stress that the specific technologies for achieving the Goals are known. What is needed is to apply them at scale. We have 10 key recommendations (pages xx–xxiii) and a more detailed set of underlying recommendations (appendix 1). More elaborate analysis and recommendations are set out in the 13 thematically oriented task force reports that underpin this plan.

Why the Goals are important

As the most broadly supported, comprehensive, and specific poverty reduction targets the world has ever established, the Millennium Development Goals are too important to fail. For the international political system, they are the fulcrum on which development policy is based. For the billion-plus people living in extreme poverty, they represent the means to a productive life. For everyone on Earth, they are a linchpin to the quest for a more secure and peaceful world.

The fulcrum of international development policy

At the Millennium Summit in September 2000 the largest gathering of world leaders in history adopted the UN Millennium Declaration, committing their

Box 1.1
The Millennium Development Goals are country goals

The UN Millennium Project interprets the Millennium Development Goals as country goals, since this is the spirit in which they are pursued the world over. While progress in China and India is to be lauded as a global-scale triumph, it would be a mistake to declare "victory" in cutting extreme poverty on the basis of progress mainly in two countries while dozens of other countries with hundreds of millions of the world's poorest people are not meeting the Goals. It is the poorest countries making the least progress that the Millennium Declaration and the Millennium Development Goals are meant to support, not the ones making the most progress even without the Goals. This is an important reason for continuing to pursue the Goals at the country level.

There are two other important reasons. First, the Goals need to be operational, and most key economic policy decisions and development assistance activities take place at the level of individual sovereign states. Virtually the entire development assistance process—including Poverty Reduction Strategy Papers, donor-country negotiations, and debt relief—is designed at country level. The UN Millennium Project stresses the need for stronger regional and international programs, to meet cross-country and global needs. But the core of development practice is likely to remain at the country level in the period to 2015.

Second, the Goals need to be applied at the country level so governments can be held accountable for signing on to them. Global or regional interpretation of the Goals diffuses responsibility and lessens the accountability of individual leaders to make every effort for poverty reduction.

A concern often voiced both in low-income countries and in donor countries is that the Goals are "unrealistic" because they require too much progress too fast. This view places a tremendous downward pressure against aspirations for the Goals and needs to be rebuffed wherever possible. It is based, implicitly, on an extrapolation of current trends, which will indeed miss the Goals in dozens of countries. Yet our analysis suggests clearly that by fulfilling the commitments at Monterrey and elsewhere, the Goals can still be met in most if not all countries of the world. Dismissing them now would be to act on hunches rather than rigorous assessment—and would be cruel to the hopes and possibilities that the Goals have engendered.

nations to a new global partnership to reduce poverty, improve health, and promote peace, human rights, gender equality, and environmental sustainability. This unprecedented joint commitment was not a one-off affair. The partnership between rich and poor countries was reaffirmed at the November 2001 launch of the Doha Round on international trade. Soon after, world leaders met again at the March 2002 International Conference on Financing for Development in Monterrey, Mexico, establishing a landmark framework for global development partnership in which developed and developing countries agreed to take joint actions for poverty reduction. Later that same year, UN member states gathered at the World Summit on Sustainable Development in Johannesburg, South Africa, where they reaffirmed the Millennium Development Goals as the world's time-bound development targets.

The framework established in the Monterrey Consensus describes the nature and importance of new global partnership as follows:

> Achieving the internationally agreed development goals, including those contained in the UN Millennium Declaration, demands a new

> partnership between developed and developing countries. We commit ourselves to sound policies, good governance at all levels, and the rule of law. We also commit ourselves to mobilizing domestic resources, attracting international flows, promoting international trade as an engine for development, increasing international financial and technical cooperation for development, sustainable debt financing and external debt relief, and enhancing the coherence and consistency of the international monetary, financial, and trading systems (UN 2002a).

While the Monterrey Consensus rightly affirmed that poverty reduction is the primary responsibility of developing countries themselves, it also set forth a balanced approach to economic growth that recognizes the interwoven nature of individual economies and how some countries need more international support (box 1.2). For instance, Monterrey outlined the need for better policies and increased assistance, for more trade and more aid. It also outlined the special development assistance needs of the least developed, African, small island, and landlocked economies. Moreover, it restated the long-standing development assistance target of 0.7 percent of gross national product (GNP). The UN Millennium Project supports these balanced principles and considers the Monterrey Consensus to be the international point of departure for detailing the specific steps to achieve the Goals.

Couched in these landmark Millennium and Monterrey pledges, the Millennium Development Goals drive a new era in international development. They are the first international goals to recognize, at the highest political levels, that poverty in the poorest countries can be dramatically reduced only if developing countries put well designed and well implemented plans in place to reduce poverty—and only if rich countries match their efforts with substantial increases in support. No well intended but impoverished country is to be left, under "business as usual," solely to its own resources. Under the auspices of the Goals, countries have agreed to hold each other to account, and citizens of both high-income and low-income countries are empowered to hold their own governments to clear standards.

Advancing the means to a productive life

For the billion-plus people still living in extreme poverty, the Millennium Development Goals are a life-and-death issue. Extreme poverty can be defined as "poverty that kills," depriving individuals of the means to stay alive in the face of hunger, disease, and environmental hazards. When individuals suffer from extreme poverty and lack the meager income needed even to cover basic needs, a single episode of disease, or a drought, or a pest that destroys a harvest can be the difference between life and death. In households suffering from extreme poverty, life expectancy is often around half that in the high-income world, 40 years instead of 80. It is common that of every 1,000 children born,

Box 1.2

The Monterrey Consensus as a framework for global partnership

Source: UN 2002a.

The Monterrey Consensus offers a valuable framework for international action, though many of its key commitments remain unfulfilled. Some important but often overlooked highlights are as follows.

First, the world committed to a broad-based development agenda, not a narrow one, taking into account not only growth but also poverty reduction and environmental sustainability:

Our goal is to eradicate poverty, achieve sustained economic growth and promote sustainable development as we advance to a fully inclusive and equitable global economic system (paragraph 1).

Second, the Consensus distinguished between developing countries that have adequate infrastructure and human capital to attract private investment (mainly middle-income countries) and those that must rely on official development assistance to build up infrastructure and human capital (mainly low-income and especially Least Developed Countries):

Official development assistance (ODA) plays an essential role as a complement to other sources of financing for development, especially in those countries with the least capacity to attract private direct investment. ODA can help a country to reach adequate levels of domestic resource mobilization over an appropriate time horizon, while human capital, productive, and export capacities are enhanced. ODA can be critical for improving the environment for private sector activity and can thus pave the way for robust growth. ODA is also a crucial instrument for supporting education, health, public infrastructure development, agriculture and rural development, and to enhance food security (paragraph 39).

Third, the Consensus noted that trade is a critical engine of growth and that low-income countries need two kinds of help to improve trade: improved market access and financial resources to remove supply-side constraints through investments in trade infrastructure, technology, and institutions:

In cooperation with the interested governments and their financial institutions and to further support national efforts to benefit from trade opportunities and effectively integrate into the multilateral trading system, we invite multilateral and bilateral financial and development institutions to expand and coordinate their efforts, with increased resources, for gradually removing supply-side constraints; improve trade infrastructure; diversify export capacity and support an increase in the technological content of exports; strengthen institutional development and enhance overall productivity and competitiveness (paragraph 36).

Fourth, the Consensus identified several regions where ODA is particularly necessary to meet the Goals:

For many countries in Africa, Least Developed Countries, small island developing states and landlocked developing countries, ODA is still the largest source of external financing and is critical to the achievement of the development goals and targets of the UN Millennium Declaration and other internationally agreed development targets (paragraph 39).

Fifth, the Consensus recognized that significant increases in aid would therefore be needed, and the donor countries committed to provide those additional resources, including the long-standing target of 0.7 percent of GNP:

We recognize that a substantial increase in ODA and other resources will be required if developing countries are to achieve the internationally agreed development goals and objectives, including those contained in the UN Millennium Declaration. To build support for ODA, we will cooperate to further improve policies and development strategies, both nationally and internationally, to enhance aid effectiveness (paragraph 41).

In that context, we urge developed countries that have not done so to make concrete efforts towards the target of 0.7 percent of gross national product (GNP) as ODA to developing countries (paragraph 42).

more than 100 die before their fifth birthday, compared with fewer than 10 in the high-income world. An infant born in Sub-Saharan Africa today has only a one-in-three chance of surviving to age 65.

For people living in extreme poverty, the Goals are ends unto themselves, directly representing the ambition for a longer, healthier, and more fulfilling life. But they are also "capital inputs"—the means to a productive life, to economic growth, and to further development in the future (chapter 3). Extreme poverty is found all over the developing world—not only in low-income countries but also in middle-income countries that have "pockets of poverty," such as remote regions and ethnic or racial minorities. So, the Goals are relevant wherever poverty exists.

Consider a typical village of subsistence farm households in a poor country, such as Afghanistan, Bhutan, Bolivia, Burkina Faso, Ethiopia, Nicaragua, or Papua New Guinea. The village lacks access to a paved road and motor transport. Also lacking electricity, its energy needs are met by extracting wood from the diminished secondary forests and woodlands. Drinking water is unsafe, and latrines regularly serve as a reservoir of infection through contamination of food and the local water supply. The children are sick from diarrhea, pneumonia, or malaria.

In an African village, adults are dying of AIDS and TB, without hope of treatment. Farmers toil but do not even produce enough food to feed their families. The soils were long ago depleted of nutrients, especially nitrogen. The rains fail and there is no backup of irrigation.

In these settings, women carry a triple burden, caring for children, the elderly, and the sick, spending long hours to gather water and fuelwood, to process and produce food, and working on farms or in family enterprises for little or no income. Impoverished families have more children than they desire because of poor access to education, contraception, decent employment opportunities, and sexual and reproductive health information and services. Education seems at best a luxury to most citizens. And since there is no emergency obstetric care, mothers die in childbirth at a hundred or more times the rate in the rich world.

Market forces alone will not rescue the village. Indeed, markets tend to bypass villages with little if any monetary income, and no ready means to earn it, given the low productivity and poor connections with the regional and world economy. The village barely lives off its own food production. Without money it cannot attract doctors, teachers, or transport firms. Without electricity or access to modern fuels it cannot run food processing equipment, irrigation pumps, computers, or electric tools for carpentry or apparel. Villagers do not have enough income to save. And since infrastructure and a skilled work force are lacking, private investors do not come. Young men and women, particularly the literate, leave the village—and the best educated, the country.

The same downward spiral applies to many urban areas. On arrival, migrants from rural areas might find employment, though informal and insecure, and they are faced with inaccessible and unaffordable housing. They take refuge in ill-serviced and overcrowded informal settlements. Many of the largest urban agglomerations in the low-income world are like extended villages, and rapidly growing cities in middle-income countries are often very poorly planned, with large areas bereft of functioning infrastructure, employment, and environmental management.

A generation or more of migrants from the countryside, combined with rapid natural population growth, results in a sprawl of densely settled humanity lacking the basics of healthcare, education, electricity, water supply, sanitation, solid waste disposal, and access to transport. People living in slums are largely excluded from enjoying their political, social, and economic rights. Some slums are so densely populated that it is not even possible to drive an ambulance into them. Diseases like TB spread like wildfire. HIV is often rampant.

Without basic infrastructure and human capital, countries are condemned to export a narrow range of low-margin primary commodities based on natural (physical) endowments, rather than a diversified set of exports based on technology, skills, and capital investments. In such circumstances, globalization can have significant adverse effects—including brain drain, environmental degradation, capital flight, and terms-of-trade declines—rather than bring benefits through increased foreign direct investment inflows and technological advances.

Yet practical steps can be taken to turn the tide. Both the villages and the cities can become part of global economic growth if they are empowered with the infrastructure and human capital to do so. If every village has a road, transport, a clinic, electricity, safe drinking water, education, and other essential inputs, the villagers in very poor countries will show the same determination and entrepreneurial zeal of people all over the world. If every city has a reliable electricity grid, competitive telecommunications, access to transport, accessible and affordable housing for the poor, a water and sanitation system, and access to global markets through modern ports or roads, jobs and foreign investment will flow in—rather than educated workers flowing out.

Investing in core infrastructure, human capital, and good governance thus accomplishes several things:

- It converts subsistence farming to market-oriented farming.
- It establishes the basis for private sector–led diversified exports and economic growth.
- It enables a country to join the global division of labor in a productive way.
- It sets the stage for technological advance and eventually for an innovation-based economy.

At a deeper level, achieving the Goals is about making core investments in infrastructure and human capital that enable poor people to join the global economy, while empowering poor people with the economic, political, and social rights that will enable them to make full use of infrastructure and human capital, wherever they choose to live (box 1.3).

Infrastructure, human capital, and human rights are vital complements to a healthy private sector. In a market-oriented economy, as long as individuals and businesses have the tools offered by infrastructure and human capital, the private sector can develop rapidly. Private sector–led growth in agriculture, industry, and services will then generate jobs and incomes, which reduce poverty and the future dependency on foreign aid. The goal, then, is to combine the critical public investments in infrastructure and human capital with market-oriented economic policies to ensure the dynamism of private sector growth. As economies grow richer, the private sector can also provide an increasing share of core infrastructure services.

The Goals are critical for global security

The Goals not only reflect economic targets, global justice, and human rights—they also are vital to international and national security and stability, as emphasized by the High-Level Panel on Threats, Challenges, and Change (UN 2004a). Poor and hungry societies are much more likely than high-income societies to fall into conflict over scarce vital resources, such as watering holes

Box 1.3
The means to a productive life

The key elements of adequate human capital include:

- Basic nutrition.
- A health system that enables people to live a long and healthy life.
- Sexual and reproductive health.
- Literacy, numeracy, and marketable skills for twenty-first century jobs.
- Technical and entrepreneurial skills to adopt existing but underused technologies and scientific expertise to advance new knowledge.

The essential infrastructure services include:

- Safe drinking water and basic sanitation.
- A sustainably managed and conserved natural environment.
- Farm inputs, including soil nutrients, reliable water for agriculture, and improved seed varieties, plus vaccines, veterinary pharmaceuticals, and feed and fodder for livestock.
- Energy, including electricity and safe cooking fuels.
- Paved roads and transport services that are safe and reliable, including nonmotorized options.
- Modern information and communications technology.

The core political, social, and economic rights include:

- Equal rights, including reproductive rights, for women and girls.
- Freedom from violence, especially for girls and women.
- A political voice for every citizen, including through civil society organizations.
- Equal access to public services.
- Security of tenure and property rights for shelter, businesses, and other assets.

and arable land—and over scarce natural resources, such as oil, diamonds, and timber. Many world leaders in recent years have rightly stressed the powerful relationship between poverty reduction and global security (box 1.4). Achieving the Millennium Development Goals should therefore be placed centrally in international efforts to end violent conflict, instability, and terrorism. As the High-Level Panel recommends, countries that aspire to global leadership through permanent membership on the UN Security Council have a special responsibility to promote the Goals and to fulfill international commitments to official development assistance and other kinds of support vital for achieving them. We endorse the Panel's recommended criterion of 0.7 percent of GNP in official development assistance for developed countries aspiring to permanent membership.

Poverty increases the risks of conflict through multiple paths. Poor countries are more likely to have weak governments, making it easier for would-be rebels to grab land and vital resources. Resource scarcity can provoke population migrations that result in conflicts between social groups, as in Darfur, Sudan, in the wake of diminishing rainfall. Without productive alternatives, young people may turn to violence for material gain, or feel a sense of hopelessness, despair, and rage. Poor farmers who lack basic infrastructure and access to agricultural markets may turn in desperation to narcotics production and trade, such as growing poppy in Afghanistan or coca in the Andes. Many slums are controlled by gangs of drug traffickers and traders, who create vicious circles of insecurity and poverty. The lack of economically viable options other than criminal activity creates the seedbed of instability and increases the potential for violence.

While violent conflicts surely result from a combination of factors, research suggests a strong causal impact of poverty and adverse income shocks on the onset of conflict. The risk of civil conflict declines steadily as national incomes increase. Negative economic growth shocks increase the risk of civil conflict dramatically (chapters 3 and 12). The implications are twofold: investing in development is especially important to reduce the probabilities of conflict, and development strategies should take into consideration their possible effects on reducing (or inadvertently increasing) the risks of conflict.

Structure of the report

In proposing a global framework that will enable all countries to achieve the Goals, this report is structured in four parts. The rest of part 1 presents an overview of worldwide progress, highlighting regions and Goals that are particularly off track (chapter 2). It then presents an analytical framework for understanding why parts of the world are falling short of achieving the Goals. It also describes the important relationship between the Goals and economic growth, and the centrality of public investments to achieve the Goals in the poorest countries (chapter 3).

Box 1.4

Poverty reduction and global security

Source: Abdullah II 2004; Blair 2004; Bush 2002; Chirac 2004; da Silva 2004; Koizumi 2004; Mkapa 2003; Obasanjo 2004; Schröder 2001.

Many world leaders have stressed the fact that the fight for global security—to stop war, internal violence, terror, and other ills of profound instability—requires success in the battle against poverty as well. Here are some of their statements, emphasizing the broad range of agreement on this vital point.

King Abdullah of Jordan, January 23, 2004

"Opportunity is a powerful force in giving people a stake in a peaceful future. It is in our hands to create a global growth economy, access to education and technology, and, most important, justice, to show young people, that ours is a world of fairness, openness, and hope. The Millennium Development Goals need to be reinforced with new benchmarks for assessing progress, for ensuring better and fairer trade, and for forging new global links."

Prime Minister Tony Blair of the United Kingdom, October 7, 2004

"The rest of the world cannot stand by—because we cannot afford to, because what happens in Africa affects and will affect the rest of the world. Poverty and instability leads to weak states which can become havens for terrorists and other criminals."

President George W. Bush of the United States, March 14, 2002

"Poverty doesn't cause terrorism. Being poor doesn't make you a murderer. Most of the plotters of September 11th were raised in comfort. Yet persistent poverty and oppression can lead to hopelessness and despair. And when governments fail to meet the most basic needs of their people, these failed states can become havens for terror.

"Poverty prevents governments from controlling their borders, policing their territory, and enforcing their laws. Development provides the resources to build hope and prosperity, and security.... Successful development also requires citizens who are literate, who are healthy, and prepared and able to work. Development assistance can help poor nations meet these education and health care needs."

President Jacques Chirac of France, May 26, 2004

"The world economy as a whole is held back when the lack of development condemns entire regions to poverty and a seeming lack of prospects. It is also a political necessity, because the security and stability of the world are under threat from the reactions of populations that are deprived of their basic rights."

President Luiz Inácio Lula da Silva of Brazil, September 21, 2004

"The path to lasting peace must encompass a new political and economic international order, one that extends to all countries real opportunities for economic and social development."

Prime Minister Junichiro Koizumi of Japan, September 21, 2004

"The protection and empowerment of individuals and communities is the foundation of international peace and security.... There will be no stability and prosperity in the world unless the issues of Africa are resolved.... Peace and security, economic and social issues are increasingly intertwined."

President Benjamin Mkapa of Tanzania, January 10, 2003

"We should address the situations and factors that have the potential to sow terrorism, namely, poverty, denial, deprivation, oppression, and injustice."

President Olusegun Obasanjo of Nigeria, September 23, 2004

"Our quest for global peace and security will prove unsuccessful unless we intensify international cooperation for development and the reduction of poverty."

Chancellor Gerhard Schröder of Germany, 2001, Program of Action 2015

"Extreme poverty, growing inequality between countries, but also within countries themselves, are great challenges of our times, because they are a breeding ground for instability and conflict. So reducing worldwide poverty is, not least, essential to safeguarding peace and security."

Part 2 presents the UN Millennium Project's central recommendations for operationalizing the Goals in developing countries. Chapter 4 outlines the content and processes of the core framework for pursuing the Goals: MDG-based poverty reduction strategies that are designed, owned, and implemented at the country level and anchored in a 10-year framework that maps out priorities by working back from the 2015 targets. Chapter 5 synthesizes the core recommendations of the UN Millennium Project's task forces for the basic interventions to be included in national MDG-based poverty reduction strategies. Chapter 6 describes the key elements for building systems to scale up these interventions.

Chapter 7 outlines the key domestic governance issues to address in a MDG-based poverty reduction strategy. It emphasizes the practical steps in building the various components of a governance system relevant to pursuing the Goals, including the advance of public sector management, human rights, civil society organizations, and private sector growth. Chapter 8 describes the important contributions of civil society organizations to national and international efforts to achieve the Goals. Chapter 9 describes the critical role of the private sector.

Chapter 10 identifies the priorities—particularly the investment priorities—in Sub-Saharan Africa, clarifying the frequent misunderstanding that African countries are stuck in a governance crisis, when they tend more to be stuck in a poverty trap. Chapter 11 follows by evaluating priorities in the other regions of the world, with special emphasis on the most vulnerable countries. Chapter 12 examines the special conditions for countries in or emerging from conflict, recommending that any international or national strategy to achieve the Goals include a focus on conflict and conflict prevention.

Part 3 discusses the implications of the MDG-based poverty reduction strategy approach for the international system. Chapter 13 outlines how the donor system of development partnership is not yet up to the task of the Goals and recommends specific steps for reform. Chapter 14 discusses the role of trade reform for the Goals, making specific recommendations for the current Doha round of trade negotiations. It emphasizes the need to distinguish effects of trade reform by country. It also addresses the supply side of developing countries' export competitiveness in addition to market access issues. Chapter 15 outlines the regional and global goods that need to be addressed in national poverty reduction strategies and through regional and global strategies and institutions. Chapter 16 then presents steps for bold and immediate action in 2005 to inaugurate 10 years of MDG success. It outlines ways to put the Goals on the fast-track they require and deserve.

Part 4 concludes the report by outlining the estimated costs and benefits of a decade of bold ambition through 2015. Chapter 17 evaluates the cost implications of country-level MDG investment strategies, and presents a Monterrey-based cofinancing approach to supporting these strategies through increased domestic resource mobilization and increased official development assistance. Chapter 18 concludes with an estimate of the benefits of achieving the Goals and an assessment of the possibilities for the world in 2015. The Goals represent a mid-station en route to ending poverty within a generation.

Chapter 2

Where we stand with only a decade to go

Economic development lifted millions of people out of poverty in the last decade. While the population of developing countries rose from about 4 billion people to 5 billion, average per capita incomes rose by more than 21 percent (table 2.1). With 130 million fewer people in extreme poverty in 2001 than a decade before, the proportion of people living on less than $1 a day declined by 7 percentage points, from 28 to 21 percent (World Bank 2004c; Chen and Ravallion 2004). The rate of undernourishment declined by 3 percentage points, and the under-five mortality rate dropped from 103 deaths per 1,000 births to 88. Life expectancy rose from 63 years to nearly 65 years (FAO 2003a; World Bank 2004c). An additional 8 percent of the developing world's population gained access to improved drinking water supply, and 15 percent more to basic sanitation services (WHO and UNICEF 2004). Of course the story has not been all good. Perhaps most notably, the spread of AIDS has been catastrophic, with more than 20 million lives lost since the first case was detected in 1981 (UNAIDS 2004).

Evaluating poverty and progress around the world

General developing world trends obscure vast differences across and within regions and countries. Some regions have made little progress or even experienced reversals in several areas (table 2.2). Many countries have seen economic growth while others have experienced stagnation. And many of the poorest countries have seen gradual economic growth, but at rates grossly inadequate to yield a dramatic reduction in poverty. From 1990 to 2002, for example, the heavily indebted poor countries saw their incomes rise only from $298 per capita to $337 in 1995 dollars (World Bank 2004c).

Much of the progress toward poverty reduction in the last decade has been driven by advances in East Asia and South Asia, home to China and India, the

Table 2.1

Measures of average progress in the developing world, 1990–2002 (population-weighted)

Indicator	1990	2002
GDP per capita (1995 US$)	1,071	1,299
Headcount poverty (percent)[a]	28	21
Undernourishment prevalence (percent)[b]	20	17
Under-five mortality (per 1,000 live births)	103	88
Life expectancy at birth (years)	63	65
HIV prevalence (percent)	0.5	1.6
Access to improved drinking water supply (percent)	71	79
Access to improved sanitation facilities (percent)	34	49

a. The poverty headcount ratio is the proportion of the national population with incomes below $1.08 a day. 2002 data unavailable; 2001 data used as a proxy.

b. Does not include CIS countries in 1990.

Source: GDP, under-five mortality, and life expectancy data from World Bank 2004c. Headcount poverty data from Chen and Ravallion 2004. Undernourishment data from FAO 2003a. HIV prevalence data from UNAIDS and WHO 2004. Water and sanitation data from WHO and UNICEF 2004.

world's most populous countries. With more than 2.3 billion people in these two countries alone, their major advances in poverty reduction drive developing world averages (table 2.3). The poverty rate in China dropped from 33 percent to 17 percent between 1990 and 2001, and in India, from 42 percent to 35 percent (Chen and Ravallion 2004) and even more by some estimates (Bhalla 2002; Deaton 2003). China's low population growth rate and rapid reduction in poverty rates have decreased its poverty headcount by nearly 165 million people since 1990.[1] By contrast, India's declining poverty rates have been offset by population growth, so the number of absolute poor there remains unchanged at approximately 360 million people (Chen and Ravallion 2004).

Other parts of Asia have also seen strong progress, with economic growth helping to reduce the rest of the region's extreme poverty by nearly 70 million people since 1990. Yet the sheer numbers of poor people in Asia remain vast, with more than 270 million in East Asia and 430 million in South Asia, all vulnerable to droughts, natural disasters, and other shocks (Chen and Ravallion 2004). East Asia and South Asia together are still home to the greatest number of absolute poor people in the world.

In sharp contrast to Asia's progress, most of Sub-Saharan Africa faces significant challenges in meeting the Millennium Development Goals on almost every dimension of poverty, with many countries falling behind. Between 1990 and 2001 the number of people living on less than $1 a day rose from 227 million to 313 million, and the poverty rate rose from 45 percent of the population to 46 percent (Chen and Ravallion 2004). In the 33 countries of

Table 2.2
Major trends in the Goals, by region

	Africa		Asia						Commonwealth of Independent States	
	Northern	Sub-Saharan	Eastern	South-eastern	Southern	Western	**Oceania**	**Latin America & Caribbean**	Europe	Asia
Goal 1 Eradicate extreme poverty and hunger										
Reduce extreme poverty by half	on track	high, no change	met	on track	on track	increasing	no data	low, minimal improvement	increasing	increasing
Reduce hunger by half	high, no change	very high, little change	progress but lagging	progress but lagging	progress but lagging	increasing	moderate, no change	on track	low, no change	increasing
Goal 2 Achieve universal primary education										
Universal primary schooling[a]	on track	progress but lagging	on track	lagging	progress but lagging	high but no change	progress but lagging	on track	declining	on track
Goal 3 Promote gender equality and empower women										
Girls' equal enrollment in primary school	on track	progress but lagging	met	on track	progress but lagging	progress but lagging	on track	on track	met	on track
Girls' equal enrollment in secondary school	met	progress but lagging	no data	met	progress but lagging	little change	progress but lagging	on track	met	met
Literacy parity between young women and men	lagging	lagging	met	met	lagging	lagging	lagging	met	met	met
Women's equal representation in national parliaments	progress but lagging	progress but lagging	declining	progress but lagging	very low, some progress	very low, no change	progress but lagging	progress but lagging	recent progress	declining
Goal 4 Reduce child mortality										
Reduce mortality of under-five-year-olds by two-thirds	on track	very high, no change	progress but lagging	on track	progress but lagging	moderate, no change	moderate, no change	on track	low, no change	increasing
Measles immunization	met	low, no change	no data	on track	progress but lagging	on track	declining	met	met	met
Goal 5 Improve maternal health										
Reduce maternal mortality by three-quarters	moderate	very high	low	high	very high	moderate	high	moderate	low	low
Goal 6 Combat HIV/AIDS, malaria, and other diseases										
Halt and reverse spread of HIV/AIDS	no data	stable	increasing	stable	increasing	no data	increasing	stable	increasing	increasing
Halt and reverse spread of malaria	low	high	moderate	moderate	moderate	low	low	moderate	low	low
Halt and reverse spread of TB	low, declining	high, increasing	moderate, declining	high, declining	high, declining	low, declining	high, increasing	low, declining	moderate, increasing	moderate, increasing
Goal 7 Ensure environmental sustainability										
Reverse loss of forests	less than 1% forest	declining	met	declining	small decline	less than 1% forest	declining	declining except Caribbean	met	met
Halve proportion without improved drinking water in urban areas	met	no change	declining access	high access, no change	met	met	high access, no change	met	met	met
Halve proportion without improved drinking water in rural areas	high access, little change	progress but lagging	progress but lagging	progress but lagging	on track	progress but lagging	low access, no change	progress but lagging	high access, limited change	high access, limited change
Halve proportion without sanitation in urban areas	on track	low access, no change	progress but lagging	on track	on track	met	high access, no change	high access, no change	high access, no change	high access, no change
Halve proportion without sanitation in rural areas	progress but lagging	no change	progress but lagging	progress but lagging	progress but lagging	no change	no change	progress but lagging	little change	little change
Improve the lives of slum dwellers	on track	rising numbers	progress but lagging	on track	some progress	rising numbers	no data	progress but lagging	low but no change	low but no change
Goal 8 A global partnership for development										
Youth unemployment	high, no change	high, no change	low, increasing	rapidly increasing	low, increasing	high, increasing	low, increasing	increasing	low, rapidly increasing	low, rapidly increasing

met or on track **progress, but too slow** **no or negative change** **no data**

a. Results based on measurements of enrollment rate. Results may change if based on measurements of primary completion rates. For example, estimates of completion rates in Latin America show that 8–10 percent of the school-age population will not complete primary school, which implies that the region is off track for reaching the goal of universal primary education.

Source: UN Statistics Division, UNDESA 2004.

Table 2.3

Population living below the poverty line, by developing region

a. Poverty lines set in 1993 US$ adjusted for purchasing power parity.

b. Calculated as rural poverty rate × (100 – urbanization rate) / national poverty rate. Note that published poverty rates often underreport urban poverty.

c. Where 2001 data are not available, uses most recent year available.

Source: Columns 1–4 and 7–10: Chen and Ravallion 2004. Columns 5–6: Calculated from World Bank 2004c.

$1.08 a day poverty line[a]

	Millions of people		Share of total population (%)		Share of poor people living in rural areas[b] (%)	Rural population as share of total (%)
Region	**1990**	**2001**	**1990**	**2001**	**2001[c]**	**2001**
East Asia	472	271	30	15	80	63
Eastern Europe and Central Asia	2	17	1	4	53	37
Latin America and Caribbean	49	50	11	10	42	24
Middle East and North Africa	6	7	2	2	63	42
South Asia	462	431	41	31	77	72
Sub-Saharan Africa	227	313	45	46	73	67

$2.15 a day poverty line[a]

	Millions of people		Share of total population (%)	
Region	**1990**	**2001**	**1990**	**2001**
East Asia	1,116	865	70	47
Eastern Europe and Central Asia	23	93	5	20
Latin America and Caribbean	125	128	28	25
Middle East and North Africa	51	70	21	23
South Asia	958	1,064	86	77
Sub-Saharan Africa	382	516	75	77

tropical Sub-Saharan Africa, the average GDP per person is only $270 a year, a mere 71 cents a day (World Bank 2004c).[2]

The Middle East and North Africa saw a consistently low poverty rate, at roughly 2 percent, while the number of poor rose slightly, from 6 million to 7 million. Latin America and the Caribbean saw fairly stagnant poverty rates while Eastern Europe and Central Asia saw a stark increase in poverty over the period. Note that measurements that use a $1 a day standard understate the real extent of poverty in regions where the cost of living is higher. For example, a $2 a day standard is more appropriate in Latin America and the Caribbean or the transition countries of Europe.

All of these regional trends mask country-level variation. Some African countries such as Mozambique have recorded substantial growth over the last decade. Asia has poor performers as well as strong ones. Variation within countries can also be very high. For example, China and India have displayed strong aggregate growth, but they have wide subnational variations in development. Similarly, Brazil and Mexico have experienced wide regional variations in poverty reduction.

Subnational variations in poverty are important because they may reflect geographical, social, or other determinants of exclusion that require special strategies or investments. The UN Millennium Project has attempted to identify this variation by developing a global map of absolute poverty. But rather than using income or consumption to measure poverty, two indicators that are notoriously difficult to compare across countries, we used data on infant mortality and malnutrition, two core measures of human poverty commonly collected at a subnational level around the world (map 2.1).[3] The map shows that most extreme poverty is in Sub-Saharan Africa, South Asia, Central America, and the Andean region of South America.

Two other important dimensions for understanding poverty across regions are the rate of urbanization and the proportion of extreme poor living in rural areas. Although the $1 a day standard underestimates the extent of urban poverty (Satterthwaite 2004), in the poorest regions—particularly in Asia and Sub-Saharan Africa—available data show that the vast majority of the population and the majority of extremely poor people live in rural areas. While urban populations are growing quickly in all regions and the number of urban poor is rising rapidly, roughly three-quarters of the poorest people in Africa and Asia still live in rural areas. This contrasts significantly with Latin America and the Caribbean, where three-quarters of the population, and at least 60 percent of the extreme poor, are estimated to live in urban areas. Not only does the

Map 2.1
Absolute poverty: infant mortality and malnutrition, 2000
Infant mortality rate (IMR) per 1,000 live births and percentage of children underweight
Source: CIESIN 2005a, b.

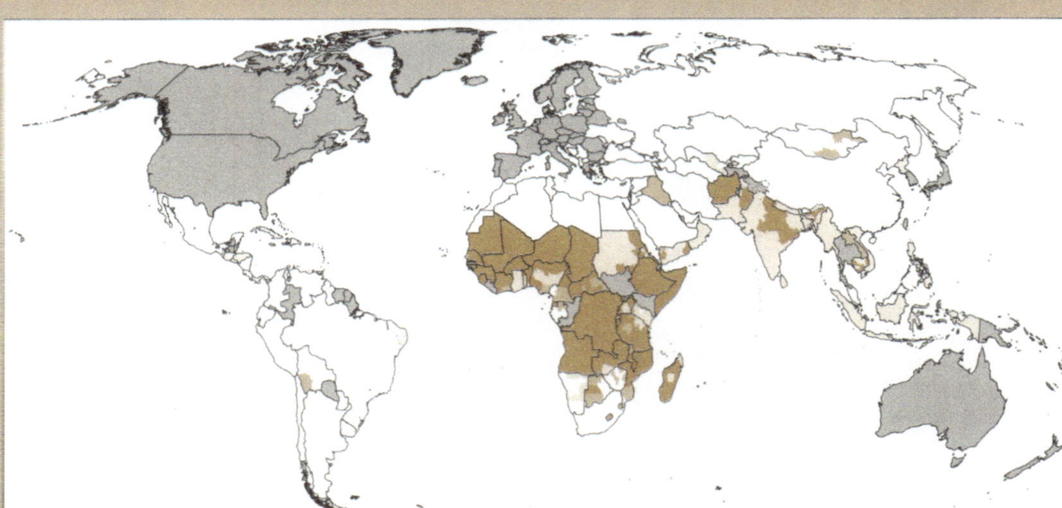

prevalence of extreme poverty differ by region, but its concentration in rural and urban areas differs markedly as well.

How regions are progressing toward the MDGs

Though there is considerable country-level variation in progress toward the Millennium Development Goals, regional trends often reflect important conditions and challenges common to many countries. This section describes progress and remaining challenges related to the pursuit of the Goals in each of the major developing regions (tables 2.3 and 2.4).[4]

North Africa

Moving in the right direction on every indicator, North Africa needs to accelerate progress to achieve the Goals. It has seen modest economic growth since 1990 and is on track to reach the target of halving the poverty headcount rate. Levels of undernourishment have been virtually stagnant at 5 percent since 1990. Gender inequality remains a challenge, reflected in the low share of women in wage employment in the nonagricultural sectors and the low numbers of women representatives in parliaments. The prevalence of HIV, TB, and malaria is low, but greater progress needs to be made on maternal mortality and rural access to safe water and sanitation. Chronic water shortages and unsustainable use of natural resources fuel growing environmental problems, particularly desertification and soil salinization. The region has also been slow to adapt to scientific and technological developments and making investments in information and communication technologies.

Table 2.4

Tracking developing world progress by region since 1990

	Eastern Asia		Southeastern Asia		Southern Asia		Western Asia	
	1990	2001/02	1990	2001/02	1990	2001/02	1990	2001/02
GDP per capita (constant 1995 US$)[a]	351	943	1,012	1,421	379	547	2,868	3,070
Undernourishment prevalence (%)[b]	16	11	17	13	25	22	7	10
Net enrollment in primary education (%)[c]	98	92	92	91	73	80	81	83
Ratio of girls to boys in secondary education[d]	—	—	0.97	0.98	0.74	0.77	0.76	0.79
Under-five mortality rate (per 1,000)[a]	48	38	78	48	126	93	68	61
Maternal mortality (per 100,000 live births)[e]	—	55	—	210	—	520	—	190
Share of land area covered by forest (%)[f]	15	17	54	49	14	13	3	3
Access to improved water supply (%)[a]	72	78	73	79	71	84	83	88
Access to improved sanitation (%)[a]	24	45	48	61	20	37	79	79
Share of urban population living in slums (%)[a]	41	36	37	28	64	59	34	35
Telephone lines and cellular subscribers (per 100 population)	2	38	1	16	1	5	10	42

— Not available.

a. Compares indicator levels between 1990 and 2002.

b. Compares indicator levels between 1990–92 and 1999–2001.

c. Compares indicator levels between 1990–91 and 2001–02.

d. Compares indicator levels between 1998–99 and 2001–02.

e. Indicator level as of 2000.

f. Compares indicator levels between 1990 and 2000.

g. Represents Latin America only.

Source: Based on data from UNDESA 2004 presented in UN 2004b; GDP data from World Bank 2004c.

Table 2.4

Tracking developing world progress by region since 1990

(continued)

	Northern Africa		Sub-Saharan Africa		Latin America and Caribbean		Oceania	
	1990	2001/02	1990	2001/02	1990	2001/02	1990	2001/02
GDP per capita (constant 1995 US$)[a]	1,263	1,484	595	581	3,290	3,735	1,017	1,112
Undernourishment prevalence (%)[b]	5	4	35	33	13	10	25	27
Net enrollment in primary education (%)[c]	82	92	54	62	86	96	74	79
Ratio of girls to boys in secondary education[d]	0.94	0.96	0.81	0.79	1.09	1.07	0.89	0.93
Under-five mortality rate (per 1,000)[a]	87	41	186	174	54	34	86	78
Maternal mortality (per 100,000 live births)[e]	—	130	—	920	—	190	—	240
Share of land area covered by forest (%)[f]	1	1	29	27	50[g]	48[g]	68	66
Access to improved water supply (%)[a]	88	90	49	58	83	89	51	52
Access to improved sanitation (%)[a]	65	73	32	36	69	75	58	55
Share of urban population living in slums (%)[a]	38	28	72	72	35	32	25	24
Telephone lines and cellular subscribers (per 100 population)	3	17	1	5	6	36	3	9

	CIS (Europe)		CIS (Asia)		CIS (Total)	
	1990	2001/02	1990	2001/02	1990	2001/02
GDP per capita (constant 1995 US$)[a]	3,553	2,617	1,223	938	—	—
Undernourishment prevalence (%)[b]	4	4	18	27	—	—
Net enrollment in primary education (%)[c]	91	87	85	94	—	—
Ratio of girls to boys in secondary education[d]	—	1.01	—	0.97	—	—
Under-five mortality rate (per 1,000)[a]	—	—	—	—	41	44
Maternal mortality (per 100,000 live births)[e]	—	—	—	—	—	—
Share of land area covered by forest (%)[f]	49	49	5	6	—	—
Access to improved water supply (%)[a]	—	—	—	—	92	93
Access to improved sanitation (%)[a]	—	—	—	—	84	83
Share of urban population living in slums (%)[a]	6	6	6	6	—	—
Telephone lines and cellular subscribers (per 100 population)	—	—	—	—	13	29

Sub-Saharan Africa

The region is off track to meet every Millennium Development Goal. It has the highest rate of undernourishment, with one-third of the population below the minimum level of dietary energy consumption. Sub-Saharan Africa has the lowest primary enrollment rates of all regions. Despite recent progress, gender disparity at the primary level is 0.86, the lowest of all regions (UN Millennium Project 2005j). The HIV/AIDS crisis is devastating much of the continent, destroying lives and livelihoods. Women are disproportionately affected, with 13 infected women for every 10 infected men (UNAIDS 2004). The region also has the highest TB incidence in the world and the highest maternal and child mortality ratios (maternal mortality ratios are 46 times higher than in the developed world).

Progress in access to safe drinking water, though more promising, is still too slow to achieve the MDG targets. More than 160 million people live in slum-like conditions where they lack security of tenure, and safe housing. Most of the region lacks access to information and communication technology, with just 5.3 telephone subscribers per 100 inhabitants. Rates of deforestation are among the highest in the world, illustrating the continent's environmental crisis. Without sustained support, Sub-Saharan Africa is unlikely to meet any of the Goals. (We focus on Africa's special needs in chapter 10.)

East Asia

The region has seen rapid falls in income poverty and in hunger, and improvements have also been recorded in gender equality, education, and child survival. It has invested heavily in infrastructure, with 37.8 phone lines per 100 people, comparing favorably with other developing regions. But it continues to suffer from pockets of extreme poverty, fairly high TB rates, and persistently low access to safe drinking water and sanitation. HIV prevalence rates are low but increasing for high-risk groups. UNAIDS estimates that without an effective response, as many as 10 million people in China may become infected by 2010 (UNAIDS 2004). Nearly 200 million people in East Asia live in slums, lacking access to secure housing and essential services. China's gender inequality remains high, with far fewer girls enrolled in school than boys. A major challenge is environmental degradation, including pollution from rapid industrialization and agricultural intensification.

Southeast Asia

The region is on track to meet the Goals for income poverty, hunger, child mortality, and gender equality. But progress toward other Goals has been mixed, with Thailand and Viet Nam making rapid progress in many areas while, for example, Cambodia and Lao PDR struggle to advance. School enrollment rates have stagnated and need to increase more quickly to achieve the universal primary education target. Other obstacles to achieving the Goals include rising numbers of people infected with HIV and TB, high maternal mortality, rapid deforestation and destruction of coastal and marine environments, and low rural access to water supply and sanitation.

South Asia

The region has made some major strides in overall poverty reduction thanks largely to rapid economic growth in India. Although aggregate income poverty is falling rapidly, South Asia is still home to more poor people than any other region and remains off track for meeting many of the Goals. Primary enrollment and gender equality indicators are lagging. Child health is improving but not quickly enough to meet the targets, while maternal mortality rates remain high, and, without urgent action, HIV is poised to spread. Severe undernourishment afflicts large parts of the population. More than 250 million people live in slum-like conditions, with insecure tenure, inadequate housing, and poor access to essential services. Most of South Asia lacks access to modern technologies and services, with approximately 5 telephone subscribers per 100 inhabitants. Parts of the region suffer from serious problems of water quality and scarcity, and access to sanitation is low throughout the region. On the positive side, access to safe water has increased rapidly in both urban and rural areas.

West Asia

This region, which includes many countries typically classified as part of the Middle East, is off track for a majority of the Goals. Both income poverty and hunger are increasing, and progress toward gender equality has been slow. Primary enrollments increased only from 81 percent in 1990 to 83 percent in 2001, and under-five mortality fell only slightly from 68 per 1,000 live births to 61 in the same period. Maternal mortality remains high, and infectious diseases such as TB are still a threat. While urban areas are on track to meet the water and sanitation Goal, rural areas are lagging behind. Youth unemployment is a significant concern in the region.

Oceania

With about 8 million people, Oceania comprises mostly small island developing states. The region is off track for nearly every Goal, and falling back in some areas. The share of undernourished people increased from 25 percent to 27 percent between 1990–92 and 1999–2001. Net primary enrollment rates remain below 80 percent. Measles immunization coverage dropped from 70 percent to 57 percent between 1990 and 2003. HIV and TB infection rates are rising, and maternal mortality remains high. Even where there is progress, it is too slow to achieve the Goals. Degradation of coastal and marine environments threatens island ecosystems and economies. Most of Oceania also lacks access to modern information and communication technologies, with fewer than 10 telephone subscribers per 100 inhabitants. Only Sub-Saharan Africa is off track on more indicators than Oceania.

Latin America and the Caribbean

The region has experienced little economic growth since 1990, yielding stagnant poverty headcounts and persistently high inequality. However, countries there are doing relatively well on the Goals for hunger, education, gender equality, and child health. The Goal for water access has been met in urban areas, but rural areas have seen little improvement. Access to sanitation lags in both rural and urban areas. Accelerating deforestation remains a major issue throughout the region. Maternal mortality is relatively high at 190 per

100,000 live births, approximately 10 times the average in rich countries. More than 125 million people live in slum-like conditions. The greatest challenges lie in the Central American and Andean countries, where the concentration of poverty is highest. The Caribbean countries, as small island developing states, face special challenges and concerns (chapter 11).

CIS countries in Europe

In the 1990s, and especially in the first half, the CIS countries had their economies collapse, with significant increases in poverty and hunger. Most are now making progress, but a few have not yet achieved pre-reform levels of per capita income. Youth unemployment remains very high. Net primary enrollment rates have increased from postindependence lows, but they still remain below those in 1990. Some of these countries may be off track for meeting the health Goals, due to alarming increases in the number of people infected with TB and HIV and to high maternal mortality. Serious environmental challenges include access to clean water and sanitation and the high levels of industrial pollution.

CIS countries in Central Asia

Central Asian countries experienced an increase in poverty levels after the collapse of the Soviet economy, and their geographic isolation compounds the challenges of post-Soviet economic development. Since 1990 the poverty headcount rates have risen significantly in several countries, as have undernourishment and child mortality rates. While overall primary enrollment rates remain fairly high, gender inequality in education remains significant in some countries. Health indicators are deteriorating across the region, and the prevalence of HIV and TB is increasing. Most countries have already met or are on track to achieve the target for urban drinking water, but there has been little progress in rural access. Desertification and water scarcity pose serious threats to agriculture and environmental sustainability. Meanwhile, access to sanitation and other forms of infrastructure is stagnant or in decline.

* * *

Each region's prospects for progress toward the Goals are affected by its demographic conditions. Sub-Saharan Africa is confronted by continuing high population growth and a large adolescent population. Western Asia has the second fastest growing population. Less severe demographic constraints affect South Asia and Southeast Asia, but the unmet need for family planning and other sexual and reproductive health services remains high. Prospects in Latin America and the Caribbean are affected by the dramatic inequality of access to family planning and safe motherhood services between wealthier and poorer social groups. Several European Commonwealth of Independent States (CIS) countries face population declines because of low fertility and migration. Countries nearing the end of their demographic transitions will need to pay special attention to the emerging needs of aging populations and migrants.

Each region will require tailored strategies and interventions to achieve the Millennium Development Goals. Chapter 11 describes investment priorities for each of these regions and for groups of countries that share special concerns, such as landlocked developing countries, Least Developed Countries, small island developing states, and countries vulnerable to natural hazards.

Summary of progress toward each Goal

In addition to high variation across regions, progress has been uneven across the Millennium Development Goals.

Poverty and hunger

Between 1990 and 2001 the percentage of the population living on less than $1 a day fell significantly in East Asia, South Asia, Southeast Asia, and North Africa. The percentage remained stagnant in Sub-Saharan Africa and Latin America and the Caribbean, and it increased in West Asia and the Commonwealth of Independent States. In 2001 poverty rates were highest in Sub-Saharan Africa, at 46 percent of the population, and in South Asia, at 31 percent.

Hunger, still high in several regions of the world, is rising in a few. A third of the population in Sub-Saharan Africa and 27 percent of the population in Oceania and CIS countries in Asia are undernourished. Undernourishment is rising in West Asia and CIS Asia. And though malnutrition rates are falling on average in Sub-Saharan Africa and South Asia, they are rising in some African countries.

Primary education

Global and regional primary school completion rates have improved since 1990, but many regions are far off track for meeting the Goal. Latin America and the Caribbean, Middle East and North Africa, and South Asia have all seen increases in primary completion rates between 1990 and 2002, though overall levels are not very high. Those rates in East Asia and the CIS countries of Europe and Asia remained more or less constant over the same period, albeit at high levels. The greatest challenges are in Sub-Saharan Africa, where average primary completion rates hovered at around 50 percent between 1990 and 2002 (UN Millennium Project 2005k).

Gender equality

Progress on gender equality targets is limited and uneven. The world is still far away from achieving gender parity and will miss the education parity target for 2005—with the ratio of girls to boys in secondary education just 0.77 in South Asia and 0.79 in West Asia and Sub-Saharan Africa in 2001. North Africa and East Asia have seen strong progress on gender parity in gross enrollments. The ratio of literate women to men is still low around the world, and trends suggest that South Asia, Oceania, and West Asia are especially off track. Meanwhile, the share of women in wage employment in the nonagricultural sector increased in 93 of 131 countries measured. The share of women in national parliaments increased significantly in Latin America and the Caribbean between 1990 and 2004. It has also risen in other regions of the world, though at a slower rate. But overall levels remain very low in Sub-Saharan Africa, South Asia, and the Middle East and North Africa (UN Millennium Project 2005j).

Child mortality

Child mortality rates fell in every region except the CIS countries—even in places that did not achieve much economic growth or reduction in poverty. But child mortality remains extremely high at 174 per 1,000 live births in Sub-Saharan Africa and 93 in South Asia. In every developing region, child mortality is still many times higher than in the developed world. At current rates, many regions are unlikely to meet the target (map 2.2). Only North Africa, Southeast Asia, and Latin America and the Caribbean appear to be on track.

Maternal mortality

Maternal mortality remains shockingly high in every developing region of the world, reflecting the low priority for women's needs and inadequate access to emergency obstetric care (map 2.3). Maternal mortality ratios in East Asia,

Map 2.2
Child mortality rate, 2002
Under-five mortality rate (per 1,000 live births)
Source: World Bank 2004c.

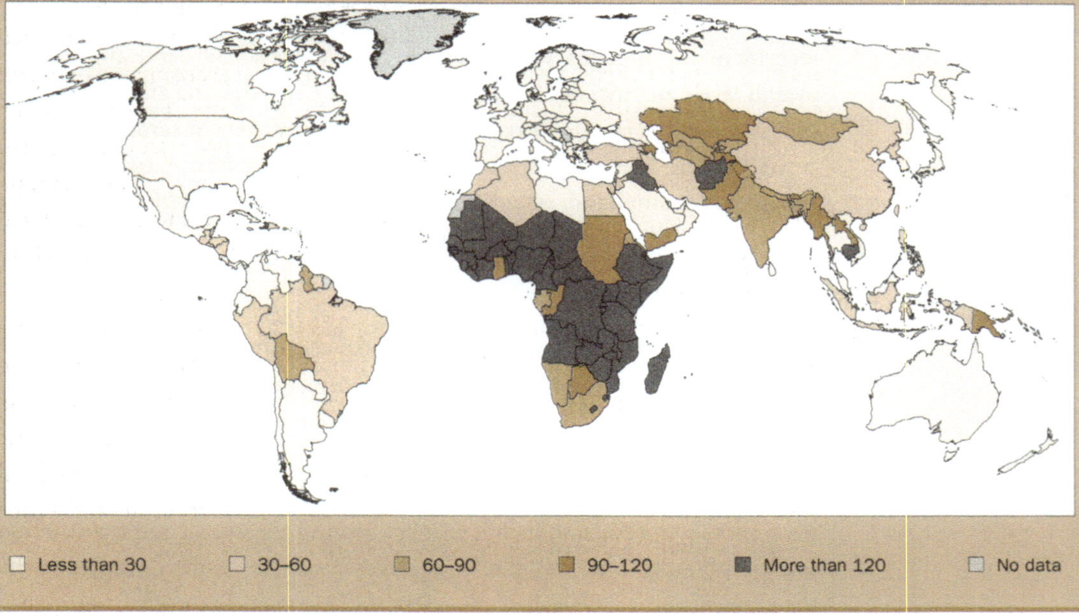

which has made significant progress on many of the Goals, are still approximately twice the ratios in the developed world. While data on maternal mortality are unreliable and do not permit time-series analysis, the best available evidence suggests that Sub-Saharan Africa, Southeast Asia, South Asia, and Oceania are unlikely to meet the targets on current trends.

HIV/AIDS, TB, malaria, and other infectious disease

HIV, now affecting about 40 million people, is pandemic in parts of Sub-Saharan Africa and poses a serious threat in other developing regions. The Caribbean has the second highest HIV prevalence rate, and India has the second highest number of HIV-infected people in the world after South Africa. Many countries are struggling to contain new infections and to treat people already infected. Infections are increasing in East Asia, South Asia, Oceania, and the CIS countries of Asia and Europe.

The incidence of TB remains extremely high around the world, increasing as an opportunistic infection associated with HIV/AIDS. It is increasing most precipitously in Sub-Saharan Africa, Oceania, and the CIS countries of Asia and Europe. Meanwhile malaria, an ecologically based parasite, kills well over a million people a year, the vast majority of them children. It remains a significant

Map 2.3
Maternal mortality ratio, 2000
Per 100,000 live births (adjusted)

Source: UNDP 2004b.

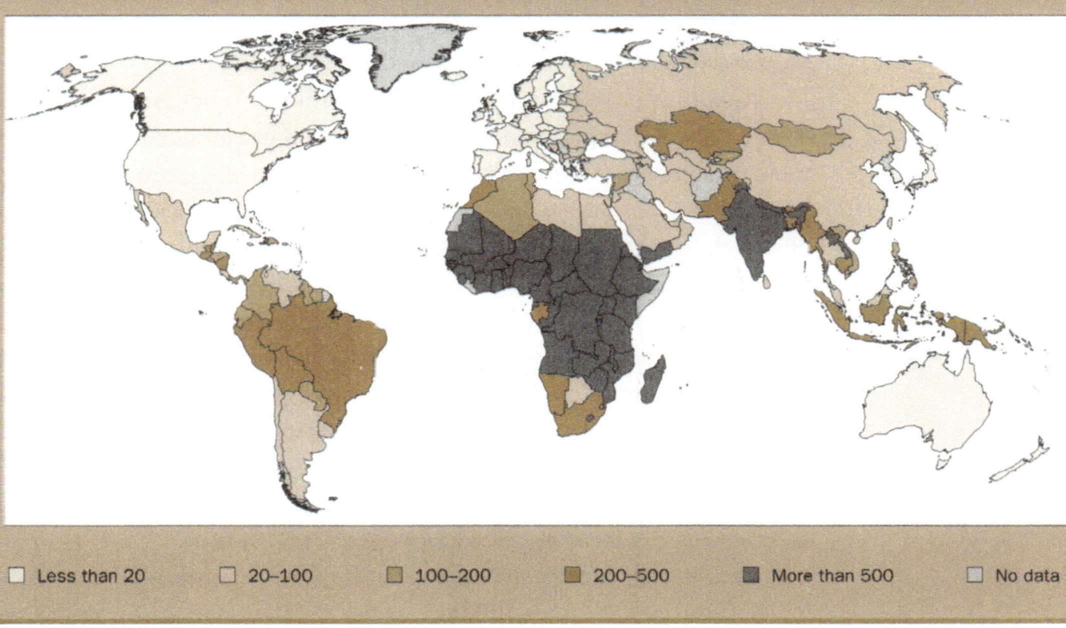

threat to human health and productivity in many tropical regions, especially in Sub-Saharan Africa, which accounts for 90 percent of all malaria deaths.

Environmental sustainability

Environmental degradation is eroding the natural resource base that many economies depend on. Few reliable indicators exist to measure environmental sustainability, and data coverage tends to be poor. As a result, environmental degradation does not show up in many official statistics. But the data on forest cover show that deforestation is increasing rapidly and threatening biodiversity in Sub-Saharan Africa, Southeast Asia, Oceania, and Latin America. Other natural resources have come under strain, with damage to marine and coastal ecosystems worsening rapidly in Latin America and the Caribbean, Southeast Asia, and Oceania. Per capita water availability is reaching critical thresholds in many parts of the world. Desertification and soil deterioration have become critical issues, especially in Asia and Africa.

Water supply and sanitation

Urban access to drinking water is relatively high in most regions, with the exception of Sub-Saharan Africa. But rural access to improved water supply remains limited in most regions, with Sub-Saharan Africa, the CIS countries, and Oceania especially off track. Access to sanitation improved in many regions but remained stagnant in West Asia and the CIS countries, and fell in Oceania. Coverage in Sub-Saharan Africa and South Asia remains extremely low, at 36 and 37 percent, respectively, contributing to widespread and preventable diarrheal disease. The problem is particularly severe in rural areas, where coverage rates are much lower than in urban areas.

Improving the lives of slum dwellers

The number of people living in slums and slum-like conditions in the world's cities is growing. Rapid rural-to-urban migration has produced massive slums in many developing country cities, where inhabitants lack secure tenure to their land and may not have access to basic water and sanitation services. Between 1990 and 2001 the slum population grew in every region except North Africa and the CIS countries of Europe. An estimated 900 million people live in slum-like conditions, more than 250 million of them in South Asia, where roughly 60 percent of the urban population lacks secure tenure. In Sub-Saharan Africa more than 70 percent of the urban population is estimated to live in slums (map 2.4). The problem is also severe in Latin America, where roughly a third of the urban population lives in slums.

Access to information and communication technologies

Information and communications technologies are critical inputs for economic development. Since 1990 access has been increasing in every region,

Map 2.4
Share of urban population living in slums
Percent

Source: UN-HABITAT 2003.

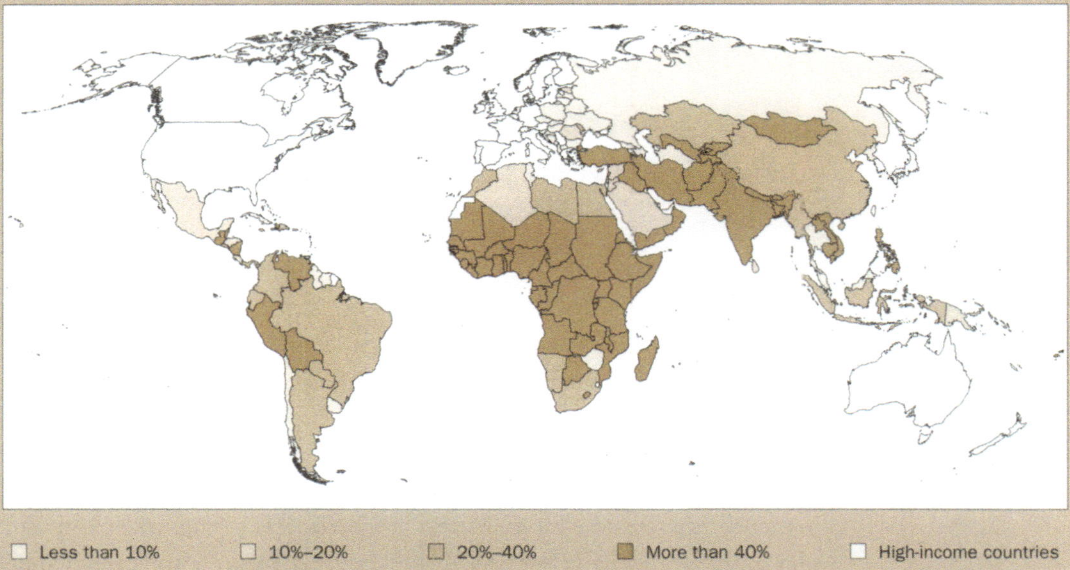

but it remains low in most. For example, the number of telephone lines and cellular subscriptions increased everywhere, most dramatically in East Asia, where it grew from 2.4 to 38 per 100 people, and Southeast Asia where it grew from 1.4 to 16 per 100 people. Access also increased significantly in Oceania, South Asia, and Sub-Saharan Africa over the period, but each of these regions still has low connectivity, at fewer than 10 subscriptions per 100 people.

Chapter 3

Why the world is falling short of the Goals

Long-term poverty reduction requires sustained economic growth, which in turn depends on technological advance and capital accumulation. The Millennium Development Goals play two roles in the growth process. First, the Goals are ends in themselves, in that reduced hunger, gender equality, improved health and education, and broader access to safe water and sanitation are direct goals of society. Second, the Goals are also "capital inputs" to economic growth and further development. A healthier worker is a more productive worker, as is a better educated worker. Improved water and sanitation infrastructure raises output per capita through various channels, such as reduced illness. So, many of the Goals are a part of capital accumulation, defined broadly, as well as desirable in their own right. In this chapter we outline the basic processes underlying economic development and progress toward the Goals, some major reasons why progress often falls short, and priorities for public action to address these shortfalls.

The links between capital accumulation, economic growth, and the Millennium Development Goals are captured in figure 3.1. The Goals for hunger and disease are part of the "human capital" box. The Goals for water and sanitation and slum dwellers are part of the "infrastructure" box. The Goal for technological innovation and diffusion are part of the "knowledge capital" box. And the Goal for income poverty is part of the "household income" box. Because meeting the Goals for hunger, education, gender equality, and health is vital for overall economic growth and development, it is a mistake to talk simply about the level of economic growth needed to achieve the Goals in a country. It is more helpful, particularly for the poorest countries caught in a poverty trap, to think about the kinds of investments that will achieve the many Goals and thus also support overall economic growth. Some important investments in human capital and infrastructure are not covered by the Goals

Figure 3.1
Capital accumulation, economic growth, and the MDGs

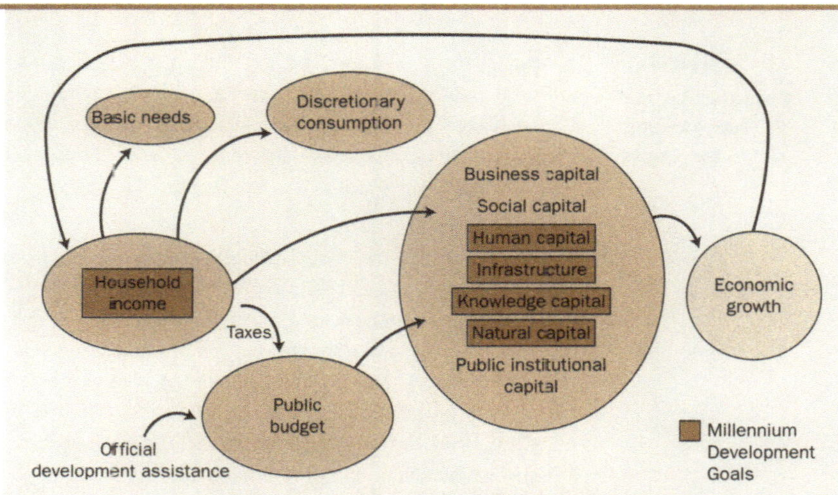

but are crucial for achieving the Goals and for spurring economic growth (box 3.1).

Various forms of capital contribute to the accumulation of other forms of capital. Human capital in the form of good health, for instance, also contributes to human capital in the form of education and skills. Water and sanitation infrastructure contributes directly to good health. Natural capital has similar feedback effects. Fish stocks, soil nutrients, and clean air all contribute to good health.

All the forms of capital are required to support long-term economic growth. Capital grows as a product of investment, with investment coming from private household savings or from public investments drawn from government revenue, savings from abroad, and other sources of income (foreign assistance, borrowing). When the process of capital accumulation breaks down, economic growth and poverty reduction break down.

Four reasons for shortfalls in achieving the Goals

There is no one-size-fits-all explanation for failure or success in achieving the Goals. Each region and each Goal requires a careful analysis. We can, however, identify four overarching reasons why the Goals are not being achieved. Sometimes the problem is poor governance, marked by corruption, poor economic policy choices, and denial of human rights. Sometimes the problem is a poverty trap, with local and national economies too poor to make the needed investments. Sometimes progress is made in one part of the country but not in others, so that sizable pockets of poverty persist. Even when overall governance is adequate, there are often areas of specific policy neglect that can have a monumental effect on their citizens' well-being. Sometimes these factors occur together, making individual problems all the more challenging to resolve.

Box 3.1
Essential inputs for reaching the Goals

Although the Millennium Development Goals were created to measure and provide targets for the most vital aspects of development, some areas important for development—and for achieving the Goals—are not included in the formal Goals framework. Energy services, sexual and reproductive health, and transport services are each vital to enabling and facilitating the achievement of the Goals.

Energy services

Improved energy services—including modern cooking fuels, access to electricity, and motive power—are necessary for meeting almost all the Goals. They can reduce child mortality rates and improve maternal health by lowering indoor air pollution. They can reduce the time and transport burden of women and young girls by reducing the need to collect biomass. And they can lessen the pressure on fragile ecosystems. Electricity is critical for providing basic social services, including health and education, and for powering machines that support income-generating opportunities, such as food processing, apparel production, and light manufacturing.

The UN Millennium Project proposes that countries adopt the following specific targets for energy services to help achieve the Goals by 2015:

- Reduce the number of people without effective access to modern cooking fuels by 50 percent and make improved cook-stoves widely available.
- Provide access to electricity for all schools, health facilities, and other key community facilities.
- Ensure access to motive power in each community.
- Provide access to electricity and modern energy services for all urban and peri-urban poor.

Sexual and reproductive health

Parts of comprehensive programs for sexual and reproductive health are included in the framework of the Goals (under Goals 4, 5, and 6). Yet sexual and reproductive health services are also essential for reducing extreme poverty and hunger, ensuring educational opportunities and gender equality, and attaining environmental sustainability (see box 5.5). These services affect the allocation of resources within the family, the prospects for household savings, the household choices about education and health investments, the exercise of the right to choose the number, timing, and spacing of one's children, and the capacities for women's social and economic participation and other practical life decisions.

At the macro-level these services affect population dynamics. A demographic transition to lower fertility and mortality (including that from HIV/AIDS) creates an opportunity to escape poverty traps and to accelerate economic and social development, a "demographic bonus" that can be realized through appropriate policies, governance, and investment. The UN Millennium Project calls for sexual and reproductive health issues to be included in national, regional, and international poverty reduction efforts.

Transport services

Transport services, such as road, rail, shipping, and air, are required to provide effective access to social services, such as emergency obstetric care, and to reduce the household transport burden and time poverty, especially of women and young girls. In addition to expanding transport infrastructure, countries need to invest in improving access to low-cost means of transport. Transport services also make many direct contributions to economic growth. They reduce the cost of agricultural inputs and raise producer prices for

Box 3.1
Essential inputs for reaching the Goals
(continued)

market produce. They facilitate the creation of export-based manufacturing and service industries, including tourism. And they increase market reach for the local private sector by lowering transport costs. Improved transport infrastructure is essential for promoting private sector development and trade, as argued by the World Bank's recent *World Development Report 2005* (World Bank 2004d and UNCTAD 2004).

Governance failures

Economic development stalls when governments do not uphold the rule of law, pursue sound economic policy, make appropriate public investments, manage a public administration, protect basic human rights, and support civil society organizations—including those representing poor people—in national decisionmaking.

The rule of law involves security in private property and tenure rights, safety from violence and physical abuse, honesty and transparency in government functions, and predictability of government behavior according to law. Too many countries fail to achieve these basic standards, sometimes due to authoritarian rulers who use violence and corruption to hold on to power—but often because upholding the rule of law requires institutions for government accountability, and those institutions are missing.

Political and social rights should ensure equality before the law and fairness in society across groups. These rights must be substantive and not merely formal. The poor must have a meaningful say in the decisions that affect their lives. Women and girls must be assured freedom from violence and from legal, economic, and social discrimination. In many places, access to public goods and services is restricted for certain groups. Minority groups, for their language, religion, or race, suffer discrimination at the hands of more powerful groups.

Sound economic policies involve a rational balance of responsibilities between the private sector and the public sector to secure sustained and widespread economic progress. The private sector is the engine of growth in production. The public sector establishes the framework and enabling environment for growth by setting sound macroeconomic policies and providing such public goods as infrastructure, healthcare and education, and support for science and technology.

Public investments are crucial for a "private-based" market economy. Every successful economy relies heavily on public spending in critical areas including health, education, infrastructure (electricity grid, roads, seaports), environmental management (national parks and protected reserves, water and sanitation), information and communications, scientific research, and land for affordable housing.

Accountable and efficient public administration requires transparency and administrators who are qualified, motivated, and adequately paid. It also requires efficient management systems, to disburse and track large investments, and monitoring and evaluation systems. Many poor countries without adequate resources for decent salaries—or the checks on political abuse that provide the incentives for performance and the ability to weed out the inept and corrupt—are unable to afford an effective public sector, so they end up suffering from large-scale inefficiencies and wasted resources.

Strong civil society engagement and participation are crucial to effective governance because they bring important actors to the fore, ensure the relevance of public investments, lead to decisions that best address the people's needs as they perceive them, and serve as watchdogs for the development and implementation of government policies.

Achieving the Goals requires that all these areas of governance be properly addressed. There is no excuse for any country, no matter how poor, to abuse its citizens, deny them the equal protection of the law, or leave them victims of corruption, mismanagement, and economic irrationality. Some improvements in governance do not cost much money, if any, and some actually save money (by cutting corruption or granting land tenure, for example). Some improvements in economic outcomes are thus available at low cost, and such opportunities must not be squandered. We describe the strategies for investing in governance in chapters 6 and 7.

Poverty traps

A second reason why many countries are not making progress on the Goals is that they are too poor to make progress and stuck in a poverty trap. To understand why countries get stuck in such a trap, it is useful to think of economic development as climbing a ladder of development. It is important that countries have strategies for moving up the ladder—that is, for achieving long-term growth. All countries face very specific challenges and thus need to tailor their national strategies to local conditions. But there are general principles of development for countries to follow as they move up the ladder.

The ladder of economic development. At the bottom of the ladder are the poorest countries, which for the most part have similar profiles.

- Most of the population lives in rural areas. Rural poverty is high, and the productivity of rural smallholder farmers is very low. The rural population is increasing rapidly, with some of the population moving to cities in search of jobs. Infrastructure is very poor, with shortages of roads, electricity, water, and sanitation. Women and girls bear much of the brunt of the poverty, with heavy labor in farming and in collecting fuelwood and water. Children are "economic assets" on the farm, and many of them, especially girls, do not attend school because they are home performing household work.
- Most of the urban population operates in the informal economy, without security of tenure and without formal employment. Cities are strongly

divided into pockets of affluence with good public services, solid infrastructure, and high-quality housing—and large squatter settlements with precarious property rights and a lack of public services. Roads, electricity, and ports tend to be congested and poorly maintained. Power failures are rampant. Foreign direct investment tends to be scarce and hard to attract. Employment is heavily informal, in services and small workshops, and in domestic food processing. Exports tend to be mostly primary commodities, subject to price volatility and long-term declines in prices.

- The population is afflicted by low human capital. Life expectancy is less than 50 years (as opposed to 80 years in high-income countries), and child mortality is 100 per 1,000 live births or higher. A significant proportion of children, especially girls, do not finish primary school. Fertility rates are high, particularly among poor people, and there is a considerable unmet demand for family planning and modern contraception. Infectious diseases are rife. Depending on climate, malaria may be year round or seasonal. TB afflicts densely populated slums. HIV/AIDS is uncontrolled among vulnerable groups (migrant laborers, truck drivers, commercial sex workers, injecting drug users) and has perhaps spread to more of the population.

In these circumstances, it is possible to envision what a successful development strategy would entail. First, it would target a rise in rural productivity, a Green Revolution to raise food output. This would accomplish several important objectives and trigger a structural change in the economy. It would enable farmers to feed their families. It would provide low-cost food for the rest of the economy. It would accelerate the transition to commercial agriculture and to urbanization (as fewer households are engaged in food production). The urbanization and movement of human resources into nonagricultural productive sectors would diversify the economy and the export base.

Almost every successful development experience has been based on a Green Revolution at an early stage. This Green Revolution could be made environmentally sustainable through thoughtful investments at the farm and village level, in soil health, water harvesting, improved seed varieties, feeder roads from farms to trunk roads, electrification, improved water sources, sanitation, and modern cooking fuels to replace fuelwood.

Second, and simultaneously, the strategy would help cities foster internationally competitive industries and services, while meeting the basic needs of all urban residents. Industrial parks, export processing zones, special economic zones, science parks, and the like would be developed as locations for internationally competitive urban industries, both in manufacturing and in services. Port services, electricity, transport services, and roads would be upgraded to support private industry. Slum dwellers would be given security of tenure, and perhaps negotiated options for relocation on a voluntary basis. Increased investments in solid waste disposal, clean air, and wastewater treatment would improve urban environmental health.

The strategy must aim to diversify the country's exports away from dependence on primary commodities toward manufactures and services. Countries with diversified exports have experienced superior growth, especially since dependence on primary commodity exports exposes the economy to volatility and long-term price decline of commodities. This transition toward diversified exports requires special attention for landlocked countries and inland economies, which face high transport costs, as well as for very small countries, which lack the scale to diversify into many sectors.

Third, these changes would be supported by massive investments in nutrition, healthcare, education, and family planning. Human capital would rise over time. The adult labor force would become literate and healthy. Infectious diseases would be brought under control through targeted disease control programs delivered through a strong health system.

Fourth, these investments in human capital and rural and urban productivity would be supported by three more overarching areas of investment. Public management systems would be upgraded, through training and retention of skilled managers and greatly expanded use of information technology. Extensive capacity building at the local level would permit effective decentralization of public investments, down to the city, town, and village. Scientific capacity would be expanded through investments in the major universities, national laboratories, and national science advisory units. And cross-border investments with neighboring countries would improve linkages in roads, electricity, environmental management, rails, and telecommunications.

History shows that investments in each of these areas can be scaled up very rapidly, in the course of a few years. Food production could double or even triple in Africa in a decade, if policymakers and donors invest in a Twenty-first Century African Green Revolution. Urban labor-intensive sectors such as garments can develop very rapidly, as Bangladesh has shown. Healthcare investments can lead to dramatic reductions in child mortality rates in just a few years. Fertility rates can fall sharply in a decade if there is a coordinated national effort to improve access to reproductive health services, including voluntary family planning. In short, a massive scaling up of both public and private investments is possible.

Why poverty traps happen. Many reasonably well governed countries are too poor to make the investments to climb the first steps of the ladder. They lack the fiscal resources to invest in infrastructure, social services, and even the public administration necessary to improve governance. Without roads, transport, soil nutrients, electricity, safe cooking fuels, clinics, and schools, the populations are chronically hungry, disease-burdened, and unable to save. Without adequate public sector salaries and information technologies, public management is chronically weak.

These countries cannot attract private investment flows or retain their skilled workers. And dozens of heavily indebted poor and middle-income countries are forced by creditor governments to spend large parts of their limited tax receipts on debt service, undermining their ability to finance investments in human capital and infrastructure. In a pointless and debilitating churning of resources, the creditors provide development assistance with one hand and then withdraw it in debt servicing with the other.

Under these severe resource constraints, countries are facing a crushing array of problems:

- *Low saving rates.* Poor households use all their income to stay alive, and so cannot save for the future. With low domestic saving there are limited possibilities for indigenous private investment. The few who can afford to save often have no access to formal banking.
- *Low tax revenues.* Governments lack the budgetary resources for public investments and public administrations using qualified managers and modern information systems.
- *Low foreign investment.* Foreign investors stay away from economies without basic infrastructure—those with costly and unreliable roads, ports, communication systems, and electricity.
- *Violent conflict.* Resource scarcity can often fuel latent tensions among competing groups.
- *Brain drain.* Skilled workers leave the country because of low salaries and little hope for the future.
- *Unwanted or ill-timed births and rapid population growth.* Impoverished people living in rural areas have the highest fertility rates and the largest families. Rapid population growth and shrinking farm sizes make rural poverty worse. Poor people (in rural and urban areas) have less access to information and services to space or limit their pregnancies in accord with their preferences.
- *Environmental degradation.* People in poverty lack the means to invest in the environment and the political power to limit damage to local resources, resulting in soil nutrient depletion, deforestation, overfishing, and other environmental damage. These degraded conditions undermine rural incomes and contribute to poor health and rural-urban migration, leading to new settlement in environmentally fragile periurban areas.
- *Low innovation.* Poor countries with limited education systems cannot afford to invest in science and technology, hindering their chances of reaching the sustained growth enjoyed by knowledge-based and innovation-based economies.

One of the many problems of being extremely poor is that almost all of a country's income must be devoted to current income rather than saving. Each household has to spend its income on food, clothing, shelter, and other basic needs, with little or nothing left over to save for the future. When income is very low, so too is the saving rate. With a low saving rate, the amount of capital per person declines, and this leads to economic decline and even more poverty (figure 3.2). The finding that saving rates are low in impoverished countries and rise with per capita income is well established (Sachs and others 2004). The situation with low savings is even worse than it looks, however, because the national income accounts data almost surely, and substantially, overestimate the true saving rate of the poorest countries. To a significant extent, these countries are living off their natural capital but counting resource depletion as income.

Many countries have been cutting down their rainforests to make room for new farmland and to provide fuelwood and timber. Farmers have been depleting the soils of nutrients by growing crops without fertilizers. But the deforestation and loss of soil nutrients are not counted as a loss of capital. Countries depleting their mineral reserves are counted as earning income rather than converting one form of capital, mineral deposits, to another, financial assets. Sooner or later these forms of resource depletion will have to come to an end. The productive capital stock in these countries is falling even faster than suspected once we take into account, even imperfectly, the decline of several forms of natural capital.

There is no readily accepted methodology for correcting measured savings rates for resource depletion effects. Figure 3.3 uses the World Bank's valiant but very preliminary attempt, and adds a calculation for soil nutrient depletion. The figure shows measured rates of national saving, augmented by spending on education (which is counted as consumption in national accounts but which should count as investment in human capital) and reduced according to estimates of the economic costs of deforestation, energy depletion, mineral depletion, and soil nutrient depletion. This corrected saving rate shows that measured saving rates may be seriously overestimating true saving rates in the economy. Those rates might be as low as 1 percent for tropical Sub-Saharan Africa.

We can now see the essence of the poverty trap. The poorest countries save too little to achieve economic growth, and aid is too low to compensate for the low domestic saving rates. Detailed data on actual saving, investment, aid, and growth rates differ greatly by region and by income level (table 3.1).

Figure 3.2
The classic poverty trap

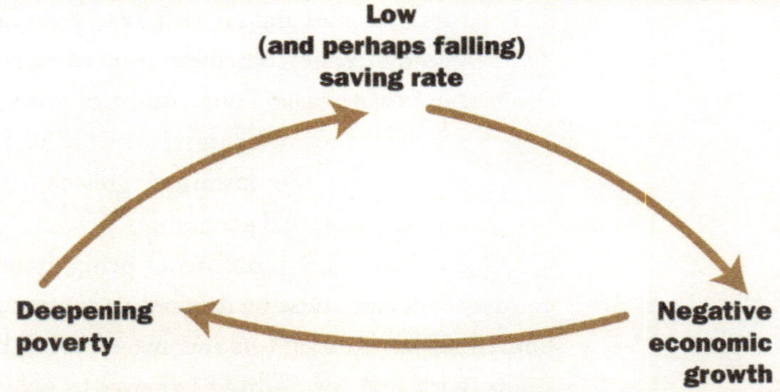

Figure 3.3

Saving rates, by developing region

Share of GNI (%)

Note: Adjusted saving is equal to gross national saving plus education expenditure and minus energy depletion, mineral depletion, and net forest depletion. "Tropical Sub-Saharan Africa" refers to a 33-country sample defined in Sachs and others (2004).

a. We use nutrient depletion indicators and fertilizer prices to calculate tropical Sub-Saharan Africa's soil depletion to be around 2% of GDP, which would reduce adjusted saving to 1.5%.

Source: World Bank 2003a. Soil nutrient depletion for 1999 Sub-Saharan Africa from Henao and Baanante (1999) and Stoorvogel, Smaling, and Janssen (1993). GDP data from World Bank (2003a); prices from African Agricultural Market Information Network (2004).

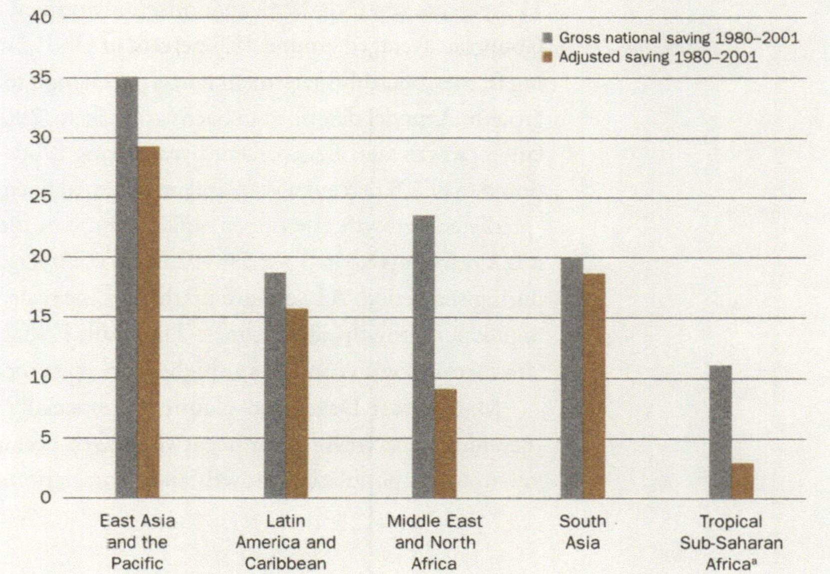

Table 3.1

Economic growth predicted from domestic saving, population growth, and capital depreciation

By developing region and by income level, 1980–2000 (%)

Note: Does not include high-income countries, countries of the former Soviet Union, or countries with populations below 1 million in 1980. All categories are annual averages across countries and years, weighted by population.

a. The measured consumption of fixed capital divided by the assumed capital-output ratio of 3.

b. Calculated by: (domestic saving / 3) – (growth in population + rate of depreciation).

c. Where 1980 data are not available, nearest available year is used to calculate the average growth rate.

Source: World Bank 2004c.

	Gross domestic saving as share of gross national income	Growth in population	Estimated rate of depreciation[a]	Annual growth in output per capita	
				Predicted[b]	Actual[c]
Central Asia	15.5	1.2	4.4	–0.4	–3.2
East Asia and Pacific	34.0	1.4	3.0	6.9	6.7
Eastern Europe	24.5	0.4	3.7	4.0	–1.2
Latin America and the Caribbean	20.1	1.8	3.3	1.6	0.4
Middle East and North Africa	19.2	2.6	3.1	0.7	1.0
South Asia	17.8	2.0	2.9	1.1	3.3
Sub-Saharan Africa	10.9	2.7	3.1	–2.2	–0.7
Least Developed Countries	6.7	2.5	2.8	–3.1	0.3
Non-LDC low-income countries	20.2	2.1	2.8	1.9	2.9
Lower-middle-income countries	31.4	1.5	3.2	5.8	5.5
Upper-middle-income countries	23.7	2.0	3.5	2.4	0.7

The Least Developed Countries show the lowest saving rate, just 6.7 percent of GNP. This very low level would result in a sharply negative growth rate of per capita income if not offset in part by official development assistance equal to around

11.2 percent of GNP. With that offset, investment rates in the Least Developed Countries averaged around 16.5 percent of GNP. From a simple growth-accounting framework, this investment rate is not enough to achieve significant economic growth. A model described in Sachs and others (2004) suggests that with a population growth rate of 2.5 percent a year during 1980–2000, an estimated depreciation rate of 2.8 percent a year, and an assumed fixed capital-output ratio of 3, the "predicted" growth rate of per capita income in the Least Developed Countries is 0.2 percent (16.5%/3 – 2.5% – 2.8% = 0.2%), exactly the average growth rate during the period. As we move up the income scale, the saving rate rises, and the population growth rate declines. The result is that predicted economic growth also increases with countries at higher per capita incomes.

Many Least Developed Countries, especially in Sub-Saharan Africa, are therefore stuck with low or negative growth because their saving rates are too low to offset population growth and depreciation (box 3.2). The result of low

Box 3.2

The poverty-demography trap

The link between extreme poverty and high fertility is strong for several interconnected reasons:

- Infant mortality rates are high when there are inadequate health services, so high fertility provides "insurance" for a surviving child.
- Children are often perceived as economic assets who provide supplementary labor for the household, especially in rural areas.
- Poor and illiterate women have few job opportunities away from the farm, and so place a low value on the opportunity (time) costs of raising children.
- Poor families in poor communities are less likely to be aware of changes in mortality and in employment opportunities for the educated and thus miss signals of the benefits of investing in child quality rather than quantity.
- Women are frequently unaware of their reproductive rights (including the right to plan their families) and lack access to reproductive health information, services, and facilities, leading to high rates of unmet demand for contraception in low-income countries and among poorer members of all developing countries.
- Poor households lack the income to purchase contraceptives and family planning services.
- Governments lack the resources to provide extensive access to reproductive health services and counseling.

Because of these multiple channels linking poverty and high fertility, an effective voluntary approach to reducing fertility rates should focus on several policy fronts:

- Investing in child survival to give parents the confidence to have fewer children.
- Investing in rural infrastructure (water, cooking fuels, roads) so that children can go to school rather than spend their days collecting water and fuelwood.
- Empowering women with skills, literacy, numeracy, and economic rights to engage in off-farm employment.
- Empowering women to gain access to family planning services and modern forms of contraception, including eliminating information gaps and provider biases (based on wealth, ethnicity, or age) that restrict use.
- Making contraceptive services available free to low-income households.
- Investing in comprehensive sexual and reproductive health services as part of scaling up public health facilities and services.

saving rates—unless offset by foreign assistance or foreign investments—is a chronic decline in capital per person and a consequent chronic decline in income per capita.

Breaking out of the poverty trap. The key to overcoming the poverty trap is to raise the economy's capital stock—in infrastructure, human capital, and public administration—to the point where the downward spiral ends and self-sustaining economic growth takes over. This requires a "big push" of basic investments between now and 2015 in key infrastructure (roads, electricity, ports, water and sanitation, accessible land for affordable housing, environmental management), human capital (nutrition, disease control, education), and public administration. This process is helped by a voluntary reduction in fertility, which promotes greater investments in the health, nutrition, and education of each child. We thus strongly support programs that promote sexual and reproductive health and rights, including voluntary family planning.

Critical to overall success in economic growth and poverty reduction, these investments help countries meet the Goals, freeing them from the poverty trap and their dependence on aid. Some countries, such as India, can probably graduate from foreign assistance by 2015. In more difficult circumstances, such as Ethiopia, graduating from aid will take longer (chapter 17).

In an important recent policy initiative, the U.S. government established a set of transparent indicators to identify 17 poor but reasonably well governed countries that qualify for funding from its new Millennium Challenge Account. The list of countries includes Bolivia, Ghana, Mali, and Mozambique. Despite significant efforts and real progress, these countries, and many like them, pass the governance test but still fail to make adequate progress toward the Goals. The reasons are clear. They lack the basic infrastructure, human capital, and public administration—the foundations for economic development.

The Goals create a solid framework for identifying investments that need to be made. They point to practical targets of public investment—water, sanitation, slum upgrading, education, health, environmental management, and basic infrastructure—that reduce income poverty and gender inequalities, improve human capital, and protect the environment. By achieving the Goals, poor countries will establish an adequate base of infrastructure and human capital that will enable them to escape from the poverty trap.

Geographical conditions make poverty traps more likely. Some countries and regions are more vulnerable than others to falling into a poverty trap. While a history of violence or colonial rule or poor governance can leave any country bereft of basic infrastructure and human capital, physical geography plays special havoc with certain regions (box 3.3). Some regions need more basic infrastructure than others simply to compensate for a difficult physical environment. Here are some of the barriers that must be offset by investments:

Box 3.3

Geographical obstacles and economic growth

Source: Calculated from World Bank 2004c; PRS Group 2003; CIESIN 2002; Kiszewski and others 2004.

Difficult geography can make it more likely for countries to fall into a poverty trap—requiring increased investments in infrastructure to reduce transport costs or curb tropical diseases. We construct indexes to estimate each country's risks from adverse agronomic conditions, adverse transport conditions due to location, and malaria ecology (see table 3.2). Using these indexes, the regressions reported in the table demonstrate the importance of geographical factors in economic growth.

Risk index regressions

Independent variable	I Growth 1980–2000 ($n = 76$, $r^2 = 0.56$)	II Growth 1980–2000 ($n = 56$, $r^2 = 0.52$)
Income (log of per capita GDP in 1980 PPP US$)	–1.27* (–5.01)	–1.26* (–3.46)
Governance, 1982 (International Country Risk Guide)	–3.18* (–3.57)	–3.61* (–2.96)
Agriculture risk (irrigation, subhumid, fertilizer)	–3.57* (–3.09)	–3.96* (–2.56)
Transport risk (coastal, low density, elevation, roads 1990)	–3.93* (–3.68)	–4.42* (–3.35)
Malaria ecology	–2.19* (–2.16)	–2.02** (–1.73)

* Significant at the 95 percent level.

** Significant at the 90 percent level.

Note: Regression II excludes high-income countries. Dependent variable is real average annual per capita GDP growth, 1980–2000. All variables except per capita GDP are a 0–1 index, where higher values indicate higher risk. Regressions include a constant that is not reported. Former Soviet countries and countries with populations below 1 million in 1980 are excluded.

Each index is constructed on a scale of 0 to 1, with 1 the maximum risk. To study the effects on economic growth from 1980 to 2000, the underlying indicators for each index use values for 1980. The regressions control for the initial level of income (since conventional economic theory expects that, all else equal, poor countries grow faster) and for the quality of governance in each country, using a similar 0–1 index (where a higher value indicates poor governance) constructed from a commonly used International Country Risk Guide governance rating for the beginning of the period.

The regression results show that the three geographical risks (agriculture, transport, and malaria) are significantly linked to lower economic growth per capita, even when controlling for initial income levels and governance quality (both of which are also significant). Indeed, the results show that countries with highest risk in either the agriculture index or the transport index grew more slowly (by almost four percentage points a year!) than the countries with lowest risk, even when controlling for governance quality. Similarly, countries with ecologies most conducive to endemic malaria tended to grow two percentage points more slowly than countries with ecologies that do not support malaria.

These results illustrate the important relationship between economic growth and geographical factors affecting agriculture, transport costs, and disease ecology. Countries suffering from these geographic vulnerabilities will need targeted investments in infrastructure, agriculture, and health to minimize the risks of falling into a poverty trap or to start the climb out of it.

Adverse transport conditions:

- Landlocked economies.
- Small island economies far from major markets.
- Inland populations far from coasts and navigable rivers.
- Populations living in mountains.
- Long distances from major world markets.
- Very low population densities.

Adverse agroclimatic conditions:

- Low and highly variable rainfall.
- Lack of suitable conditions for irrigation.
- Nutrient-poor and nutrient-depleted soils.
- Vulnerability to pests and other postharvest losses.
- Susceptibility to the effects of climate change.

Adverse health conditions:

- High ecological vulnerability to malaria and other tropical diseases.
- High AIDS prevalence.

Other adverse conditions:

- Lack of domestic energy resources (fossil fuels, geothermal or hydropower potential).
- Small internal market and lack of regional integration.
- Vulnerability to natural hazards (tropical storms, earthquakes, volcanoes).
- Artificial borders that cut across cultural and ethnic groups.
- Proximity to countries in conflict (box 3.4).

Sub-Saharan Africa is especially burdened by poor geographical endowments (table 3.2 and map 3.1). Africa has the highest agriculture risk (tied with South Asia), the highest transport risk, and by far the highest malaria risk. Africa is also uniquely vulnerable to drought conditions. High vulnerability as of 1980 was inversely correlated with economic growth during 1980–2000 (see box 3.3). A recent major statistical analysis of economic growth corroborates the importance of geographical variables.[1]

Africa's vulnerability is very high but not insurmountable. Indeed, our message is that geographical vulnerabilities can and need to be offset by targeted investments in infrastructure, agriculture, and health. Countries far from markets can be brought closer by adequate investments in roads and railways. Countries with nutrient-depleted soils and inadequate rainfall can be helped by special programs for soil nutrient replenishment and water control (such as irrigation and water harvesting). Countries suffering from malaria and other endemic diseases can battle these diseases with appropriate programs of disease prevention and control. Yet such investments are costly—too costly for the poorest countries to bear on their own—and so require much greater help from the donor countries.[2]

Box 3.4

The poverty-conflict nexus

Source: Humphreys and Varshney 2004; Fearon and Laitin 2003; Miguel, Satyanath, and Sergenti 2004; Homer-Dixon 1994; Reno 1995.

In the 1990s up to a quarter of all countries were affected by conflicts and more than a billion people were living in conflict-afflicted countries. Importantly, the statistical relationships between poverty and violent conflict are very strong. They have been found to be robust to variation over time and to variation in the definition of what constitutes a civil war. A country with a civil war within its borders typically has only one-third the per capita income of a country with similar characteristics but at peace. Poor countries are also more likely to experience new conflicts. A country with a per capita income of $500 is about twice as likely to have a major conflict within five years as a country with an income of about $4,000 per capita (see figure).

Similar trends hold for a broader class of internal conflicts defined as "internal intermediate armed conflicts" or "internal wars." The poorest countries have the highest risks of new conflict and there is a systematic decline in risk as incomes grow.

Independent of income levels, low economic growth rates are associated with higher risks of new conflict. An important recent study by Miguel, Satyanath, and Sergenti (2004) also shows that a negative growth shock of five percentage points increases the risk of civil war by 50 percent in the following year—and that economic conditions are likely the most important determinants of civil conflict in Sub-Saharan Africa.

While there is a broad consensus among researchers on the strong bidirectional links between poverty and conflict, there is disagreement about why the relationship holds. Perhaps most obviously, the relationship arises because violent conflict destroys wealth. It results in the destruction of physical and human capital, massive dissavings, and interruptions of economic activity. But research indicates clearly that the relationship is due not simply to conflicts resulting in income losses—although this certainly occurs—but also to the fact that poverty makes countries more vulnerable to conflict. In aiming to identify the specific channels of mutual causality, the following reasons figure prominently:

- *Poor state capacity.* Poor countries are more likely to have weak states, so they are vulnerable to forcible takeover and attack by armed groups. They are also less able to resolve local disputes peaceably and more vulnerable to manipulation and control by third parties. This reduces their freedom to react to threats of conflict and makes them more prone to the spread of predatory forms of financing and asset stripping, leading in turn to greater levels of frustration toward the government in power.
- *Scarcity and inequality.* While poverty affects state capacity, it also affects the incentives of citizens to engage in violence. Scarcity, including that of environmental resources, can lead to migrations that result in conflicts between identity groups over resources. Without productive alternatives, youths, especially, may turn to violence out of frustration or for material gain. Such behavior is especially likely to occur not just when countries are poor but when there is also inequality between segments of society—when pockets of poverty persist within national economies.
- *Demography and social structures.* Poorer countries are more likely to have demographic regimens marked by high fertility and high mortality, resulting in high child-to-adult ratios. Such demographic profiles are also associated with greater conflict risks. Indeed since 1945 almost every instance of massive one-sided violence (genocide or politicide) has occurred in countries with more than a two-to-one child-adult ratio.

Other risk factors include a highly unequal distribution of wealth—especially when this wealth is unevenly distributed across political groups, such as regional, ethnic, or religious groups—dependence on high-value natural resource exports (particularly

Box 3.4
The poverty-conflict nexus
(continued)

diamonds, drugs, and oil), sudden and sharp political or economic transitions, weak and instable political regimes and institutions, and political tensions drawn from historical ethnic tensions and identity clashes. In many instances conflicts have also been initiated or exacerbated through external involvement, including governments and corporations.

Rising national incomes reduce the risk of civil war

Predicted probability of observing a new conflict within five years (%)

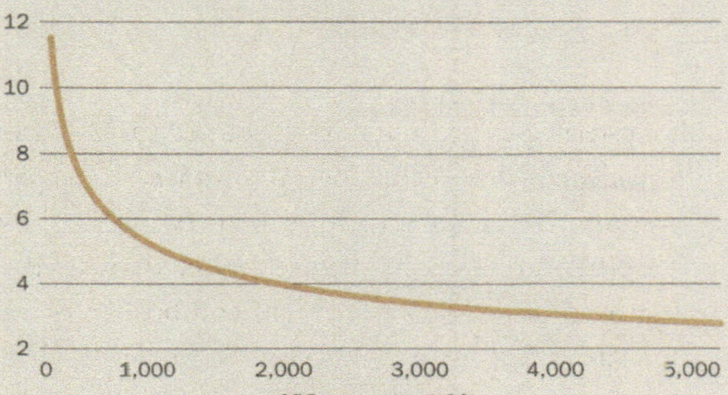

Note: Estimated probabilities are derived from the relationship between GDP per capita (constant 1985 US$) and civil war onset. The figure denotes only average relationships identified across countries and over time and does not imply that for any income levels conflict risks are the same in all places.

Pockets of poverty

The next step up the ladder of economic development occurs when countries have made the transition from subsistence agriculture to commercial agriculture and from commodity exports to urban-based exports, with a large proportion of the population living in urban areas. Most of Southeast Asia achieved that level of development a generation ago. Yet most economies have considerable variation in household incomes, so even middle-income countries may have large numbers of extremely poor households, especially large countries with sophisticated market structures with considerable regional and ethnic diversity. Economic development often leaves some parts of an economy, or some groups in society, far behind. This occurs both in lagging regions and in cities, where a growing proportion of the poor live in slums. In many countries there are cities within cities—a dual reality of haves and have-nots in close proximity. In many cases, geographical disadvantages (distance from markets) are worsened by the political disempowerment of minority groups.

The major policy implication for middle-income countries is to ensure that critical investments—in infrastructure, human capital, and public administration—get channeled to lagging regions, including slums, and to social groups excluded from the political process and economic benefits. Some notable lagging regions include:

- Western China, burdened by great distance from the eastern coast.
- Southern Mexico, burdened by tropical diseases, agronomic risks, great distances from the U.S. market, and political marginalization of the indigenous peasant populations.

Table 3.2
Agriculture risk, transport risk, and malaria risk, by region

Region	Agriculture risk[a]	Transport risk[b]	Malaria risk[c]	Human vulnerability index[d]
Central Asia	0.31	0.41	0.00	**0.24**
East Asia and the Pacific	0.68	0.27	0.04	**0.33**
Europe	0.38	0.27	0.00	**0.22**
Latin America and the Caribbean	0.76	0.36	0.03	**0.39**
Middle East and North Africa	0.71	0.36	0.02	**0.36**
North America	0.51	0.23	0.00	**0.25**
South Asia	0.86	0.26	0.02	**0.38**
Sub-Saharan Africa	0.86	0.52	0.42	**0.60**

Note: Indexes range from 0 to 1, with a higher value indicating higher risk. Country averages are weighted by population.
a. Averages indexes for share of cropland under irrigation in 1980, fertilizer use per capita in 1980, and share of population living in subhumid ecological zones.
b. Averages indexes for share of population living near the coast, share of population living in low-density areas, share of population living above 800 meters elevation, and paved roads per capita in 1990 (earliest available data).
c. A 0–1 index for malaria ecology.
d. Averages agriculture risk, transport risk, and malaria risk.

Source: Calculated from World Bank 2004c; CIESIN 2002; Kiszewski and others 2004.

Map 3.1
Human vulnerability index, 1980
1 = highest risk

Source: Calculated from World Bank 2004c, CIESIN 2002, and Kiszewski and others 2004.

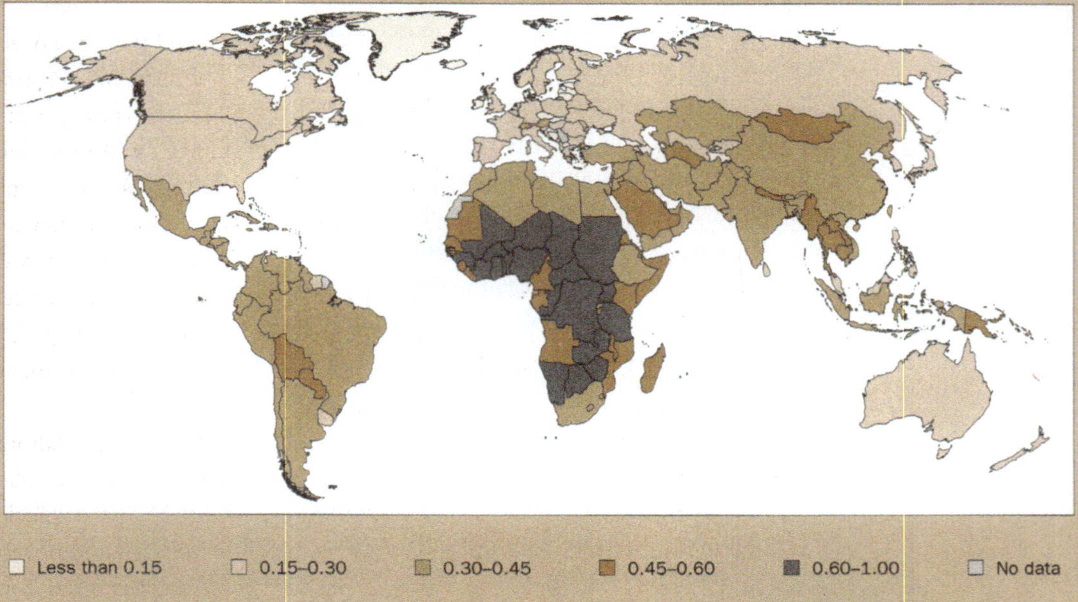

- Northeastern Brazil, burdened by vulnerability to drought and a long history of heavily concentrated land ownership.
- The Gangetic states in India, burdened by low-productivity agriculture, long distances to coastal trade, and a large landless population.

Concurrently, to continue climbing the ladder of economic development, middle-income countries need to focus critically on building an improved business environment that will deepen integration with the world economy in a widening range of manufactures and services. They also need a growing capacity in science and technology, to make the transition from a pure technology importer to a country that can innovate and commercialize technology on its own.

The Republic of Korea and Taiwan (China) are key examples of economies that made the transition from technology importer to technology innovator and exporter in the past couple of decades. The investment strategies of middle-income countries need to focus on sophisticated infrastructure (such as state-of-the-art container ports and intermodal transport systems) and on innovation systems comprising national laboratories, research universities, and public-private R&D partnerships.

Areas of specific policy neglect

A fourth reason why some Goals are not being met is simply that policymakers are unaware of the challenges, unaware of what to do, or neglectful of core public issues. Environmental policy is often grossly neglected because of politically weak environment ministries, even weaker law enforcement, and considerable deficiencies in information and in the capacity to act on that information. Few governments currently have the capacity to assess the deep links between ecosystem services (hydrology, biodiversity, natural hazard reduction) and poverty reduction. The environment is much too often taken as given, or taken for granted, or regarded as a resource to be exploited in the short term. This approach is now failing around the world, as population densities increase and human impacts on the environment increase markedly. Environmental sustainability must become a centerpiece of public investments.

Also common are gender biases in public investment and social and economic policies, maternal health, and sexual and reproductive health. Adolescents are also widely underserved for life skills, nutrition information, education and employment opportunities, and sexual and reproductive health information and services. Throughout the developing world and even in middle-income countries, maternal mortality ratios remain appallingly high. High maternal mortality and morbidity have a specific major remedy: access to emergency obstetric care. Despite its life-saving potential, there has been a pervasive underinvestment in this service and in the health systems to deliver it. For a long time, high maternal mortality was addressed through skilled birth attendants, an approach that has not proved sufficient. Yet investments in the capacity of local hospitals to perform Ceasarean sections, for example, can have huge benefits in saving the lives of women in prolonged labor.

More generally, policymakers and civil society must take the opportunity to identify the key areas of public policy and public investment that have been left behind in countries falling short of the Goals. All regions and policy areas will need to be part of any national strategy to achieve the Goals.

Private and public investments to meet the MDGs

The public and the private sectors both have a role in almost every form of investment needed for the Goals. In some areas the private sector is predominant—as for business growth, generating employment, raising incomes, and raising productivity. In others the government is predominant—as for governance and a regulatory framework to foster the private sector. In still others there is a mix of responsibilities—as for human capital, infrastructure, science and technology, and environmental sustainability. Public and private investments, when well designed, tend to be complementary, not rivals or substitutes. It is therefore a huge mistake to be dogmatic about public versus private investments. Both are needed.

The limits to private investment

A common assessment for countries stuck in extreme poverty is that they simply need more private investment (including foreign capital inflows) to stimulate market growth. Too simplistic, this view mischaracterizes the challenges of promoting private investment in low-income countries. Private investment in general, and foreign investment in particular, require that certain threshold conditions be met. When infrastructure and human capital are inadequate, potential investors will stay away completely. One of the roles of the public sector is to ensure that infrastructure is adequate to push the economy across the threshold, so that private investors can earn at least the minimum return they need to invest.

The key variables of interest to private investors include:

- Adequate infrastructure (roads, ports, electricity, water).
- Physical safety, including peace and security.
- Reliable sites for operations, such as industrial zones close to seaports and airports.
- The rule of law for predictable contracts and relations between business and government.
- A healthy and skilled labor force.
- A safe and accommodating location for expatriate managers and their families.
- Favorable tax treatment.

The key reason that foreign direct investment (FDI) can rarely "lead" the development of impoverished countries is that these preconditions are not met. A chicken-and-egg problem arises: growth depends on private investment, and especially on FDI (since it brings technology along with capital), but private investment depends on a sound infrastructure and adequate human capital, both of which depend on an adequate level of national saving. Here is where aid-financed infrastructure development and human capital investments can make a vast difference in promoting investment, particularly FDI.

It is no accident that a mere 12 countries account for nearly 85 percent of non-oil FDI in the developing world (table 3.3). These are countries with especially favorable coastal locations, large domestic markets or proximity to large markets, and reasonably salubrious climatic and agronomic conditions. On the other side, the Sub-Saharan countries receive only 4.3 percent of the world's FDI flows and the Least Developed Countries only 2.5 percent.

In addition to the business environment, we believe that favorable tax treatment is important to attract the "early mover" investors to a newly emerging market. There has been a long debate about the role of taxes in attracting FDI, a debate that in our view has been widely misunderstood. It is true that tax breaks, by themselves, will not be decisive. Without adequate infrastructure, property rights, human capital, and physical safety, a tax holiday will not make much difference. Yet this does not mean that the tax system is unimportant or that tax holidays and other promotional instruments can be ignored. Once the other preconditions are met, tax rates become a pivotal concern of major foreign investors. This is why the IMF–World Bank advice over the years to avoid tax concessions and other promotional tools has been in error.

A recent study powerfully makes the point that it is precisely in the low-income countries competing for export-oriented FDI (as for apparel, electronics

Table 3.3
Concentration of non-oil FDI among developing countries, 2002

	FDI received as share of FDI to non-oil producing developing countries (%)
China	36.8
Brazil	12.4
Mexico	10.9
Czech Republic	7.0
Poland	3.1
Slovak Republic	3.0
Malaysia	2.4
India	2.3
Peru	1.8
Colombia	1.5
Chile	1.3
Viet Nam	1.0
Total	**83.5**
For comparison	
All Least Developed Countries	2.5
All Sub-Saharan Africa	4.3

Note: Oil-producing countries are counted as those with yearly crude oil production of more than $100 per capita and where manufactured exports account for less than 50 percent of total exports. Excludes countries receiving less than 1 percent of global FDI.

Source: World Bank 2004c and U.S. Department of Energy 2002.

assembly, and other highly mobile international sectors) where taxes have a huge effect:

> The results confirm and strengthen conclusions drawn from earlier research, that taxes do influence the real operations of MNCs [multinational companies], not simply where they attempt to shift profits or financial assets. The particular tax elasticities reported here suggest that in the circumstances where the greatest responsiveness occurs (non-OECD countries with open trade policies and firms that produce for export markets), the elasticity can be as high as 3. In such circumstances, FDI-promotion programs based on tax holidays, government provision of key infrastructure such as land, designation of convenient industrial parks, and the like, can make a very big difference (Mutti 2003, p. 68).

Natural resource–based FDI. Resource-based FDI, such as that for oil and gas deposits, happens under very different circumstances. The ability to attract FDI depends much less on good governance and taxation than on the resource. Indeed, oil companies have proved repeatedly that they will operate in the worst conditions, even war zones, and pay bribes or other suspect payments, to gain concessions over resource deposits. Perhaps it is for this reason among others that resource-based FDI has proved to be a double-edged sword. It can provide critical revenues to the state for undertaking broad-based investments in infrastructure, human capital, natural capital, and public capital. Too often, however, it has resulted in plunder and corruption, rather than development, a phenomenon known as the "natural resource curse."

Turning natural resource–based FDI into economic development is thus a major challenge in selected parts of the low-income world, notably in West Africa today, with its large reserves of offshore oil and gas. Such investments should be guided by the following considerations:

- Transparency in bidding and concessions.
- Transparency in use of the resulting revenues.
- Use of a sufficient part of the oil and gas for the development of the domestic energy infrastructure.
- An overall public investment plan to explain how the natural resource rents will be invested on behalf of broad-based development.

Remittances. When economies are not highly productive, workers naturally try to migrate to other countries, especially to rich ones. Impoverished countries thus benefit from remittances sent home by migrant workers. These remittances can increase household income and saving, especially if they can be channeled through a formal financial system, as is now beginning to happen in some countries in Latin America and Asia. But global migration from the poorest countries to wealthier countries remains limited, so most countries in poverty traps receive only limited benefits from remittances.

Remittances are also a double-edged sword. Much international migration is by high-skilled workers, such as doctors and scientists, who have more opportunities to cross borders, since the rich countries have made it easier for them to obtain visas and work. The result is a brain drain—and sharply rising costs of providing skilled services in the low-income countries. In other words, migration often depletes a country's knowledge capital and public capital. Doctors in parts of Africa, for example, command salaries of $25,000 or above, often nearly 100 times the average per capita income, because they have alternative work opportunities in the high-income economies.

This outmigration of skilled workers such as doctors and nurses requires two kinds of responses:

- Higher public sector salaries in low-income countries, with donor support, to compete with world markets.
- New kinds of skilled workers, such as community health workers with one or two years of (likely postsecondary) training, who can provide some services without being subject to the same intensity of brain drain (see Joint Learning Initiative 2004).

The need for public investment

Without public-led investments in infrastructure and human capital, the private sector simply stays away. But this pattern, observed repeatedly in modern economic history, begs an important question. Why not let the private sector simply take the lead in infrastructure, health, and education if those investments are indeed so important as preconditions for other kinds of private investment? There are four kinds of answers, with the circumstances determining their relative significance.

First, many of the key preconditions for growth—such as roads, infectious disease control, and education—are public goods, meaning in shorthand that the social returns to providing them are much higher than the private returns. Such investments are characterized by increasing returns to scale, network externalities, and other kinds of positive spillovers, so that the private supply of such investments is far below the social optimum. So, even if the savings are available in the economy, it is better to mobilize it through public spending rather than private spending. Thus, even when roads can be financed through tolls, it is often highly advantageous to foster free access rather than toll-based access. Similarly, the social returns to immunization are higher than the private returns, so that mass immunization based on public financing is more desirable than private financing with limited public uptake of immunization services.

Second, even though these infrastructure and human capital investments are preconditions for long-term growth, the private rate of return on these investments is very low. In some countries, there simply are few or no investments that currently meet the international market test. Building roads and railways in impoverished landlocked countries such as Chad, Mongolia, or

Tajikistan may be necessary for long-term growth, but the returns on those investments are very low. In this case, foreign aid rather than private capital is essential to break the deadlock.

Third, some of the key investments—such as public health and primary education—are merit goods, meaning that universal access to such goods is a goal in itself. The Millennium Development Goals identify several such merit goods: primary education, maternal and child survival, control of infectious diseases, access to safe drinking water and sanitation.

Fourth, an impoverished country may well be a serious credit risk, even when the return to the investment is relatively high. Enforcing international claims is very difficult, and many of the world's poorest countries are already swamped with unpaid and unpayable foreign debts. In this case, domestic savings may be unavailable and foreign private savings uninterested in entering the economy.

In general, we stress that there is an important distinction between public financing of capital and direct public provision of capital services. The public sector may pay for the capital (whether roads, or schools, or clinics) but arrange for it to be provided by private companies or choose to organize public delivery itself. Either way, the public sector has the lead in arranging financing and in the regulatory structure, but it may try to foster private provision. This is the essence of public-private partnerships, in which the public sector provides some or all of the financing for a project or program, and the private sector provides the services on a contractual basis. The key question is whether the private provider is likely to be a more efficient provider, taking into account the complicated contracting costs in which the public sector pays but the private sector delivers the service. There is no single answer to that question. The answer varies by sector and by country.

Aid to help break out of the poverty trap

The primary responsibility for development lies with countries themselves. As an indispensable condition for defeating poverty, each country must recommit to pursuing the national institutions and policies conducive to dynamic and sustainable economic growth. But many low-income countries, including many fairly well governed ones, cannot afford the public investments in basic infrastructure, human capital, and public administration that are foundations for private sector growth and economic development. Many Least Developed Countries, especially in Sub-Saharan Africa, are stuck with low or negative growth. Why? Because their saving rates are too low to offset population growth and depreciation, and they are unable to attract the needed investments from abroad.

The core idea of official development assistance, therefore, is to push the elements of the capital stock—infrastructure, human capital, public administration and so forth as in figure 3.1—above the threshold needed for self-sustaining economic growth. ODA should not be a safety net (except for

Box 3.5
Large-scale aid works—when done properly

Criticisms of aid come in many forms. Some critics charge that aid is inherently flawed because it strengthens governments, often corrupt governments, at the expense of the private sector. This is the famous argument of the late British economist Peter Bauer. Some charge that aid is not needed, since private saving and investment can and should be the backbone of economic growth. Some have taken the middle ground that aid works when it is channeled to well governed countries. This is the conclusion of the highly influential study by Burnside and Dollar (2000).

Our view, explained in the text, is that aid is most useful if channeled to the countries that truly need it (mainly those stuck in a poverty trap) and channeled to the right sectors (mainly infrastructure and human capital). It works best when delivered to well governed countries. And aid used to support public investment complements private saving and investment, rather than competing with private capital.

Many negative conclusions about the link between aid and economic growth have come from cross-country regressions of economic growth on aid volumes (and other variables). The volume of aid is often found to be statistically insignificant as a determinant of economic growth, leading some authors to conclude that "aid is ineffective" in promoting economic growth. An important weakness in such studies is that they tend to examine the links of growth to overall volumes of aid without paying attention to how the aid is actually delivered. Specifically, much aid comes in the form of technical assistance (for consultants from the donor country), administrative costs of running bilateral and multilateral agencies, and emergency food aid. It is not really surprising that such aid is not correlated with economic growth in the recipient country. Food aid, especially, is given in the midst of deep crises. So a regression of economic growth on food aid would tend to prove (erroneously) that aid causes output to decline, instead of the correct conclusion that an output decline (caused by drought, for example) causes emergency aid to rise!

In an important new study, Clemens, Radelet, and Bhavnani (2004) correct for this typical shortcoming by considering only aid volumes that effectively support investments and services on the ground in the recipient country, taking out emergency aid, technical assistance, and other kinds of aid that do not translate into growth-promoting investments and services. They find that aid, when measured properly, contributes significantly to economic growth. This suggests that aid is effective, if it is well targeted and administered as direct support for country-level investments. Of course, a minimum adequacy of governance is required for a country to be able to channel aid into investments.

Aid can and must be disbursed in ways that align the incentives of donors and recipients to support positive development outcomes. As this report argues, elements of a successful disbursement strategy include aid in the form of budget support for national poverty reduction strategies based on the Millennium Development Goals. While there have been real problems with the way that aid has been distributed in the past, governments in rich and poor countries alike are learning from their mistakes to design more effective ways of delivering financial assistance to those who need it most.

In sum, foreign aid can play a hugely positive part in growth and poverty reduction when properly targeted and administered toward vital infrastructure and human capital. This finding is underlined by the recent experience of Mozambique, Tanzania, and Uganda, which all experienced substantial social sector improvements financed largely through development assistance. Mozambique is a particular success story over the past decade, having averaged real per capita economic growth rates of 5 percent while receiving aid ranging from 20 percent to 60 percent of GNP every year since 1993.

humanitarian relief). It is an investment in economic development to help countries begin climbing the ladder of self-sustaining economic development. The MDG-based investment program described in the following chapters is fundamentally an investment program for self-sustaining growth, not a program for increased dependence on welfare handouts.

Any aid strategy must confront many complexities, and the effects of aid on economic growth have been debated. When poorly designed or used for purposes other than real development—such as when it has been used to support foreign policy clients of developed countries—aid can create perverse incentives in recipient countries that are unhelpful for development.[3] But recent studies indicate that when aid is properly measured (that is, subtracting things counted as aid but which actually do not reach the recipient country in a form available for investments), evidence suggests that it greatly benefits economic growth (Clemens, Radelet, and Bhavnani 2004) (box 3.5). In addition, studies have shown that aid is especially effective in developing countries with good fiscal, monetary, and trade policies—and less effective if policies are poor (Burnside and Dollar 2000). The evidence is thus quite compelling: foreign aid with good policies in the recipient country helps to create economic growth, if the aid is delivered so that it becomes investment "on the ground."

To be adequate for a country in a poverty trap development assistance needs to support proper investments at a level sufficient to get capital accumulation ahead of population growth and depreciation. A big push of aid-supported investment that puts the country on a path of increased savings and self-propelling growth is far more efficient than low quantities of aid that do not change the fundamental growth potential of the economy. The key insight is that it will be much cheaper for the donors to frontload their aid over 2005–15 to raise each low-income country to the point of self-sustaining growth as rapidly as possible—rather than to continue to dribble out aid in small measures for several decades. If aid—even well targeted aid—is so small that the country's infrastructure and human capital are persistently insufficient, growth will never take off in a self-sustaining manner, and aid will remain a handout rather than a solution to the poverty trap.

2

Country-level processes to achieve the Millennium Development Goals

Chapter 4

MDG-based poverty reduction strategies

To enable all countries to achieve the MDGs, the world must treat them not as abstract ambitions but as practical policy objectives. The Goals are essential for transparency and accountability, so it is important that they be taken literally since the pressures in development policy push overwhelmingly for lower expectations rather than higher. National governments and international donors not wanting to be held accountable for their role in poverty reduction will always want to water down the Goals—particularly if achieving them requires increased budgetary commitments or major policy changes. In many countries the Goals are deemed "unrealistic" because they would require dramatic progress. Such statements should generally be met with skepticism. The practical steps to achieve the Goals in each country can and should be diagnosed, planned, and implemented with the proper focus and actions, combined with suitable support from the international community.

Designing a national strategy to achieve the Goals

In every country that wants to achieve the Goals, particularly those with basic conditions of stability and good governance, the starting assumption should be that they are feasible unless technically proven otherwise. In many of the poorest countries, the Goals are indeed ambitious, but in most or even all countries they can still be achieved by 2015 if there are intensive efforts by all parties—to improve governance, actively engage and empower civil society, promote entrepreneurship and the private sector, mobilize domestic resources, substantially increase aid in countries that need it to support MDG-based priority investments, and make suitable policy reforms at the global level, such as those in trade.

It is crucial that technical constraints to meeting the Goals not be confused with financial constraints. Although poverty reduction is the primary responsibility of developing countries, as this report shows (chapter 17), achieving the Goals in the poorest countries—those that genuinely aspire to the MDG targets—will require significant increases in official development assistance to break the poverty trap. We urge all low-income countries to increase their own resource mobilization for the Goals by devoting more budget revenues to priority investments. And in countries where governance is adequate but domestic resources are not, we call on donors to follow through on their long-standing commitments to increase aid significantly. In short, we call for co-financing the scaling up of MDG-based investments. The rich countries must no longer delay on their side of the bargain.

To implement the core element of partnership outlined in the Monterrey Consensus, the international system requires a baseline approach to encourage all developing countries to outline their specific and systematic strategies to achieve the Goals and, where needed, the path of co-finance required. To that end, our core operational recommendation is that each developing country with extreme poverty—including middle-income countries with pockets of poverty or areas of specific policy neglect—should adopt and implement a national development strategy ambitious enough to achieve the Goals. The country's international development partners—including bilateral donors, UN agencies, regional development banks, and the Bretton Woods institutions—should give all the support needed to implement the country's *MDG-based poverty reduction strategy*. Official development assistance should be generous enough to fill the financing needs, assuming that governance limitations are not the binding constraint and that the recipient countries are making their own reasonable efforts at domestic resource mobilization (see chapter 17 for more discussion on resources for the Goals). Donors must commit credibly to make sufficient funds available with actual disbursements being allocated to high-quality MDG-based poverty reduction strategies. Where the Goals are already within reach and greater progress is sought, we suggest that countries adopt an "MDG-plus" strategy, with more ambitious targets.

Working back from the 2015 targets and timelines

Serious implementation of the MDG targets and timelines implies a major shift in development practice. Low-income countries and their development partners now plan around modest incremental expansions of social services and infrastructure. We recommend instead a bold, needs-based, MDG-oriented framework over 10 years—aimed at achieving the quantitative targets set out in the Goals. Rather than strategies to "accelerate progress toward the Goals," countries need strategies to "achieve the Goals."

Underlying this point is a fundamentally new approach for development policy. Instead of asking the typical question, "How close can a country come to achieving the Goals under current constraints?" we strongly recommend asking, "Given the urgency of the Goals and the repeated international commitments to achieve them, what sequence of investments and policies is required and what constraints, financial and otherwise, need to be overcome?" The guiding element of this approach is that it works backward from the MDG targets to focus on the policies and investment needed between 2005 and 2015 (figure 4.1).

We recommend a four-step approach.

- First, countries need to map the key dimensions and underlying dynamics of extreme poverty—by region, locality, and gender—as best as possible with available data.
- Second, consistent with the poverty maps, countries should undertake a needs assessment to identify the specific public investments necessary to achieve the Goals, including faster overall economic growth supported by major public investments in infrastructure and private sector promotion (chapter 17).
- Third, the needs assessment should be converted into a 10-year framework for action, including public investment, public management, and financing.

Figure 4.1

An MDG-based poverty reduction strategy

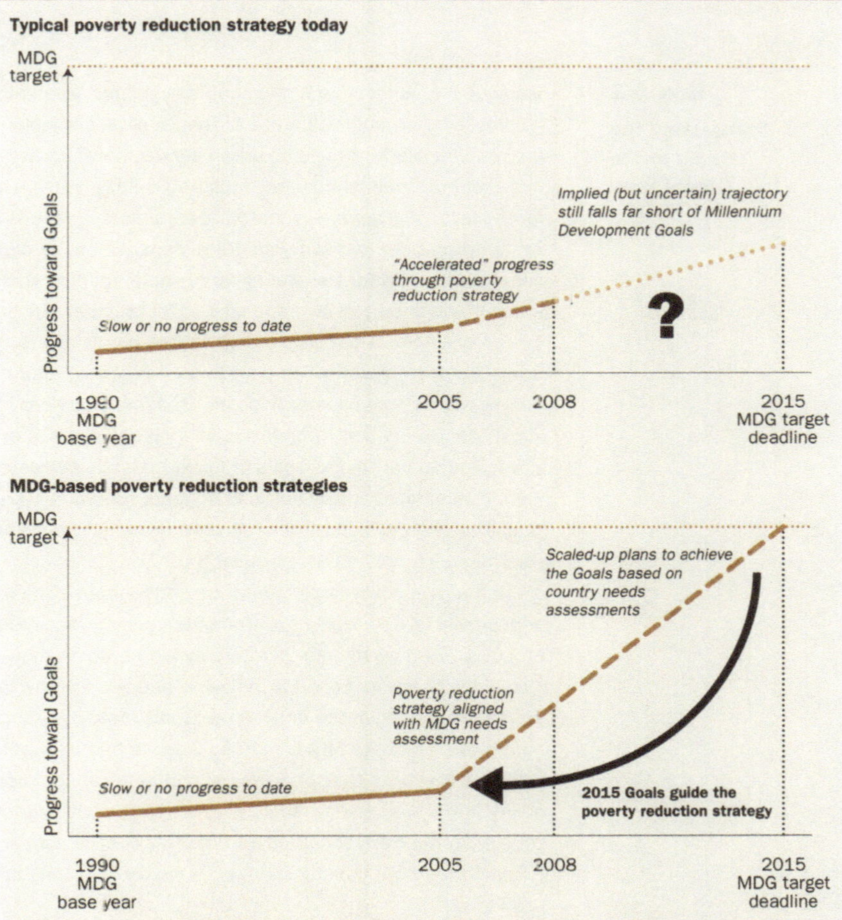

- Fourth, a 3-to-5-year MDG-based poverty reduction strategy should be elaborated within the 10-year framework. The poverty reduction strategy is a more detailed, operational document, linked to a medium-term expenditure framework that translates the strategy into budgetary terms.

Crucially, the 10-year framework and 3-to-5-year poverty reduction strategy should each include a public sector management strategy—with a key focus on transparency, accountability, human rights, benchmarking, and results-based management. They should also include a clear strategy for decentralizing target-setting, decisionmaking, budgeting, and implementation responsibilities at the level of local governments (box 4.1). They should encourage service delivery with the participation and oversight of local communities and nongovernmental organizations. And there should be a clear private sector strategy to promote economic growth and have countries "graduate" from donor assistance in the longer term.

Importantly, we are not advocating new development processes or policy vehicles. Instead, we are simply recommending that the current processes be truly MDG-oriented. In many low-income countries, the national poverty

Box 4.1

Translating the Goals to the local level

Many of the services and investments required to meet the Goals need to be delivered by provincial or local authorities. This applies particularly to cities where municipalities are responsible for providing urban services and infrastructure and upgrading slums. To this end the Goals should be "localized"—that is, translated into operational objectives for the level of government that will bear primary responsibility for their achievement. This is important both for linking program management as closely as possible to the intended beneficiaries and for translating large-scale national goals into more manageable pieces that communities can own and pursue for themselves.

For example, a city like Nairobi might be invited to define its "Nairobi Development Goals," setting quantitative targets for education, health, housing, water supply, sanitation services, solid waste disposal, transport services, and so forth. Similarly a village could identify its own targets for schools, clinics, water supply, sanitation, energy services, and so on. On the basis of localized goals consistent with the Millennium Development Goals, local authorities and regional governments should develop their own strategies that align with the national poverty reduction strategy and that partner with local community groups in service delivery.

To manage policy implementation and responsibilities for localized MDG targets, local and provincial (or state) governments require real resources. An MDG-based poverty reduction strategy should therefore include fiscal frameworks to provide these resources, including transfers from the national budget. For low-income countries with domestic resources insufficient to finance the Goals, poverty reduction strategies need to map out the systems for transfers of external finance to the lower levels of government. While such resource transfers to subsovereign entities raise complex questions of financial accountability and transparency, scaling up requires consolidated mechanisms in each country. Otherwise the transaction costs between donors and large numbers of local authorities would be too high, making impossible the rapid scaling up of essential investments at the local level.

reduction strategy is embodied in a Poverty Reduction Strategy Paper (PRSP), used as a basis for programs with the IMF and the World Bank. Those PRSPs urgently require revision to align them to the Millennium Development Goals. Very few PRSPs are ambitious enough to achieve the Goals, largely because they have been prepared in a context of insufficient donor assistance. Even when the PRSPs claim to aim for the Goals, they rarely identify the path of public investments that would be needed to achieve them.

We strongly support the PRSP as a powerful tool for achieving the Goals, but it needs to be deployed for that purpose, including both targets and time horizons. In our view, the World Bank, as the international development institution with the most direct financing engagement with the PRSP process, should work energetically to help countries prepare PRSPs that are MDG-based. Most PRSPs have a three-year time horizon, so they need to be embedded within countries' longer-term plans, many of which already exist. In addition, the process of preparing the PRSP needs to become more inclusive.

In countries that do not use a PRSP to guide policymaking, we similarly recommend that their processes be made MDG-based, with full support from the international community. There are five criteria for evaluating whether a national strategy is genuinely consistent with the Millennium Development Goals (box 4.2).

In the MDG-based policymaking process, we recommend that the government lay out a 10-year MDG framework for scaling up critical interventions over the entire period and sketch out 10-year financing needs as they appear as of 2005. Countries have often already prepared such strategies within individual sectors, but those strategies typically have not received support for implementation. Nor are they linked to other sectors in a consolidated framework. In many countries, such bold plans sit in the drawers or on the shelves without public discussion, since the donors have told the countries that sufficient financing is not available for such bold programs. The MDG-based process would enable such strategies to be empowered with financing commitments, implementation strategies, and alignment across sectors.

The 10-year MDG framework should then become the basis for elaborating the detailed budget and plan of action in the MDG-based poverty reduction strategy (or PRSP, if appropriate) for 2005–08, the time horizon typical of IMF and World Bank–supported programs. To link the MDG-based poverty reduction strategy with actual budgetary processes, the international community—including the international financial institutions and the bilateral donors—should support each country as it develops its medium-term expenditure framework to finance it.

This differs starkly from the prevailing practice in developing countries, which is to formulate investment strategies after the macroeconomic framework, official development assistance, and overall budgetary ceilings have been set independent of needs (Oxfam 2004). For many developing countries to achieve

Box 4.2

How to know if the content of a poverty reduction strategy is up to the task of meeting the Millennium Development Goals

For a poverty reduction strategy to be truly MDG-based, it needs to do much more than mention the Goals as aspirations. It needs to be linked systematically with the MDG targets and timelines—and be based on a detailed assessment of the public investment strategies needed to achieve the Goals. As a general rule, one can apply the following five-point checklist to review whether a poverty reduction strategy is really MDG-based.

Ambition: Are the targets aligned with the Millennium Development Goals?
The simplest thing to evaluate is whether the poverty reduction strategy's targets are aligned with the Goals. Are they equal to the Goals, more ambitious, or less ambitious?

Scope: Is the poverty reduction strategy aligned with all of the Goals?
Many PRSPs refer to the Goals in name but actually discuss only a few of them. For example, even if a poverty reduction strategy sets MDG-consistent targets for education, it also needs to set MDG-consistent targets for health, hunger, and the rest of the Goals. Moreover, issues like the environment, gender equality, and urbanization need to be addressed in an integrated manner.

Rigor: Are the targets substantiated with solid analysis of the needed inputs?
Many poverty reduction strategies set lofty targets, often much more ambitious than the Goals, without a clear plan on how to achieve them. For instance, a country could aim to cut its child mortality rate by 80 percent within 10 years, but not have a clear intervention-based strategy on how to achieve that target. Outcome targets are crucial, but input targets are also crucial for achieving outcomes. It is further important that all necessary investments be included in an MDG-based poverty reduction strategy. For example, all too often major investments in transport and energy are not included in existing strategy documents, even though the Goals cannot be met without them.

Timeframe: Is the strategy grounded in a long-term assessment of needs?
Most poverty reduction strategies cover only a three- to five-year period and are not grounded in an assessment of long-term investment and policy needs. Critical capacity constraints are typically not addressed since they would require long-term investments in training human resources—such as doctors and nurses who require many years of training—or infrastructure development plans. To be MDG-based, a poverty reduction strategy needs to be embedded in a decade-long needs assessment and action plan that work backward from the Goals to identify the required sequence of investments and policies.

Financing: Is the budget consistent with the level of inputs needed to achieve the Goals?
Often poverty reduction strategies do not have budgets or expenditure frameworks that are linked to the Goals. Where they exist, macroeconomic frameworks are typically set prior to designing sectoral investment strategies, regardless of the investments needed to reach the Goals (Oxfam 2004). Instead MDG-based budgets should be set from a careful assessment of how they will meet a population's needs. For example, the WHO Commission on Macroeconomics and Health outlined that the baseline costs of a scaled-up functioning health system are at least $30–$40 per capita (WHO 2001). The UN Millennium Project has identified similar benchmarks for other sectors (chapter 17).

If one sees a $4 per capita annual public health budget as part of a strategy to achieve the Goals, one knows that the budget has not been properly linked to a full needs assessment. At a deeper level of detail, particularly if cost estimates are conspicuously low, one needs to inquire whether the budget includes full costs of service delivery—for instance, both capital and operating costs. The latter are very often overlooked, consigning strategies to a low probability of sustainable implementation.

Box 4.2

How to know if the content of a poverty reduction strategy is up to the task of meeting the Millennium Development Goals
(continued)

* * *

These five questions can guide the evaluation of an MDG-based poverty reduction strategy or PRSP. Note that they are still separate from questions of implementation. Even the best poverty reduction strategy needs to be systematically implemented and managed through benchmarking, results-based management, and a medium-term expenditure framework.

the Goals, particularly low-income countries, this process needs to be reversed so that it starts with an assessment of the actual MDG investment needs—followed by the design of a supportive macroeconomic framework, including a viable financing strategy that includes increased development assistance.

A transparent, integrated, and consultative process

The process of developing an MDG-based poverty reduction strategy needs to be open and consultative, including all key stakeholders, domestic and foreign. Each country should convene an MDG strategy group chaired by the national government—but also including bilateral and multilateral donors, UN agencies, provincial and local authorities, and domestic civil society leaders, including women's organizations. The MDG strategy group can then organize a series of thematic working groups, each with broad participation, to develop the strategies to scale up in such areas as health, rural infrastructure, and agricultural productivity. Many countries already have similar working groups in place, so these can be used—with enhanced membership, as needed—to focus on the Goals. The MDG strategy group and each of the thematic groups should include gender expertise to ensure that actions for gender equality are fully integrated. Likewise, adequate environmental expertise should be available to all groups to ensure sector strategies are aligned with environmental objectives.

Development partners and civil society leaders should designate focal points for each of these government-led groups, contributing to the local MDG problem-solving process at the outset rather than midstream or after the fact. Developing multisector 10-year scale-up plans will naturally require much technical work—and considerable ingenuity. The UN agencies—including such technical agencies as FAO, UNICEF, UNIDO, UNFPA, and WHO, and the international financial institutions (the IMF, World Bank, and regional development banks)—should play an important advisory role in supporting this work as the host government requests. Middle-income countries requiring less external financial support will still often benefit from this technical support.

An MDG coordinator, typically in the ministry of planning or economy, or in the office of the head of government, should ensure that the work of the various working groups is properly integrated.[1] Importantly, the MDG strategy group will need to work closely with the ministry of finance to ensure the strategy is linked to the government's operational budget and not left to float as an irrelevant document. Meanwhile, line ministries such as health and water often complain that they have been excluded from the planning process, so they need to take leading roles in the thematic working groups. This process requires sustained executive leadership from the head of state and senior decisionmakers in each country.

In 2004 the UN Millennium Project began advising a selected number of UN Country Teams as they provide real-time support to governments developing MDG-based poverty reduction strategies. In each of these countries the UN Millennium Project and UN Country Team are working with government, local partners, and multilateral organizations to identify the best ways to integrate MDG targets and the time horizons into ongoing national policy processes. Some early lessons from these countries are discussed in chapter 13.

While this MDG-based framework is the UN Millennium Project's recommended norm for countries wanting to achieve the Goals, we recognize that in some countries—particularly those in conflict or with highly corrupt governments—the contents of a national strategy, the domestic process for preparation and implementation, and the international partners' financial support will each require adjustment to the local situation. These difficult circumstances require careful case-by-case strategies. We discuss the need for differentiated priorities and approaches to countries with weak governance and fragile states in chapters 7, 11, 12, and 13.

Chapter 5

Public investments to empower poor people

A central finding of the UN Millennium Project is that the world has the practical knowledge, tools, and means to reach the Millennium Development Goals. Development can be achieved through specific actions on the ground. We know how to prevent mothers from dying in labor. We know how to encourage girls to complete sixth grade so that they have more choices than their mothers. We know how to vastly increase maize yields to feed villages. We know how to make sure that hospitals have uninterrupted electricity. We know how to plan cities to avoid the misery of slums and how to connect remote villages to markets and schools. We know how to combat violence against women and girls. We know what it takes to make sure all citizens have the equal right and opportunity to make choices. Even if we don't know everything about such challenges, we know enough to achieve the Goals. Moreover, the necessary interventions are utterly affordable (chapter 17).

To achieve the Millennium Development Goals, huge new investments and, in many cases, better policies and institutions are needed to implement the practical measures that we know to work. The tools have existed in the rich world for decades. The UN Millennium Project task force reports, essential accompaniments to this report, describe these required investments and policies in detail (UN Millennium Project 2005a–m). This chapter summarizes their recommended investments and policies.

In the poorest countries, especially those caught in a poverty trap, the process should start with an increase in public investments—not only to meet certain critical needs but also to trigger a rise in private investment that is held back by the lack of infrastructure, effective service delivery, and a healthy and skilled labor force (chapter 3). Such ambitious investment strategies typically are not pursued in the developing countries because of extreme resource constraints.

For all developing countries, but especially those stuck in a poverty trap, we recommend that the MDG-based frameworks to meet the 2015 targets (as described in chapter 4) should be designed around seven broad "clusters" of public investments and policies:

1. Promoting vibrant rural communities, by increasing food productivity of smallholder farmers, raising rural incomes, and expanding rural access to essential public services and infrastructure.
2. Promoting vibrant urban areas, by encouraging job creation in internationally competitive manufactures and services, upgrading slums, and providing alternatives to slum formation.
3. Ensuring universal access to essential health services in a well functioning health system.[1]
4. Ensuring universal enrollment and completion of primary education and greatly expanded access to postprimary and higher education.
5. Overcoming pervasive gender bias.
6. Improving environmental management.
7. Building national capacities in science, technology, and innovation.

In designing the recommended investment and policy clusters, we have given careful consideration to gender equality. In addition to women's various productive roles—as farmers and wage workers—women are the main caregivers and household managers in most countries around the world. Though women's reproductive roles affect their participation in various social, economic, and political activities, they often lack the information and services needed to determine the number, timing, and spacing of their children. They often perform a disproportionate amount of physical labor every day merely to keep their families alive. In addition to the direct investments in fighting gender discrimination detailed below, many of the other recommended investments—such as improved access to water supply, modern cooking fuels, enhanced transport services, and improved soil nutrients—can reduce gender inequalities and empower women.

Environmental sustainability is also essential to any effort to improve the well-being and health of the world's poorest people. A degraded environment has dramatic and harmful effects on health, education, gender equality, and economic development. People cannot work and study if they are frequently ill from drinking and bathing in polluted water or if they are malnourished because of soil erosion and desertification. The oft-heard argument—that the poor should wait until their incomes have risen before investing in better management of the environment—is misplaced. Why? Because in many parts of the developing world environmental degradation already places a binding constraint on development. A successful MDG strategy must thus be created and implemented in an environmentally sustainable manner right from the start.

As a general point for every country's MDG-based poverty reduction strategy, the detailed public investment plans should meet six criteria. They should:

- Be ambitious enough to achieve the Goals.

- Lay the ground for private sector–led economic growth, with particular support to the indigenous private sector.
- Be based on known interventions that can be rapidly scaled up.
- Describe a path for scaling up through 2015, beginning with investments that open up capacity bottlenecks for scaling and investments that each country prioritizes as most urgent.
- Promote community organizations and other nongovernmental organizations as vital partners in delivering services to poor people.
- Ensure that women's specific needs are taken into account and that women and girls benefit equally with men and boys.

In each of the seven investment clusters, national and local governments, in partnership with civil society, should take the lead in setting the "rules of the game" for households and private enterprises. They should set the policies and institutions to ensure effective, equitable, and sustainable delivery of key services. They also have the primary responsibility for planning and financing key investments and services at scale, even if some are delivered by the private sector. Importantly, the seven national investment areas must be complemented by investments in regional cooperation and infrastructure services and in the global goods described in chapter 15 (including environmental management and international science and technology).

Here, we identify the priority investment packages and enabling policies for their implementation, giving examples of successful scale-ups. In chapter 17, we identify the costs of the national-scale investment packages—and how those costs can be met through greater domestic resource mobilization combined with increased aid for countries that cannot finance them with domestic resources alone.

Before proceeding with the range of investment priorities below, it is important to recognize that the full range of clusters might seem daunting for countries, particularly those with limited capacities. For certain, development processes are complex, so building up human resources and infrastructure in an integrated manner across sectors will require intensive long-term investment programs over the coming decade, with a particular focus on strengthening capacities (chapter 6). But there are many actions and investments that can be implemented in a much shorter term, using existing capacity and infrastructure. These actions, which we call "Quick Wins," provide immediate opportunities to save and improve large numbers of lives, starting in 2005 (box 5.1).

Rural development: increasing food output and rural incomes

The global epicenter of extreme poverty is the smallholder farm. Of the roughly 850 million people living in chronic hunger, smallholder farmers constitute half (FAO 2004). Remember that these farmers also constitute the bulk of private sector economic activity in many developing countries, so improving their economic lot will make a huge difference to their countries' prospects for

Box 5.1

Quick Wins—solutions to implement now

Implementing a full set of MDG-based investments and policies will take time and in many cases complex institutional change. But by implementing the interventions set out below, large-scale progress can begin immediately in 2005 and see major results within three or fewer years. We call these Quick Win interventions for the Goals, ones that can and should be implemented right away if the world is serious in its commitment. Although far from comprehensive, the Quick Wins could bring vital gains in well-being to millions of people and start developing countries on the path to the Goals.

- Eliminate school and uniform fees to ensure that all children, especially girls, are not out of school because of their families' poverty. Lost revenues should be replaced with more equitable and efficient sources of finance, including donor assistance.
- Provide impoverished farmers in Sub-Saharan Africa with affordable replenishments of soil nitrogen and other soil nutrients.
- Provide free school meals for all children using locally produced foods with take-home rations.
- Design community nutrition programs that support breastfeeding and provide access to locally produced complementary foods and, where needed, micronutrient (especially zinc and vitamin A) supplementation for pregnant and lactating women and children under five.
- Provide regular annual deworming to all schoolchildren in affected areas to improve health and educational outcomes.
- Train large numbers of village workers in health, farming, and infrastructure (in one-year programs) to provide basic expertise and services to rural communities.
- Distribute free, long-lasting, insecticide-treated bed-nets to all children in malaria-endemic zones to cut decisively the burden of malaria.
- Eliminate user fees for basic health services in all developing countries, financed by increased domestic and donor resources for health.
- Expand access to sexual and reproductive health information and services, including family planning and contraceptive information and services, and close existing funding gaps for supplies and logistics.
- Expand the use of proven effective drug combinations for AIDS, TB, and malaria. For AIDS, this includes successfully completing the 3 by 5 initiative to bring antiretrovirals to 3 million people by 2005.
- Set up funding to finance community-based slum upgrading and earmark idle public land for low-cost housing.
- Provide access to electricity, water, sanitation, and the Internet for all hospitals, schools, and other social service institutions using off-grid diesel generators, solar panels, or other appropriate technologies.
- Reform and enforce legislation guaranteeing women and girls property and inheritance rights.
- Launch national campaigns to reduce violence against women.
- Establish, in each country, an office of science advisor to the president or prime minister to consolidate the role of science in national policymaking.
- Empower women to play a central role in formulating and monitoring MDG-based poverty reduction strategies and other critical policy reform processes, particularly at the level of local governments.
- Provide community-level support to plant trees to provide soil nutrients, fuelwood, shade, fodder, watershed protection, windbreak, and timber.

Box 5.1
Quick Wins—solutions to implement now
(continued)

Of course, these Quick Wins alone will not achieve the Goals. They need to be matched by investment strategies with longer timeframes, such as those for transport infrastructure, energy services, and teacher and nurse training. But they represent a feasible and immediate set of actions that could begin today and could save and improve millions of lives around the developing world.

These strategies do not require complex systems or extensive infrastructure for their delivery. Time-tested, they have proven their effectiveness in the poorest of countries. To succeed, they need to be customized and implemented by developing countries and supported by immediate and adequate financial and technical assistance from the rich countries.

Each of these interventions is discussed at greater length in the reports of the UN Millennium Project task forces. Once again, they are not the only interventions required to reach the Goals—just ones with very high impact in the short term. The world cannot afford to let another year go by without investing in these simple and proven strategies.

long-term economic growth. Impoverished smallholder farmers scratch out an existence that is brutally difficult, living on the edge of survival and often falling off the edge. They live in communities that are geographically isolated and burdened by disease, climatic shocks, environmental degradation, and social exclusion and violence. They not only suffer—they pass on their suffering to the next generation.

Poor rural families tend to have many children, because they fear for their children's survival, because they regard their children as social security, and because they lack access to safe and effective modern contraception and family planning. Women bear a disproportionate burden of rural poverty, living lives of great physical labor as well as rearing children and giving care to the elderly and sick. Making matters worse, smallholder subsistence farming is often highly destructive of the environment, especially as population pressures and low productivity lead to deforestation, soil nutrient depletion, and soil erosion. Addressing rural poverty requires making farms more productive, raising farmers' incomes, improving the lives of the rural landless, and expanding essential services.

Making farms more productive

Agricultural productivity (of crops, livestock, forests, and fisheries) remains low in many countries for many reasons. Among the most important are that soils are widely depleted of nutrients, water resources are not adequately managed, farmers do not have the means of using improved plant varieties and animal breeds suitable to their forms of agriculture, and extension services do not adequately teach the findings of modern agricultural research. Rising water scarcity for agriculture coupled with poor management of water resources is a major challenge in many rural parts of the developing world, and water scarcity in some regions is also being exacerbated by long-term climate change.

A Green Revolution has preceded virtually every economic takeoff in modern history, from Britain's original Industrial Revolution—preceded by its agricultural revolution—to Asia's boom in the 1970s and onward, fueled by the modern Green Revolution, based on high-yield varieties of basic staple crops. On July 5, 2004, the UN Secretary-General called for a Twenty-first Century African Green Revolution that would "address the interconnected challenges of agriculture, health care, nutrition, adverse and unfair market conditions, weak infrastructure, and environmental degradation." (Annan 2004, MDG Technical Support Centre 2004). He called special attention to farming systems in areas largely disconnected from functioning markets. This applies equally to remote areas in Africa, Asia, and Latin America.

A Twenty-first Century Green Revolution in Africa and in bypassed regions in Asia and Latin America would include several components all working together (see appendix 1): improved rural infrastructure services in the form of roads and means of transport, modern energy services, and communication technologies. Central to this new Green Revolution is improving soil health through a combination of applying fertilizers—using fertilizer trees, mineral fertilizers, green manures, cover crops, and crop residues to replenish soil nutrients—and controlling soil erosion (box 5.2).

Other important on-farm investments include expanding and improving small-scale water management, postharvest storage, crop varieties, and livestock breeds. In many parts of Africa and Asia, livestock and livelihoods are closely related since livestock strengthen both nutrition and income security. Investments in vaccines, veterinary pharmaceuticals, and quality feed and fodder contribute to increasing farm animal productivity and reducing both poverty and hunger.

Public and private investments will be required to improve water storage and harvesting, as well as to increase the water efficiency of agriculture by raising the "crop per drop." (In periurban areas and in cities, appropriately treated wastewater can be used for irrigation.) To support all of these actions aimed at boosting land yields, agricultural research systems need significantly higher levels of funding.

Unlike the Asian Green Revolution, farming techniques exist today that permit an environmentally friendly intensification (such as agroforestry and small-scale water management). Extension services need to be strengthened with village-level paraprofessionals who use a strong participatory approach and have up-to-date knowledge of ecologically sound agricultural techniques.[2] In many parts of the world, especially in Africa, the majority of smallholder farmers are women. Providing them with access to improved farming techniques is most effective through well trained women extension workers. We stress that these farm-level investments will be more effective than fighting the symptoms of low productivity with food aid brought from abroad.

Box 5.2

Increasing crop yields in Africa

Source: UN Millennium Project 2005d; map data calculated from World Bank 2004c.

More than 200 million Africans are undernourished. Unlike other regions where inadequate food supply is a result of inequalities and poor distribution, food insecurity in Africa is primarily a result of the lack of adequate food production. One of the major reasons that so many Africans do not have access to enough food is that crops are grown in fields with utterly depleted soil. The land has been over-farmed and few subsistence farmers can afford to use fertilizers (see map), which in parts of Africa can cost nearly four times as much as they do in North America or Europe.

Researchers from the World Agroforestry Center, along with national research and extension services, and NGOs working with farmers, spent 10 years developing soil fertility replenishment methods with resources naturally available in Africa. Simple and affordable "green fertilizers" were in most cases developed and tested in farmer-designed trials on farmer fields. Three components can be used in combination or separately: nitrogen-fixing leguminous trees, indigenous rock phosphates in phosphorus-deficient soils, and biomass transfers of leaves of nutrient-accumulating shrubs. By adding nutrients to the soil, these natural fertilizing methods have in many cases doubled or tripled yields for farmers, meaning that in just a few years, more and better food has been made available to thousands. These agroforestry technologies also help the environment, by increasing biodiversity, sequestering carbon, and protecting watersheds.

Tens of thousands of farm families in Kenya, Malawi, Mozambique, Tanzania, Uganda, Zambia, and Zimbabwe now use various combinations of fertilizer trees, phosphorus, and biomass transfers with good and consistent results. Adoption has been almost entirely through farmer-to-farmer and village-to-village knowledge transfers and through the support of community-based organizations, national research and extension institutes, and universities.

These technologies to combat soil depletion can and must spread to reach tens of millions more families dependent on agriculture. They should be combined with the strategic use of mineral nitrogen fertilizers, the return of crop residues to the soil, and erosion control practices to restore soil health. Large and sustained investments from governments and donors are necessary to extend the remarkable success in increasing yields.

Fertilizer consumption, 2001

Metric tons per million people

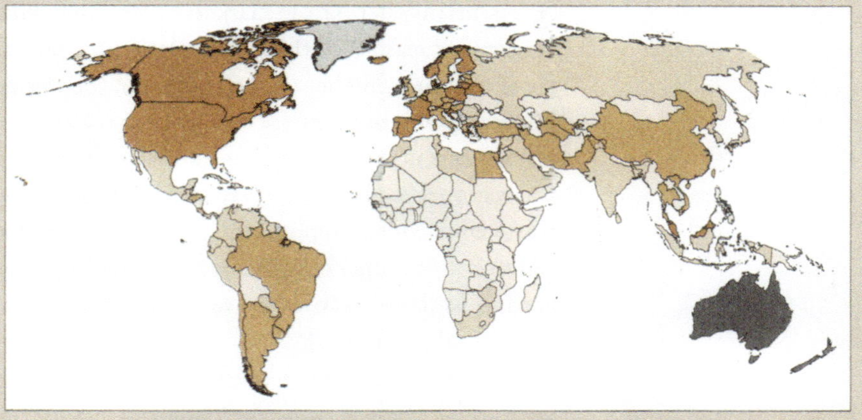

□ Less than 10,000 □ 10,000–20,000 □ 20,000–40,000 □ 40,000–80,000 ■ More than 80,000 □ No data

(continued on next page)

Box 5.2
Increasing crop yields in Africa
(continued)

The cost of such an ambitious scale-up is remarkably small. Distributing "green fertilizers" to tens of millions of African farmers will cost an estimated $100 million a year for each of the next 10 years. This is only a tenth of the amount currently spent each year to deliver food aid and contend with the effects of hunger and malnutrition on the continent.

Combining healthy soils with small-scale water management, improved crop germplasm, and targeted nutrition and market interventions, a science-based Twenty-first Century African Green Revolution—driven by technologies already developed and adopted by many African farmers—could save millions of lives and provide a sustainable future for African agriculture.

Success factors for scale-up

- Developing the technologies together with farmers. The result was scientifically sound and fit the farmers' requirements.
- Strong participation of national research institutions.
- Local NGO involvement in developing and testing.
- Local government commitment and management of modest scale-up plans in some of the countries.

"Smart" subsidies for qualifying food-insecure farmers are strongly recommended to ensure access to key farm inputs. In other words, we call for a reversal of the policies in recent years of slashing subsidies for fertilizers and other critical inputs. But we underscore that such subsidies would need to be targeted to very poor regions and translated into farm-site investments in soils, water, improved seeds, and other critical needs—with an exit strategy as rural productivity and incomes rise over time (UN Millennium Project 2005d).

Raising farmers' incomes—getting farm products to market

Today most farmers in rural Africa and remote parts of Asia and Latin America are essentially cut off from markets beyond their village. Poor transport services are a major reason for this, particularly in rural Africa, where transport costs are several times higher than elsewhere in the world. Since rural road networks in many poor countries are too small and falling into disrepair, major public investments in the construction and rehabilitation of footpaths, feeder, district, and national roads are required—using techniques that are labor intensive to create employment and minimize the adverse impact on the environment. In addition to commencing major road-building programs, adequate provisions need to be made to cover the maintenance costs through targeted road funds that are fully resourced.

In many cases major investments are required to build and strengthen institutions charged with the design, operation, and maintenance of transport infrastructure. Equally important are measures to improve access to low-cost transport services. Examples include providing access to credit, ensuring efficient transport markets through legislation, lowering entry costs into the transport market, and improving the supply of low-cost vehicles, bicycles, and other means of transport. Throughout, rural development strategies should focus on strengthening economic linkages between rural and urban areas, where the markets for farmers' products and inputs lie. To this end, transport corridors linking key agricultural regions with primary and secondary cities need to be promoted.

Improved transport infrastructure and services will have an important positive impact on reducing the time burdens of women and girls who today spend much of their day walking to obtain water and other essentials for survival. They will also allow women better access to life-saving health services, such as emergency obstetric care.

To further improve farmers' ability to market their products and access markets, national strategies can focus on building storage facilities, encouraging networks of agrodealers, and improving credit and savings facilities. All these investments will succeed when smallholder farmers and rural communities are empowered to establish their own institutions—for example, farmer field schools to gain access to new agricultural technologies, village banks to gain access to financial services, and farmers' associations to negotiate with market intermediaries.

Improving the lives of the rural landless

The foregoing interventions will help reduce hunger and increase incomes for the rural poor who farm their own land—but they are not enough. Many of the poor and hungry, especially in Asia, are the rural landless. Most of the landless poor depend on rural off-farm labor markets to earn their livelihoods. Yet these markets often function poorly in rural areas where alternative opportunities to farming may be scarce or the poor may have limited education and training. The large numbers of landless laborers in South Asia, for example, lack productive assets, marketable skills, and alternative employment opportunities. They are thus obliged to do on-farm work at low wage rates. Many more decent opportunities to earn income are needed.

A variety of interventions to increase skills can expand labor opportunities for the rural landless and promote the nonfarm economic sector more broadly. Primary education and skills training for both men and women can also create a more competitive labor market. The position and bargaining power of the poor in labor markets can be strengthened through legislation that permits labor organizations and the free movement of people in search of work, as well as through government labor agencies that serve the poor.

Expanding essential services in rural areas to meet the other Goals

Beyond improving farmers' incomes, expanding rural infrastructure and enhancing service delivery is also required to fight rural poverty, disease, and heavy work burdens, especially for women. Improving effective access to domestic water supply and sanitation, an important end in itself, is critical for meeting the health, education, gender, environment, and other Goals, as is better management of water resources. Hygiene education and awareness-building programs need to precede and accompany infrastructure provision to raise household demand for sanitation services and ensure proper use and maintenance of the infrastructure. Since low-cost sanitation technologies can often be built using locally available materials without significant outside funding, community participation and mobilization are key to ending open defecation, with its deleterious effects on human health and dignity in communities the world over.

Most rural areas also require better access to energy services. To accelerate the shift toward cleaner fuels for cooking and space heating, countries should provide efficient cookstoves and improve access to modern fuels. Any strategy for meeting the Goals should also place a strong focus on setting up the necessary infrastructure to provide schools, clinics, hospitals, and other key community facilities with low-cost electricity.[3] These can then serve as access points for basic communication services and battery charging stations to supply domestic households.

Also necessary to fight "hidden hunger" or malnutrition, which contributes to disease and lowers productivity, are nutrition interventions targeted at pregnant women and children under two, specific vulnerable groups (such as people living with HIV/AIDS), and the general population through, for example, food fortification or salt iodization. Sexual and reproductive health services can improve birth spacing and reduce nutritional and health burdens on mothers and their children. These interventions are of course equally relevant to rural and urban dwellers.

Urban development: promoting jobs, upgrading slums, and developing alternatives to new slum formation

A sharper focus on reducing urban poverty is necessary, in part because official statistics tend to underestimate urban poverty, which is very high in most developing countries (Satterthwaite 2004). Ending extreme poverty also requires ensuring a productive urban environment, improving the lives of slum dwellers, and providing alternatives to the formation of new slums. Most non-agricultural activities—industry and services—thrive best in an urban setting where the concentration of economic activity reduces transaction costs (such as transport and communication) and allows the face-to-face contact vital for a sophisticated division of labor. That is why the urban economy is generally an important center of gravity of economic life and the focus of technological advance and specialization.

In developing countries around the world, cities are struggling to function. They are home to extreme poverty and fail to create the jobs necessary for growth. The share of the population living in urban areas is rising inexorably and will continue whether rural development is successful or not. If rural development is successful, it will mean that a shrinking proportion of the population can feed the entire population. The children of farmers will therefore move to cities in search of a new life. If it is unsuccessful, then "rural refugees" will escape from intense rural poverty, shrinking farm sizes, and environmental degradation. They will come to the cities in search of jobs and services. If good jobs do not exist, the migrant workers will live in extreme poverty under slum-like conditions and swell the ranks of the informal economy.

Over the coming decades, countries in Asia and Africa will continue to urbanize rapidly, approaching the urban population shares in Latin America and the high-income countries. While rapid urbanization in poor countries poses an unprecedented challenge, it also creates an opportunity. Due to high population densities, critical social services such as education and healthcare can be more easily provided than in rural areas. Even so, these services often remain inaccessible to many urban poor. In some slums health outcomes are worse than in rural areas. If the social exclusion of people living in informal settlements or slums can be ended, urbanization can be a powerful driver for improving the lives of a country's population and for generating economic growth.

Given the pressures that urbanization imposes on cities, finding alternatives to new slum formation and improving the lives of slum dwellers, as called for in the Millennium Development Goals, are essential goods in themselves and necessary for raising urban productivity. The package of investments for urban development, summarized in appendix 1, should include five broad areas: improving security of tenure for slum dwellers, upgrading slums and improving housing, expanding citywide infrastructure and effective service delivery, creating urban jobs through local economic development, and providing alternatives to slum formation.

Improving security of tenure

Strategies for improving security of tenure are central for improving the lives of slum dwellers and land use in urban areas. They often require reform of tenure and land-use legislation, coupled with legislation to prevent forced eviction. In enhancing access to land, particular attention should go to ensuring that women have equal access to land tenure and titling rights. Throughout, improving security of tenure requires a high degree of tailoring to local needs, since preferences for and the feasibility of a particular tenure regime vary tremendously within cities, let alone countries or regions (Durand-Lasserve and Royston 2002). It is also conditional on a high degree of participation and decisionmaking by the slum dwellers themselves, whose organizations should be recognized as critical partners with local authorities.

Upgrading slums

Upgrading housing and retrofitting infrastructure for water supply, sanitation, transport, and energy services are critical for improving the lives of slum dwellers. Successful slum upgrading is best carried out by local authorities and communities working in close partnership (box 5.3). Where possible community organizations should be supported and allowed to play an active role in preparing and executing plans for slum upgrading. Moreover, upgrading must be citywide to avoid having the remaining informal settlements continue to grow by attracting new migrants. Of particular importance are investments in housing, which can often be carried out incrementally by the poor, if they have adequate security of tenure, and which can become an important means of asset accumulation. (The report of the Task Force on Improving the Lives of

Box 5.3

Improving urban sanitation in India

Source: Water Supply and Sanitation Collaborative Council 2000.

Starting in the 1970s the NGO Sulabh International developed and implemented a low-cost sanitation system in India. The Sulabh program made two main innovations: the modification of an existing low-cost technology, and community education to increase demand for services.

The technology, known as a pour-flush system, has many advantages. It is affordable, even for more economically disadvantaged segments of the population. It is never out of commission since, with the twin-pit option, one pit can always be used while the other one is being rested. The latrine can be built with locally available materials and is easy to maintain. It is also easy to upgrade, as it can be connected to a sewer system if one is introduced in the area. The toilet also has a water seal that makes it odorless and fly-free. And flushing requires only 2 liters of water, rather than the 10 needed by other flush toilets.

Despite these technical virtues, the Sulabh program would not have succeeded without improving public awareness and encouraging community participation. For populations unfamiliar with modern sanitation practices, the Sulabh International Social Service Organization undertook community-based educational campaigns, including door-to-door efforts to persuade people to convert from bucket latrines. Sulabh then constructed the twin-pit, pour-flush toilet for those who agreed to the conversion. Sulabh also educated people on the use and maintenance of their new latrine, promising to fix construction defects and solve technical problems at no cost. The program also helped local communities set up, operate, and maintain the community toilet complexes.

More than 1 million units have been constructed in private homes (or substituted for existing unhygienic latrines), and about 5,500 have been installed in pay-and-use public toilets. This has vastly improved the quality of facilities available to users. An attendant staffs the public toilets 24 hours a day, supplying powdered soap for hand washing, bathing, and laundry. Free services are offered to children, the disabled, and the poor. More than 10 million people now use the complexes every day, and some facilities have even begun providing new services, such as telephone calling plans or basic primary healthcare. As a result, some municipal governments have relinquished control of public sanitation provision to Sulabh for up to 30 years.

Success factors for scale-up

- Partnership between an NGO, local communities, and the government.
- Shift in role of central government from implementer to facilitator.
- Stepwise approach to service provision rather than all-or-nothing.
- Community involvement and awareness programs to ensure demand for services.
- Capacity building to enable communities to assist with service delivery.
- Service delivery approach adapted to local conditions.

Slum Dwellers describes successful strategies for slum upgrading in detail; UN Millennium Project 2005f.)

Compared with rural areas, slum upgrading requires a stronger focus on networked technologies, such as sewers, piped water, and electricity grids. Investments in improved sanitation should receive high priority to improve the quality of life and reduce the high burden of oral-fecal diseases in informal settlements caused by widespread open defecation. The high density of informal settlements makes sanitation particularly precarious. Where space constraints are high, low-cost communal toilet blocks have been used successfully. Effective hygiene education and awareness building programs are essential to create demand for sanitation and to ensure adequate use by all household members. Other investments required as part of slum upgrading include storm drainage, community facilities, local markets, and street lighting. Health services require investments as well. As mentioned earlier, nutrition interventions for both the general population and vulnerable groups will reduce morbidity and mortality among the urban poor. Clinics need to be more accessible, be located where needs are greatest, and be open during hours that can accommodate the schedules of the working poor.

Expanding citywide infrastructure

To complement the upgrading of individual informal settlements, citywide infrastructure and services need to be extended and upgraded. A high priority should be meeting the transport needs of the urban population through investments in transport services and infrastructure, such as footpaths, kerbing, bus lanes, roads, and mass transit systems. In many cases investments in mass transit systems do not require expensive infrastructure. Many large cities have successfully developed efficient bus-based mass transit systems that can provide transport services to the poor at a moderate cost. They are often more cost-effective than investments in large-scale road and rail-based transport infrastructure. Also important are policy changes to improve the availability of low-cost means of transport, including bicycles. In addition, effective regulation of industrial water and air pollution must complement an urban development strategy to ensure a safe urban environment. Solid waste disposal using well designed landfills and, in some cases, wastewater and sewage treatment also need to be provided.

Creating jobs

Cities must create jobs to employ their rising populations. Good infrastructure attracts domestic and foreign investment, which is necessary for large-scale job creation. Another task for urban planners is to improve industrial efficiency and attract foreign investors with industrial parks, export processing zones, or other designated areas for private sector development. Successful cities are able to link industrial zones with seaports and airports to reduce the time, cost, and hassle of shipping goods.

Equally important are measures to support the informal sector, where most of the urban poor work in low paid, low productivity, and low security jobs. To facilitate the shift into the formal sector, local authorities should adjust their laws and regulations to lower the costs and increase the benefits for people to formalize their enterprises. They should also provide assistance to small enterprises to upgrade skills and increase access to productive resources and market opportunities.

Providing alternatives to slum formation

Since cities in many developing countries will continue to grow at a fast pace, local authorities and national governments need to strengthen urban planning and citywide strategies to provide alternatives to slum formation. By making land available to the poor at affordable prices and ensuring the provision of housing, urban infrastructure, and transport services at the fringes of cities, urban planning can provide alternatives to the formation of new slums. In the past some cities—particularly in Latin America—have used land banking for this. Local authorities should also provide much of the trunk infrastructure in development areas and establish clear regulatory standards regarding minimum plot sizes, infrastructure standards, and so forth. Sound urban planning and standards also are central in averting or mitigating the impact of floods, landslides, and storms.

Making it happen—empowering city governments and the urban community

Perhaps the most important change needed in managing cities is to foster a collaborative partnership between local authorities and communities, with strong support from the national government. Local authorities are the city planners, financiers, and providers of infrastructure services. Their performance depends on good governance at city level—involving civil society, including communities living in informal settlements, and working with the urban poor as partners in making cities work, not seeing them as obstacles, as is too often the case today.

A key to productive and sustainable urban development is for city governments to have the policy autonomy and financial independence to design and implement plans and infrastructure programs. Decentralization strategies need to strengthen local authorities that are directly accountable to urban communities. Donors should ensure that their assistance to cities does not get bottled up in national capitals, but reaches and empowers local urban governments to take the lead in their own development efforts. National governments, in turn, should strengthen policies for local tax mobilization and expenditure assignment to ensure predictable and adequate financial transfers to local authorities.

Community organizations can provide a voice for the urban poor and ensure that their interests are met in slum upgrading and urban planning (chapter 8). Federations of slum dwellers have access to unique information on informal settlements—information central to successful upgrading. They should be involved as equal partners from the beginning of the planning processes. In many cities, community organizations, like the ones federated under Shackdwellers International, have led slum upgradings on a massive scale. Wherever this is possible, local authorities should support community-led initiatives financially and treat community organizations as equal partners. This is particularly important where resettlements of slum dwellers become a necessity—say, to free up critical railway lines in a city.

Without the support and participation of the poor, such resettlement programs can lead to the mere relocation of slums—or much worse. Community organizations can help mobilize the resources of the urban poor to co-finance improvements in housing and investments in basic urban services. For example, the work of the NGO Sulabh International in India showcases a successful scale-up of the provision of sanitation services (see box 5.3).

Health systems: ensuring universal access to essential health services

Health, a fundamental human right, is also a key input to economic development because it raises the productivity of the work force and increases the attractiveness of the economy for investors, domestic and foreign. Pandemic diseases such as malaria, TB, and AIDS not only increase suffering but deter investments in infrastructure, tourism, agriculture, mining, and industry. But developing countries continue to endure enormous rates of avoidable illness and premature death. Moreover, inequalities in health status and in access to healthcare are pervasive and growing, both among and within countries. Despite all this, technical interventions to prevent and treat the vast majority of health conditions affecting people in these countries are well known. The central challenge to achieving the health Goals is one of implementation—of ensuring access to these known interventions in ways that simultaneously promote the fundamental aims of development and social justice (box 5.4).

Appendix 1 summarizes the key interventions that could, if implemented broadly and equitably, allow attainment of each of the health Goals. These are discussed in detail in the reports of the individual health task forces. While many of these interventions could be delivered through disease-specific vertical programs—and in some circumstances this may be the most efficient solution—in most cases they are best provided through an integrated district health system centered on primary care and first level referral hospitals. That is the course strongly recommended by the UN Millennium Project. Vertical initiatives should in general not bypass the health system in delivering services, since this tends to undermine existing health systems by duplicating management structures and compete for scarce resources, especially skilled staff.

Box 5.4
Controlling malaria in Viet Nam

Source: WHO 2002; UN Millennium Project 2005b.

Nearly a third of the population of Viet Nam resides in malaria-endemic regions, and in 1991 the country faced an intensive malaria epidemic, with 144 outbreaks recorded and more than a million people sickened. The commonly used antimalaria drugs proved ineffective due to drug resistance.

Between 1992 and 1997 the government of Viet Nam made malaria a national priority, dedicating significant additional funding for its control. The money went for improving village health systems and coordinating malaria control. The package of specific interventions included free distribution of insecticide-treated bed-nets, adoption of new antimalaria drugs, and application of indoor residual insecticides. Services to pregnant women were expanded to include preventive treatment for malaria.

There was a major investment in training and supervision, and 400 mobile teams were set up to supervise health workers in malaria-endemic areas. Volunteer health workers were mobilized in communities to educate villagers and help them seek appropriate care.

Simultaneously the government worked with private biomedical firms to produce artemisinin—a powerful new malaria drug—locally. Widespread use of artemisinin combinations was an important ingredient in the dramatic reduction of malaria deaths. Coverage of indoor residual spraying rose from 4.3 million in 1991 to 12 million in 1998. In parallel, the number of people using bed-nets rose from 300,000 to more than 10 million.

This integrated package of interventions was evaluated over 1992–97. Mortality was reduced by 97 percent and morbidity by 60 percent. Local malaria outbreaks have been virtually eliminated.

Success factors for scale-up

- Government commitment to equitable access to health services and oversight of control efforts.
- Significantly increased funding and abolition of user fees for bed-nets.
- Simultaneous strengthening of village-level health systems.
- Adoption of a multipronged prevention and treatment approach.
- Significant increase in health workers—both formally trained and village level—and improved supervision of performance.
- Investment in disease surveillance and monitoring systems.
- Community involvement in prevention.

Disease control interventions need to be delivered in ways that strengthen stewardship, human resources, and management of services to improve general health outcomes and reverse the major epidemics. The key to successful and sustainable scale-up of these key interventions is to strengthen health systems, whose deplorable condition is a major obstacle to improving health outcomes in many countries. The added benefit of investing in health systems now is that it will create a sustainable base to enable countries to respond to the next wave of health challenges, including chronic diseases, which are an increasingly large proportion of the disease burden in developing countries.

A health system, defined by WHO as "all the activities whose primary purpose is to promote, restore, or maintain health," includes interventions in the household and community and the outreach that supports them, as well as the facility-based system (both public and private) and all categories of providers (WHO 2000). It should also be understood and addressed as a core social institution. When characterized by neglect, abuse, or exclusion of certain individuals or groups, the health system is a major contributor to social injustice. Conversely, the strengthening of health systems increases social capital within the community and fulfills the rights of individuals.

Delivering health services—from the community to the hospital

Different delivery strategies are appropriate for different interventions; reaching the health Goals will require strengthening all elements. Many of the key interventions—including exclusive breastfeeding, oral rehydration therapy, healthy and responsible sexual behavior, and appropriate use of antimalaria bed-nets—occur in households and communities. Households, in particular, are important in "producing health." They do this by practicing health-promoting behaviors and by delivering home-based interventions (Wagstaff and Claeson 2004), supported by community health workers. Women's education increases the use and effectiveness of health interventions. Enabling households to deliver appropriate interventions requires public education and health promotion, as well as the distribution of basic commodities by community health workers, but it does not necessarily depend on health facilities or clinical staff. So these outreach interventions could, in principle, be scaled up immediately.

Other critical interventions require functioning primary health clinics and district hospitals. Examples of services that community clinics can provide are treatment of malaria, sexually transmitted infections, and pneumonia and prevention interventions, such as childhood vaccinations, HIV testing and counseling, contraceptive provision, and antenatal care. More sophisticated health centers or hospitals are indispensable for emergency obstetric care, antiretroviral treatment, and safe abortion services (where permitted by law[4]) as well as for treating referrals of severe cases from lower levels of the system.

But few people in the developing world have access to facilities providing these services—because the facilities do not exist or lack basic equipment, essential medicines, or trained staff, because the lack of roads or transport prevents people from reaching them, or because people cannot afford the fees charged for even the most basic services (maps 5.1 and 5.2). And even where facilities are accessible and affordable, cultural obstacles, poor information, perceptions of poor quality (often justified), and a lack of trust may mean that they are not used. In Sub-Saharan Africa, the met need for emergency obstetric care—the proportion of women with direct obstetric complications treated in emergency obstetric care facilities—can be as low as 5 percent (Uganda Ministry of Health 2003).

Required investments and policies to support health systems

Addressing the obstacles to access and quality requires scaled-up investments in the health sector, backed by supportive health policies. Investments will certainly be required to ensure an adequate supply of essential drugs, clinic and

Map 5.1
Physicians per 1,000 people
Most recent available year

Source: World Bank 2004c.

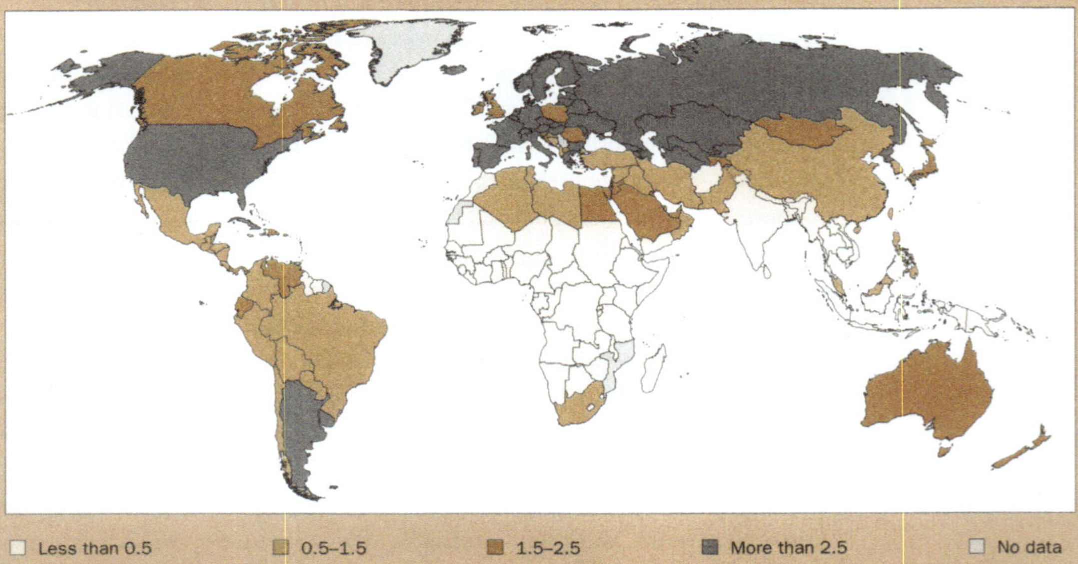

laboratory facilities, and most important, competent, motivated health workers. The latter requires high-quality training, adequate salaries, and appropriate performance rewards. Additional investments will be needed to reduce barriers to access and to improve quality of care. These include maintaining health infrastructure, improving roads, providing access to cheap transport services including ambulances, ensuring uninterrupted supplies of water and electricity, improving housing, and promoting modern communication technologies.

Better policies will often be required to complement these investments. For example, in many countries, the health worker situation is catastrophic, with extremely low absolute numbers of health workers. In many countries the available skills are inadequate or inappropriate to local needs. And the distribution of workers is strongly biased toward urban areas. Not surprisingly, motivation and productivity are low, and migration to countries offering better conditions of service is pervasive. Human resource policies are needed that align the training, deployment, supervision, and empowerment of specific cadres of workers with national requirements. Community health workers can play a vital role, one that is currently missing in many impoverished countries. Moreover, training and otherwise engaging traditional healers, who in many countries are the first point of contact for the sick, can be another important aspect of a national human resource plan (Joint Learning Initiative 2004).

Map 5.2
Antiretroviral drug coverage
Share of HIV-infected individuals in need of treatment who have access to antiretroviral drugs (%), 2003
Source: USAID and others 2004.

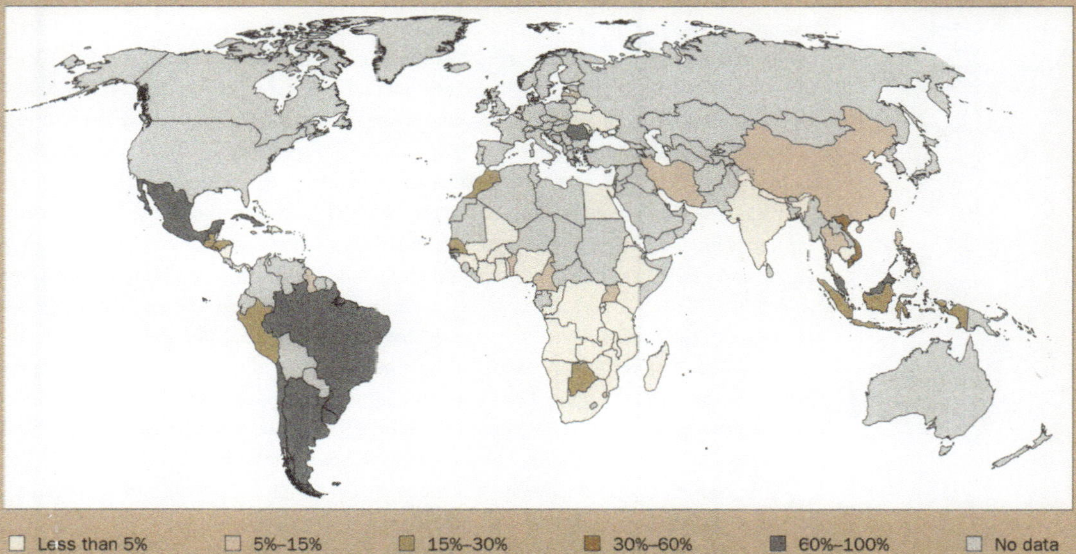

There are at least two other essential policy levers to ensure the effectiveness of investments. The first is to strengthen health system management, including planning, program implementation, financial management, supervision, and a supportive environment that fosters trust and cooperation among health workers, recognizing the vital role of human relationships in successful implementation. The second is to improve utilization through the elimination of user fees for basic health services, the adoption of culturally sensitive health education, and the involvement of communities in decisionmaking.

Changing behavior can also require changing deeply rooted community norms and understanding the economic, cultural, and gender constraints on change. Enabling real participation by the community requires investing in community capacity and developing facilitative legal frameworks and policies. Providing people with a legal right to the highest attainable standard of health may ensure more equal and equitable access to treatment by empowering groups likely to be denied such access. These challenges are highlighted when ensuring universal access to sexual and reproductive health, which is essential to the attainment of many Millennium Development Goals (box 5.5).

Box 5.5

Sexual and reproductive health—essential for reaching the Goals

Promoting sexual and reproductive health

Sexual and reproductive health is essential for reaching the Goals. It entails healthy, voluntary, and safe sexual and reproductive choices—voluntary choices of individuals and couples about family size and family formation, including early marriage and other exposures to sexual risks. Reproductive health issues thus deal with vital (and frequently sensitive) concerns including sexuality, gender roles, male and female power relations, and social and personal identity.

The current situation shows how devastating the neglect of sexual and reproductive health can be. The differences in reproductive health—between the rich and poor, both within and between countries—are larger than in many other areas of healthcare (Gwatkin and others 2003; Bernstein 2004). Maternal mortality takes some 529,000 lives a year. Of those deaths, around 68,000 are due to unsafe abortion, a sign of the need for better access to higher quality family planning services to prevent unwanted pregnancies and (where permitted by law) to safe abortion services.

Levels of unplanned or ill-timed fertility are high in many countries. The unmet need for family planning comprises women at risk of pregnancy who do not desire another birth (limiting desires) or who wish to space their birth at least two years (spacing desires) but who are not using a method of family planning. An estimated 29 percent of women in developing countries have an unmet need for modern contraception.[a] The highest proportion is in Sub-Saharan Africa, where 46 percent of women at risk of unintended pregnancy are using no method. Adding those using traditional methods to those with unmet need for modern methods brings to 63 percent the proportion of Sub-Saharan women and couples who have unmet limiting and spacing desires (Singh and others 2004).

But promoting reproductive health requires more than simply delivering services and information to prevent disease and reduce risk. It includes:

- Family planning, including access to modern contraceptives and informed and voluntary choice of family planning methods.
- Safe motherhood, including antenatal, postnatal, and normal delivery services and emergency obstetric care.
- Postabortion care and access to safe abortion, where permitted by law.
- A continuum of prevention, treatment, and care for HIV/AIDS and other sexually transmitted infections.
- Prevention, surveillance, and care for gender-based violence.
- Action against harmful traditional practices, such as female genital mutilation and early and coerced marriage.
- Information and services for underserved populations, including diverse groups of adolescents, people in emergency situations, and men (Singh and others 2004).

Each of these elements affects progress toward the Goals. Some examples:

Goal 1. *Eradicate extreme poverty and hunger.* Smaller families and longer birth intervals, a result of contraceptive use, allow families to invest more in each child's nutrition and health. That can reduce poverty and hunger for all members of a household. At the national level, voluntary reduction of birth rates may enable faster social and economic development.

Goal 2. *Achieve universal primary education.* Families with fewer children, and children spaced further apart, can afford to invest more in each child's education. This has a special benefit for girls, whose education may have lower priority than that of boys in the family. In addition, girls who have access to contraceptives are less likely than those who do not to become pregnant and drop out of school.

a. Details about the calculations and an extended discussion of the reliability, validity, and implications of the concept of unmet need can be found in Casterline and Sinding (2000).

Box 5.5

Sexual and reproductive health—essential for reaching the Goals

(continued)

Goal 3. *Promote gender equality and empower women.* Controlling whether and when to have children is a critical aspect of women's empowerment. Women who can plan the timing and number of their births also have greater opportunities for work, education, and social participation outside the home.

Goal 4. *Reduce child mortality.* Prenatal care and the ability to avoid high-risk births (such as those to very young women and those spaced closely together) help prevent infant and child deaths. Children in large families are likely to have reduced health care, and unwanted children are more likely to die than wanted ones.

Goal 5. *Improve maternal health.* Preventing unplanned and high-risk pregnancies and providing care in pregnancy, childbirth, and the postpartum period save women's lives.

Goal 6. *Combat HIV/AIDS, malaria, and other diseases.* Sexual and reproductive health care includes preventing and treating sexually transmitted infections, including HIV/AIDS. In addition, reproductive health care can bring patients into the health care system, encouraging diagnosis and treatment of other diseases and conditions.

Goal 7. *Ensure environmental sustainability.* Providing sexual and reproductive health services, and avoiding unwanted births, can help stabilize population numbers in rural areas, slow urban migration, and balance natural resource use with the needs of the population.

Goal 8. *Develop a global partnership for development.* Affordable prices for drugs to treat HIV/AIDS and a secure supply of contraceptives would greatly advance reproductive health programs in all developing countries.

Policies and interventions to address sexual and reproductive health needs

Reproductive health approaches give heightened priority to strengthening prevention efforts while recognizing the importance of a full spectrum of prevention, treatment, and care and to improving the linkages between these service systems.

Action in sexual and reproductive health also encompasses an analytic and programmatic framework, based on a human rights approach, rooted in stronger health systems, supported by communities, and accompanied by complementary interventions in other sectors. Promoting reproductive health thus requires a broad range of interventions that facilitate access to information and services, increase gender equality and empowerment, involve communities and cultural leaders, strengthen health delivery systems at multiple levels, create effective referral systems, and improve logistics and management.

There are numerous channels to integrate sexual and reproductive health services in a strengthened health system. For example, maternal and child health services can provide an opportunity for family planning information programs, referrals, and services. Integrated maternal health, family planning, and child healthcare services should add appropriate personnel and increase referral capacities. HIV prevention can be better linked with other reproductive health information and service interventions. Health system contacts for abortion, where permitted by law, and for treating postabortion complications also provide entry points for family planning services to reduce the incidence of recourse to abortion. Expanding the scale of family planning service delivery should include a range of contraceptive options to meet the needs of specific populations and accommodate choice and appropriate method-switching.

(continued on next page)

Box 5.5
Sexual and reproductive health—essential for reaching the Goals
(continued)

A large cohort of adolescents—1.3 billion, mostly in less developed countries—will require new approaches to information and service delivery. Youth-friendly services will require separate facilities addressing a complex of life skill needs—including productive and entrepreneurial skills for employment, literacy and numeracy training, and nutrition and health information, including that for reproductive health. They will also require specially trained staffs sensitive to the needs of the young. Curriculum reform must be undertaken to make age-appropriate materials, acceptable in the local context, available in school systems.

Promoting gender equality and respect for human rights together with building skills (including self-esteem, self-confidence, and negotiation skills, particularly for young women) is important for protecting young people and developing their capacities. Mass media, folk media, and other information outreach approaches must be expanded to reach the large number of out-of-school young. Working with existing institutions, parents, parent groups, and cultural leaders can make information and services more effective for young people.

Male reproductive health needs and the role of men in supporting women's reproductive health is another area for priority attention (Greene and others 2005). Local efforts have produced significant improvements in antenatal care, in using skilled birth attendants, and in contraceptive acceptance. They demonstrate the potential of scaling up these efforts. Service delivery to men (and their families) in the military and police forces has also been important in scaling up many national reproductive health programs. Additional programs addressed to men and boys are needed to encourage them to be responsible in their behaviors and to end violence against women.

Civil society and nongovernmental organizations are often the main providers of reproductive health information and services, especially for the poor (chapter 8). In scaling up access to sexual and reproductive health services, governments should seek partnerships with NGOs to test new approaches, to identify culturally appropriate interventions, and to build an evidence base for scale-up. Investments to strengthen such partner organizations and their monitoring and evaluation capacities can lead to faster increases in quality and coverage.

Education: ensuring universal primary education and expanded postprimary and higher education

There are few jobs beyond subsistence for people who are illiterate and innumerate. A lack of education is thus a sentence to a lifetime of poverty. When girls and boys drop out of school before completing primary school, or leave primary school without having acquired basic skills, they cannot earn their way in a competitive world economy, and they have less capacity to rear healthy, educated children. An essential role of government is to ensure that every child in society, boy and girl, has the opportunity to complete quality primary basic schooling—and that a substantial proportion also completes secondary or some other form of postprimary education by the year 2015 (box 5.6). Among all levels of education, postprimary education has been shown to have the greatest payoff for women's empowerment. (The gender target for Goal 3 focuses on gender parity at all levels of education.)

Box 5.6

Getting every child into school in Tanzania

Source: Mkapa 2004.

Foreign aid to Tanzania was cut in the mid-1990s, and the country's education budget was reduced dramatically. The gross enrollment rate—98 percent in the early 1980s—fell to 78 percent in 2000. The net enrollment rate also fell—from more than 80 percent to only 59 percent. The enrollment of girls fell faster, and dropouts rose higher, than among boys. Recognizing that tremendous development gains could be undone in just a few short years, the government made education a top priority. The Primary Education Development Plan, adopted in 2001, aims to achieve universal basic education by 2006, nine years ahead of the global target.

With a clear, targeted plan calling for 100 percent enrollment by 2006 and increased donor funds to support the endeavor, Tanzania took two important steps in 2001. It increased poverty reduction expenditures by 130 percent to channel more resources for basic inputs like classrooms, teacher housing, and teacher training. More important, it eliminated school fees, bringing down the cost of education for households. The results have been dramatic. In Tanzania's schools today:

- There are as many girls in school as boys.
- The number of children in primary school is up 50 percent.
- Net enrollment has risen from 59 percent in 2000 to nearly 90 percent.
- More than 30,000 new classrooms have been built.
- Pass rates for primary school examinations rose from 19 percent in 1999 to 40 percent today.
- 7,530 teachers' houses have been constructed with the direct participation of local communities.
- 17,851 new teachers have been recruited, and 14,852 sent to upgrading courses.
- More than 9,000 science-teaching kits have been supplied to schools.

Success factors for scale-up

- Government commitment and management for the scale-up plan.
- Significantly increased funding, and the abolition of school fees.
- Involvement of communities in teacher retention.
- Significant increase in new teachers and investment in upgrading of existing teachers.
- Large investment in new classrooms and other infrastructure.
- Simultaneous improvements in improving curriculum and teacher quality.

Ensuring universal primary completion and expanding access to postprimary education

Achieving universal primary completion and increasing participation in postprimary education in the developing world will require both getting out-of-school children into school and instituting changes to make schools more effective and responsive to parents and students. Depending on local conditions, specific strategies will be needed to attract out-of-school children. Abolition of school fees and special incentives to get the most marginalized groups into school can also be powerful tools. For example, cash and in-kind subsidies, such as school meals using locally produced foods, for extremely impoverished households, orphans, and girls, can promote school attendance at the primary and secondary levels. In instances where the supply of schools is the binding constraint, this will require constructing new classrooms and hiring new teachers.

Investments in primary education should be balanced with selective support to postprimary education, with particular attention to educational opportunities for girls and young women. Indeed, planning for the expansion of the postprimary level should be done in parallel with planning for achievement of universal primary education. Primary school investments can help create the pipeline for postprimary education, just as opportunities to enter into postprimary education are required to reinforce demand at the primary level. Additional interventions required to increase girls' participation in primary school (and which may also apply to the postprimary level) include making schooling more affordable by reducing costs and offering targeted scholarships; building schools close to girls' homes; making the infrastructure of schools, such as sanitary facilities, safe and "girl-friendly"; eliminating gender biases in curricula; and improving the content, quality, and relevance of education. Other important ways to promote education for girls and women include informal education opportunities such as women's literacy programs.

Countries also need to address institutional shortcomings in the education sector, many of which are linked to dysfunctional incentives for administrators and teachers. This entails:

- Developing and strengthening the constituency for a national commitment to education, with a legal and institutional framework that places high priority on public sector provision of quality education.
- Promoting mechanisms for local control of education, with an explicit role for parents and other citizens in holding schools and teachers accountable for delivering results.
- Improving the quality of information about education sector performance.
- Instituting systems to assess acquisition of skills and knowledge to an international standard, and ensuring transparency in the dissemination of this information at both national and local levels.
- Recognizing civil society organizations as legitimate participants in debates about the direction of the education system.

Increasing opportunities for higher education

But primary and secondary education are not enough. Low-income and middle-income countries should also scale up their investments in higher education—both to train the teachers and managers who will provide the primary and secondary education and to train the scientists and engineers who will underpin the continuing advance of technological capacities in the country (UN Millennium Project 2005g). Higher education is also necessary to train the doctors, nurses, natural resource managers, and other professionals who will implement MDG-based poverty reduction strategies. By building universities, technical institutes, and professional associations, countries can establish some of the most critical resources for their economic transformation. But they also need measures to ensure that all the opportunities for higher education are not captured by the rich alone. Inevitably, investments in higher education will have a regressive nature, but this can be moderated by merit-based and needs-based scholarships (Mkapa 2004).

Gender equality: investing to overcome pervasive gender bias

In many societies, women suffer deep and systematic discrimination and exclusion. Gender biases are often firmly set by cultural and social norms, and enshrined in laws that discriminate against women. The roles assigned to women and men often involve unequal labor burdens and unequal access to resources and opportunities, usually to the disadvantage of women. Girls and women usually receive less schooling than boys and men, have poorer access to health care, are at greater risk of contracting sexually transmitted diseases, including HIV, and are less able to start businesses, obtain credit, or enter higher-level occupations. They also have less voice in the decisionmaking of the household, community, and nation. Violence against women exists in epidemic proportions in many countries around the world. Because it has serious health and development impacts and is a gross violation of women's rights, it must be eliminated if the gender equality Goal is to be met.

Recent statistical evidence strongly supports the presumption that women bear more of the work burden than men. Data from time-use surveys show that women's total work time is greater than men's in most countries. In urban developing country areas, women's time spent on paid and unpaid activities was 7 percent higher than men's, and in rural areas 20 percent higher (UNDP 2003d). Most of women's work time is spent in nonmarket activity, while the opposite is true for men. Rural women's time, especially in Sub-Saharan Africa and South Asia, is heavily engaged in collecting water and fuelwood, farming, subsistence production, and domestic work, including rearing children. The time involved in these activities could be eased through improvements in rural infrastructure, especially affordable safe water and sanitation near the home, modern cooking fuels (such as liquid petroleum gas), and accessible and affordable modes of transport. The lack of basic goods and services imposes enormous time and work burdens on women that are not counted in the national income and product accounts.

Moreover, women and girls own far less of the world's productive assets—land and housing—than do men. Studies from around the world find that women represent one third or less of land owners in Latin America, Sub-Saharan Africa, and South Asia (UN Millennium Project 2005j). Yet, ownership and control over assets such as land and housing provide economic security, incentives for taking economic risks which lead to growth, and important economic returns including income.

Integrating gender empowerment throughout national strategies

In a modern economy—dependent on human capital, demographic transition, social mobility, and technological change—these and other forms of gender inequality form a fundamental barrier to economic development, with significant economic and social costs. Interventions to address gender inequality should therefore be an intrinsic part of the strategies for each investment cluster described in this chapter and should address systemic challenges as well. Reducing structural gender inequalities will require additional financial resources. Moreover, to ensure sustainability of public investment, women must be involved in key decisions about priorities and implementation.

The UN Millennium Project Task Force on Education and Gender Equality has identified seven strategic priorities that are the minimum necessary to empower women and alter the historical legacy of female disadvantage that remains in most societies of the world (UN Millennium Project 2005j). These include:

- Expanding girls' access to postprimary education, while simultaneously ensuring primary completion.
- Guaranteeing sexual and reproductive health and rights (see box 5.5).
- Investing in infrastructure to reduce women's and girls' time burdens.
- Guaranteeing property and inheritance rights.
- Eliminating gender inequalities in employment.
- Increasing women's representation in political bodies.
- Combating violence against women.

The first three priorities are addressed elsewhere in this chapter. Here we briefly discuss the other four.

Improving women's economic and political opportunities

To improve women's economic opportunities, governments need to guarantee women effective and independent property ownership and access to security rights, especially land and housing, both in law and in practice. Land rights can be in the form of ownership or rights of use, and can encompass differing degrees of freedom to lease out, mortgage, bequeath, or sell. Ensuring female property and inheritance rights would help empower women both economically and socially and would rectify a fundamental injustice.

Other strategies for economic opportunities include improving women's access to employment and conditions of work by offering job training, improving pay and working conditions, and providing child care. For countries with large informal economies, one of the highest priorities is social protection for the workers in that sector.[5]

To promote political opportunities, statutory or voluntary gender quotas and reservation policies can enhance women's representation in political bodies at all levels of government. National budgets should include the costs of providing women political candidates with training, information, and means of communication.

Combating violence against women

To address violence against women, multidisciplinary strategies must be implemented that combine infrastructural, legal, judicial, enforcement, educational, health, and other service-related actions aimed at prevention, protection, and punishment. Sometimes these strategies exist but budgetary resources are not available to implement and monitor them. Fully funding these strategies should be an integral part of any national poverty reduction strategy (UN Millennium Project 2005j) (box 5.7). These are described in detail in the report of the Task Force on Education and Gender Equality (UN Millennium Project 2005j).

Box 5.7
Empowering women in Rwanda

Source: Zuckerman 2001; Zuckerman and Garrett 2003.

The genocide and civil war in 1994 traumatized Rwanda's economy and social fabric. Women were systematically raped, murdered, and disfigured. Today they make up 52 percent of the population and head 34 percent of households.

Since the genocide the country has committed to gender equality and women have been pivotal in political and economic reconstruction. The country adopted a new constitution, which guarantees equal rights for women. The parliament has begun to repeal laws biased against women, including legislation prohibiting women from inheriting property. Prosecuted as a genocidal act under the International Criminal Tribunal for Rwanda, rape is now a first-degree crime in local Rwandese courts (previously, it was a third-degree crime).

Rwandan women are also becoming community leaders, entrepreneurs, and elected officials. In the multiparty presidential and parliamentary elections in 2003, women gained 49 percent of the seats in the lower house and 30 percent of the seats in the upper house, up from 17 percent in 1990 (IPU 2004). Rwanda has also created local women's councils elected by women only and a government ministry for women to ensure that policies are gender equitable.

The Ministry for Gender and Women in Development helped mainstream gender in the country's Poverty Reduction Strategy, and external evaluations note that Rwanda's PRS is one of the best on gender equality issues. Budgetary expenditures give a high priority to activities that address gender inequalities, and all sector expenditures incorporate Ministry for Gender inputs.

Although Rwanda faces enormous economic development challenges, the country's gender indicators are above the regional averages. In 2000 the girls to boys gross enrollment rate was 0.99 for primary education and 0.98 for secondary education (absolute rates for secondary education are very low) (UNESCO 2004). The country is on its way to parity at both levels in 2005 and to a reverse gap at the secondary level in 2015. The country's fertility rate has been declining steadily, from 7.1 in 1990 to 5.7 in 2002. Births attended by skilled health staff increased from 25.8 percent in 1990 to 31.3 percent in 2001, and the unmet need for contraception was also reduced for both adolescents and women of reproductive age (World Bank 2004c; ORC-Macro 2004).

Success factors for scale-up

- Presidential leadership and political will.
- Multifaceted effort to address gender rights through several channels.
- Changed economic and social context as a result of upheaval.
- A strong ministry for gender and women in development.
- Active and vocal women's organizations.
- Recognition by men of women's important contributions.

Environment: investing in improved resource management

A healthy environment underpins human life and well-being by providing food, clean water, disease control, and protection from natural disasters—and is thus necessary to achieve each Goal. But the environment is under threat in all parts of the world because of rising pollution, soil degradation (including rapid desertification), deforestation, destruction of coastal and freshwater fisheries, rising water scarcity, and declining biodiversity. Anthropogenic climate change, already causing environmental change, is projected to threaten agricultural productivity in many parts of the developing world, spread vector-borne diseases, and lead to a rise in sea levels and a higher incidence of natural disasters.

Environmental degradation and the effects of climate change are therefore major development issues. Most countries cannot wait until they reach higher incomes before investing in better environmental management. The degradation of the environment threatens the very basis of sustained economic growth, particularly where agriculture accounts for a large share of national income. Achieving environmental sustainability will require interventions at the national, regional, and international level (see chapter 15).

Unfortunately, the concept of environmental sustainability does not provide clear operational guidance for choosing policies and outcome targets. The first step is thus for countries to decide which environmental objectives they want to achieve by 2015 and beyond. As agreed at the World Summit on Sustainable Development, rich countries and rapidly growing middle-income countries need to take the lead in ending unsustainable patterns of production and consumption. Perhaps most important, urgent action is required to stabilize greenhouse gas concentration by reducing emissions and promoting carbon sequestration.

Developing countries, in turn, need to concentrate on integrating environmental strategies into all sectoral policies and more specifically on promoting direct investments in environmental management, considering the effect on the environment when designing sector strategies, promoting regulatory and market reforms to reduce environmental degradation, and improving environmental monitoring (UN Millennium Project 2005c). In each of these intervention areas, countries will need to consider the growing need to adapt to climate change. This includes changes in agricultural practices, improved disease monitoring and reporting systems, investments in local climate modeling and projections, and measures to stem the impact of rising sea levels.

Direct investments in environmental management

Examples of direct investments in environmental management include planting trees to combat deforestation, improving farming and land management practices to combat desertification, treating wastewater to reduce nutrient loads in freshwater ecosystems, curbing chemical pollution to protect human health and ecosystems, and preserving critical ecosystems to protect biodiversity.

Sector strategies to benefit the environment

Targeted sectoral investments are instrumental to improving the environment. Investing in modern cooking fuels to shift away from biomass will not only lower ambient and indoor air pollution, but also reduce pressure on fragile ecosystems. Likewise, improving access to water and sanitation will improve environmental quality. Of particular importance to the environment are improved agricultural practices and investments in soil health and sustainable water management for agriculture, which can stem soil degradation and biodiversity loss. To this end, agricultural extension workers should be trained to promote environmentally friendly practices that can raise yields while minimizing the use of environmental resources. In addition, countries should meet the Johannesburg goals by initiating the design of strategies for integrated water resources management during 2005.

Sectoral strategies, such as investments in infrastructure and agricultural intensification, also need to manage environmental tradeoffs. Some negative tradeoffs are unavoidable, but they can be mitigated by environmental impact assessments and improved scientific advice to senior decisionmakers (UN Millennium Project 2005c).

Regulatory and market reforms to reduce environmental degradation

Reforms to land tenure regimes and an improved regulatory environment to combat pollution are required to minimize the adverse impacts of sectoral policies on the environment. To this end, countries will need to invest in strengthening the capacity of environmental protection agencies or equivalent government bodies. In most cases this will require a substantial scaling up of their human resources, equipment, and operating budgets.

Environmental monitoring

No strategy for environmental sustainability can be successful without better monitoring. Yet, monitoring systems for water flows and quality, air quality, deforestation, and other land degradation are woefully inadequate in many developing countries. Sustained investments in strengthening environmental monitoring systems will thus be essential.

An important mechanism for implementing national strategies is the set of multilateral environmental agreements and conventions, such as the Convention on Biological Diversity, the UN Convention to Combat Desertification, the Ramsar Wetland Convention, and the UN Framework Convention on Climate Change. More funding and targeted technical support need to be made available to developing countries that wish to implement these agreements.

Science, technology, and innovation: building national capacities

The long-term driving force of modern economic growth has been science-based technological advance. Without modern technologies, the world would still be where it was centuries ago, with people at the edge of survival, always pressing on the margins of available food supply. Technologies allow human society to fight disease, to raise crop production, to mobilize new sources of energy, to disseminate information, to transport people and goods with greater speed and safety, to limit family size, and much more. Yet these technologies are not free. They are themselves the fruits of enormous social investments in education, scientific discovery, and targeted technological development to strengthen national systems of innovation.

Every high-income country makes special public investments in higher education and in scientific and technological capacities. Poor countries have largely been spectators, or at best users, of the technological advances produced in the high-income world. They lack large scientific communities, and their scientists are chronically underfunded, with the best and brightest moving abroad to find colleagues and support for scientific research.

Enterprises transform scientific and technological knowledge into goods and services, but governments play an important role in promoting the application of science and technology. They need to act in the four areas described here (UN Millennium Project 2005g). But national efforts alone are not sufficient. Meeting the Goals requires a special global effort to build scientific and technological capacities in the poorest countries—and to direct research and development toward specific challenges facing the poor (chapter 15).

Expanding access to science and technology education and research

Enhancing science and technology education has been one of the most critical sources of economic transformation. To build science, technology, and innovation capabilities, developing countries need to expand access to higher education. But more than simply offering more places, universities need to become more entrepreneurial and oriented toward key development challenges. They can participate in technology parks and business incubator facilities. They can introduce entrepreneurial training and internships to their curricula. And they can encourage students to take research from the university to firms. Most universities will need to change to take on these new roles. Governments should also expand and set up research centers focused on specific needs, such as agriculture or public health.

Promoting business opportunities in science and technology

Developing countries should use today's technologies to help create new business opportunities. Most developing countries still distinguish between industrial policies that emphasize building manufacturing capabilities and those that support research and development (R&D) to generate new knowledge.

Adopting a "fast follower innovation strategy," aimed at making full commercial use of existing technologies, would combine these two approaches while building a foundation for future R&D.

In promoting business opportunities, countries should focus on platform technologies that have broad applications or impacts in the economy, such as information and communications technology, biotechnology, and new materials. In addition, governments should adopt policies and invest in infrastructure that stimulates small and medium-size businesses, improves access to credit and other forms of capital, increases participation in international trade, and promotes the integration of regional markets. Attracting foreign direct investment can diffuse tacit knowledge and help enterprises learn about the world's technological frontiers.

Promoting infrastructure development as a technology learning process

Infrastructure projects can also be a valuable part of a nation's technological learning process. Every stage of an infrastructure project, from planning and design to construction and operation, involves the application of a wide range of technologies and requires deep understanding and capabilities from the many engineers, managers, and government officials. Policymakers need to recognize this dynamic role of infrastructure development in economic growth and take the initiative to acquire available technical knowledge from the international and local construction and engineering firms they contract with for such projects.

Improving science and technology advice

Governments must incorporate science and technology advice in their decisions for scientific and technological investments. They need first to set up an advisory structure, usually with a science advisor who reports directly to the president or prime minister. Whatever the structure, the function should have some statutory, legislative, or jurisdictional mandate to advise the highest levels of government—and be transparent to the public. It should have its own operating budget and a budget for funding policy research. Countries also need to strengthen the capacity of scientific and technical academies to participate in advisory activities, in cooperation with other institutions, especially judicial academies.

Interdependence of investment clusters

Many interventions are expected to have benefits for several Goals. For example, reducing gender inequality is essential for reducing hunger, containing HIV/AIDS, promoting environmental sustainability, improving settlements, and reducing child and infant mortality. Environmental management and ready access to clean water are essential for ensuring that clinics and hospitals function, for reducing women's and girls' time burdens so that they can engage in productive economic activity and attend school, and so forth. And to achieve any Goal, it is not enough to invest merely in one sector. Many interventions outside the health sector are needed to reduce child mortality. Reducing child mortality requires better treatment of diarrhea and pneumonia as well as improved access to clean water, hygienic behavior, improved sanitation, and reproductive services to improve birth spacing.

The work of the task forces shows that no "silver bullet" exists to reach any individual target, let alone the ensemble of Goals. Required instead are integrated strategies for complementary and mutually reinforcing interventions. (Appendix 2 summarizes the inputs across sectors required to reach each Goal.)

Implementing the recommendations: scale and sequencing

The last 50 years of development practice have shown that project success is not enough. The greatest technical challenge lies not in identifying the right interventions or making them work in one village—but in taking known interventions to scale. We know that providing access to domestic water supply and sanitation services to a nation requires far more than multiplying a village-level intervention by several orders of magnitude. It also requires a governance and public management system that extends from the national level right down to the communities. Similarly, bringing antiretrovirals to treat AIDS to all those who need them requires a functioning health system, including national, regional, district, and village health facilities. Small-scale replication of a successful NGO clinic is simply not enough to meet national health needs. As the next chapter discusses, nationwide scale-up needs a systematic plan for policy and planning, management, infrastructure, and human resources.

Some of the investments, such as those described in the list of Quick Wins, do not require significant infrastructure or highly trained workers for their delivery and can thus be implemented immediately. But many others will require better management systems, upgraded infrastructure, and significant new numbers of highly trained workers for successful scale-up. Many of the interventions in the seven clusters focus specifically on building capacity over the long term. But the term "capacity constraint" is often used in an unspecific manner to describe irremediable institutional weaknesses that pose a barrier even to getting started.

The current constraints to scaling up can be addressed with a long enough planning and investment horizon. Indeed, the next chapter recommends that "capacity" investments—such as improving transport services, providing electricity, training teachers and doctors, and instituting better management systems—must begin in the early years of the MDG framework. In this way, countries can build their ability to expand key services to their entire populations by 2015.

Chapter 6

Key elements for rapid scale-up

The UN Millennium Project recommends that the known interventions outlined in the previous chapter be implemented in the context of an MDG-based poverty reduction strategy, one that reaches the scale of investment needed to achieve the Goals. The implementation challenge has two main aspects. One is the sheer range of interventions that should be implemented simultaneously to achieve the Goals. The second is the need to reach large parts of the population to have a measurable impact on national outcomes. Here we describe the core elements of successful scale-up programs, citing examples that highlight the feasibility of reaching the Goals.

The need to scale up arises from the limited impact of pilot projects, or "islands of excellence" amidst a sea of inertia—small projects implemented at local or district levels without a measurable impact on national indicators (Uvin, Jain, and Brown 2000). National scale-up is the process of bringing essential services to most or all the population quickly, equitably, and lastingly (Carter and Currie-Alder n.d.). Equality and nondiscrimination, ensuring that the services reach all of the population, especially the most disadvantaged, are central.

National scale-up is a major managerial challenge for many developing countries. It is significantly more complex than planning and implementing a single project, no matter how large. Going to national scale demands an intersectoral approach and a carefully designed multiyear planning framework to ensure investments have the expected impact. For instance, expanding pre-service training is typically an immediate priority, since future implementation often cannot proceed without a dramatic expansion of human capacity to deliver services. Scaling up is of necessity a process of experimentation, requiring careful monitoring and mid-course corrections.

We stress that while governments have the primary responsibility for managing this complexity, by planning and funding the core services required to reach the Goals, the services can often be delivered by NGOs or the private sector and with real input from civil society. Reaching the Goals thus requires a working partnership among all stakeholders.

Successful service scale-up to achieve the MDGs cannot begin without political leadership and strong government commitment. This is an absolutely necessary (though far from sufficient) condition. Once the government has committed to reaching the Goals, it must lead on four specific actions:

- Setting concrete objectives and plans of work.
- Building national and local capacity in public management, human resources, and infrastructure.
- Adopting replicable and locally appropriate delivery mechanisms.
- Monitoring to measure progress and allow for mid-course corrections.

The three other requirements for success are the involvement and ownership of communities and civil society organizations (chapter 8), mobilization of the private sector (chapter 9), and long-term and predictable funding commitments and technical assistance from donors to give countries the means to proceed (chapters 13 and 17).

Political leadership

In most of the successful scale-ups described in the previous chapter, political leadership was the primary impetus for progress, often starting with support from the head of state. A recent review of 17 large-scale successes in health by the Center for Global Development confirms that political leadership was important in nearly all cases (Levine and Kinder 2004). Heads of state and other leaders must thus establish scaling up for the Millennium Development Goals as a national priority. They can set an ambitious tone and encourage a culture of results-based management in the often inertia-bound bureaucracies of line ministries. They need support from a broad coalition of interest groups and must work to obtain that support through open communication and inclusive planning. In many countries this support hinges on a serious donor commitment to long-term predictable funding and technical support.

Uganda's success in bringing down the prevalence of HIV shows the power of political will in mobilizing national action. In the mid-1980s, when the HIV prevalence in Kampala was 15 percent and rising, President Yoweri Museveni set the stage for a national response to the epidemic by stressing that fighting AIDS was a patriotic duty of every Ugandan. He directly appealed to civic leaders for strong leadership and to the public for open communication to combat the stigma of infection. With this encouragement, the media picked up the story of "Slim," as the disease was known, emphasizing effective prevention strategies.

In 1992 the Uganda AIDS Commission was formed to coordinate a multisectoral response to the epidemic, and AIDS control programs were set up in the ministries of education, gender, defense, and social affairs (USAID 2002).

The president also encouraged community leaders to take up the fight and bring the message to every village and town. Today, HIV prevalence is 4.1 percent (UNAIDS 2004), which signals both a large number of deaths among the infected and a promising decline in new cases. The Uganda AIDS Commission is coordinating with 1,000 partner agencies to continue to drive down transmission and, increasingly, to provide antiretroviral treatment to those in need (Uganda AIDS Commission 2002).

There are, of course, many other important cases of political leadership leading national scale-up. In Brazil, for instance, presidential leadership and the commitment of central and local governments has been critical for addressing root causes of slums (box 6.1). Similarly, the Indian government's recent commitment in its federal budget to extending rural infrastructure and services for the Millennium Development Goals will provide a crucial mandate for broad action.

Setting concrete objectives and sequencing work plans

To ensure that the process is MDG-oriented and stays on track, it needs concrete long-term objectives and shorter term guideposts for monitoring progress. As described in chapter 4, each MDG scale-up strategy must begin with an understanding of the status of poverty and its manifestations in the country. Each country then needs to set ambitious coverage targets for 2015 that will lead to reaching the Goals. These targets should take into account the human rights obligations of the state, including the rights to health, education, and food for all. Interim coverage targets and process indicators—such as how many children have access to effective healthcare or primary education—will help in monitoring progress. Interim coverage targets and indicators should be

Box 6.1

Transforming the lives of slum dwellers in Brazil

Source: Caixa Economica Federal 2002; UN Millennium Project 2005f.

Brazil's government has in recent years demonstrated an extraordinary commitment to ending the unchecked proliferation of squatter settlements and to encouraging sustainable urban development. In July 2001 the federal legislature passed "The City Statute," based on decades of local experimentation, to create a more equitable city. With these legislative conditions, the Secretariat for Housing and Urban Development then introduced its Slum Action Plan.

In April 2003 significant national government support for the items in the Slum Action Plan was secured when President Lula established a housing fund of $1.6 billion for financing new housing construction and upgrading favela or slum neighborhoods. The fund was also charged with providing direct credit support to families investing in home improvement. A variety of financial instruments, ranging from microcredit to assisted loans, are available to low- and middle-income families.

The Secretariat is upgrading 30 slums and has approximately 31,000 housing units in various states of construction or rehabilitation. In scaling up these programs, priority will go to special zones of social interest, 600 of which have already been defined in São Paulo's new master plan.

disaggregated by sex, ethnicity, and income to ensure that services are reaching marginal groups at a pace equal to or faster than the rest of the population.

Sequencing investments is also a critical feature of the MDG work plans. Each country will need to decide on policy and investment priorities for early implementation, depending on local circumstances. These might be based, for example, on where the need is greatest, or where interventions can have the greatest immediate impact. The early investments should, however, include some of the Quick Wins described in chapter 5 as well as needed investments in infrastructure, human resources, and management systems. Clearly, long-term investments—such as training engineers and doctors or building roads and sanitation infrastructure—must begin early to yield results by 2015. Other long-term investments that require early implementation include improving management and statistical systems, and encouraging behavior change in the population. These investments can unlock what are often described as a system's capacity constraints and so must be made early to enable national scale-up.

Today's development planning instruments, like the PRSP's typical three-year time horizon, do not encourage planning for these kinds of long-term investments. As a result, key constraints for physical infrastructure and personnel—which, if left unaddressed, would block ambitious expansions of services—are taken as given. An MDG-based poverty reduction strategy should instead guide countries to assess these capacity constraints realistically and then to develop an MDG framework for relieving them over a 10-year period. With this 10-year horizon the "capacity" discussion can focus on how many people need to be trained and how much physical infrastructure must be built—not on how existing shortages limit the feasible scope of scaling up.

The MDG strategy group recommended in chapter 4—reporting to the head of state, the planning minister, or the finance minister—would be an important mechanism for ensuring coherence and progress in planning and implementation among government agencies and national and international NGO partners. It would be responsible for a high-level work plan that specifies the key actions of each ministry and identifies any nongovernmental partners for service delivery. Each ministry would in turn prepare detailed MDG-based work plans that include both longer term activities (training human resources or building power plants) and shorter term activities (purchasing and distributing essential drugs). In many instances public service managers will need to learn to work more closely with community-based organizations, which at a minimum should participate in the design and monitoring of scale-up plans through representation in the strategy group and through regular civil society consultations. To be sure, these complicated but necessary processes will be extremely challenging for the poorest countries with limited human resources, so international partners will often need to provide important technical support.

Building national and local implementation capacity

The short time left before 2015 means that national capacity to manage scale-up—and thus to absorb additional resources—will need to be strengthened at the same time as service delivery is expanded. The UN Millennium Project emphasizes the need for simultaneous investments in direct service delivery and in building capacity, here defined as public sector management and administration, infrastructure, and human resources.[1] These capacity investments will also have the effect of improving governance and transparency.

This two-pronged strategy is critical because waiting for capacity to grow organically or for reforms to be implemented before making the necessary investments will make it impossible to meet the 2015 deadline. Over the past decade, donors have often made funding for infrastructure and service delivery contingent on capacity building and institutional reform. But in many cases the acquired skills atrophied before the investments materialized—or the "reforms" were merely cosmetic. In other cases the aid or private investment in service delivery expected to follow institutional reforms never appeared. Allowing reforms and investments to take place simultaneously can help address the tension between the desire to have reforms in place before making investments and meet the Goals by 2015. It can also ensure that needed capacity building and reforms are grounded in reality.

Public sector management

Management systems are an essential part of service delivery, if often overlooked and underfunded. Also frequently overlooked is that the quality of public sector governance depends crucially on investments in public sector management systems. Even in countries with good governance, public management tends to suffer mightily from a lack of trained managers, poor information systems, rigid civil service procedures, and inadequate budgets to address these concerns.

The situation of public sector managers and civil service workers in many low-income countries has deteriorated over the past 20 years as a result of prolonged underfinancing of the public sector. Cash-strapped governments are often forced into draconian actions such as civil service hiring freezes, or across-the-board reductions in work forces and budgets to maintain macroeconomic balance. IMF and World Bank–supervised programs sometimes include those freezes because an increase in official development assistance that could ease the fiscal austerity is simply not forthcoming from donors. Even when IMF and World Bank staff recognize the deleterious aspects of such policies on the delivery of public services, the macroeconomic margin for maneuver may be limited unless increased ODA or debt relief are made available by the country's donors. We recommend that IMF and World Bank staff use the evidence of MDG-based needs assessments to highlight these constraints more forcefully to donor governments and to promote the needed overall increases in donor assistance (see chapters 13 and 17 for more on donor assistance needs).

In the poorest countries these fiscal austerity programs have often led to a catastrophic failure to improve public management processes. Macroeconomic stability may be achieved but the victory comes at great cost, since public services are deeply compromised (or the chance for improving public services is missed). What's more, development theory has not focused on this area. As the Shanghai Conference on Scaling Up noted, "The development literature...has largely ignored the underlying processes and systems for institutions to innovate, fail along the way, learn from that failure, and continue to expand" (Malhotra 2004). This makes reinvesting in public sector management an urgent need today. Our definition of public sector management includes planning, financial systems, human resource management, reporting and accountability structures, data and information systems to inform decisionmaking, and adequate record keeping.

Managerial roles should be clearly defined and supported, and managers should be given sufficient authority—over priorities, finances, and staff—to perform effectively. Specific management bottlenecks in many poor countries include a lack of information and communications technology and overly rigid organizational structures that discourage innovation. There is also a severe shortage of managers in most line ministries, especially at the district and community levels. Again, these are all issues amenable to investment.

Underlying good management is access to high quality data. Expanding national statistical services for data collection and monitoring measurable outcomes at the country level enables results-based management. The essence of managing for results is that good policies, based on empirical evidence and a clear understanding of the development process, lead to improved outcomes consistent with national priorities and objectives.

In many countries, decentralization has made building management capacity at regional, district, and municipal levels a particularly high priority. The intent of these reforms is correct: many aspects of program design and implementation are best carried out at more local levels of government, which are closer to those who require the services and have better access to local information. But local managers have often been given new responsibilities—for setting priorities, implementing and monitoring programs, and managing financial flows—without receiving appropriate training and without corresponding increases in their administrative budgets. Moreover, in some cases processes for ensuring the prompt flow of resources from the center to the periphery need to be streamlined, as severe bottlenecks have impeded the local use of allocated funds. Building the managerial capacity of local NGOs and developing more efficient procedures for channeling funds to them are also priorities, since these organizations are often best placed to deliver certain services, such as community HIV prevention or orphan support.

Some specific public sector functions and tools that require strengthening to improve the effectiveness of public sector managers in implementing an ambitious scale-up of service delivery are listed below. As discussed further in chapter 7, these strategies are equally critical in promoting transparency and good governance.

- *Civil service planning.* Each country needs a merit-based civil service with adequate pay to attract and retain the human resources for scale-up within a global labor market. This requires careful human resource planning to eliminate redundancies and deploy civil servants in the most effective way possible.
- *Information technology and management systems.* Transparency and accountability mechanisms can ensure that civil servants at all levels of government have an incentive to perform. Setting up such systems requires political will but also increased resources for investing in information and communication technology and financial accounting systems to track implementation. In addition to improving transparency, those systems can make government processes such as budgeting, monitoring, and issuing such documents as licenses and registrations more efficient and responsive.
- *Monitoring and evaluation systems.* Monitoring and evaluation systems can ensure that different departments perform their tasks in the most effective manner possible. Such systems may need independent auditing structures and periodic reporting procedures. Civil society groups can be part of monitoring and evaluation at the local, regional, and national levels.

Human resources

People manage the systems of service delivery. And people deliver most services. In many poor countries the shortage of trained workers and managers is the binding constraint on scaling up services. To achieve the Goals, human resource needs across sectors have to be carefully assessed and strategies for recruitment and retention have to be created. Retention packages should reward high performance and include concrete incentives for service in rural areas. And preservice training (such as that in medical schools and teachers' colleges) will in most cases need to be vastly increased to scale up core services. Expanding tertiary training is expensive and time-consuming, and few donors have invested in this.

The human resource challenge has been perhaps most extensively researched in the health sector. The Joint Learning Initiative—an international effort to identify solutions to the human resource crisis in healthcare—reported a worldwide shortage of more than 4 million health workers (doctors, nurses, and midwives). The situation is particularly severe in Sub-Saharan Africa where numbers of health workers have stagnated or even fallen in the last three to four decades due to emigration, inadequate investment in training and salaries, and AIDS. The Joint Learning Initiative (2004) estimates that this region needs an additional 1 million health workers to effectively provide essential health services.

Training technical and professional workers is equally important in other service areas. Qualified teachers are in short supply in many countries, particularly in Sub-Saharan Africa. FAO estimates that there is a "critical need" for trained agricultural extension workers in developing countries (Van Crowder 1996). Training institutions for agricultural extension also need to be brought up to date, and those already working in the field need substantial upgrading to address new developments in agricultural technology and markets. Needs differ from context to context, but the general picture is the same nearly everywhere.

In addition to increasing the overall output of training programs, the curricula in many countries need to be overhauled to emphasize local priorities and solutions. For instance, medical schools in developing countries often use Western European or American curricula and textbooks that pay little attention to the tropical and infectious diseases the new doctors will spend most of their time treating. Similarly curricula may need updating and revision in other fields like education or agriculture.

In a series of recent interviews that the UN Millennium Project conducted with bilateral donor agency representatives on expanding capacity, very few mentioned assistance for preservice training as an area of focus. Training initiatives tend to focus much more on in-service programs, such as continuing education workshops for professionals. But there are early signs that this may be changing. Recently, the UK Department for International Development (DFID) announced a program to boost the supply of health workers in Malawi, a country particularly hard hit by AIDS. Among other initiatives, DFID is helping fund a 50 percent expansion of preservice training for doctors and nurses through investing in infrastructure and teaching staff at existing medical schools. The impact on the numbers of nurses will thus be seen within three years and on the number of physicians within five to seven (DFID 2004).

Another factor that contributes to the human resource crisis in many developing countries is the emigration of professional and skilled workers to countries offering a better wage and quality of life, also known as brain drain. This is particularly relevant in the health sector, since the global labor market for doctors and nurses has seen many low-income countries invest significantly in training outstanding young people for careers in healthcare, only to lose them to rich countries facing health professional shortages. While some migration will inevitably accompany globalization, rich countries have a responsibility not to fill their human resource gaps by draining the professional work forces of developing countries through aggressive recruiting. This is a priority for the ILO, which is seeking to develop recommendations that put more responsibility onto developed countries to train more staff of their own and to control recruitment from countries at risk of significant impact from brain drain (Lowell and Findlay 2001).

Some developed countries, like the United Kingdom, have made important strides to reduce brain drain. In 2001 the UK's National Health Service adopted a code of practice that bans active recruitment of developing world physicians and nurses and directs NHS employers not to use employment agencies that recruit in those countries (UK Department of Health 2004). In addition to such voluntary restrictions on recruiting in rich countries the Joint Learning Initiative recommends establishing a global educational reinvestment fund to support expansion and enhancement of training opportunities in developing countries (Joint Learning Initiative 2004).

Developing countries, for their part, should be creative in filling major human resource capacity gaps by delegating activities to lower level providers, such as nurses and clinical officers in health, and training additional cadres of frontline workers. In many cases this will require revising regulations restricting delegation. For instance, in many developing countries, nurses are taking on a much greater role in delivering antiretroviral therapy. In addition to skill delegation, countries can train special cadres of lower level providers, such as clinical officers, community health workers, and pharmacy technicians. These workers require less training than physicians or nurses and can provide important services, especially in rural and remote communities.

In many instances a basic level of competence can be achieved with only one or two years of postsecondary training. This is what the Ethiopian government is doing right now in training 20,000 secondary-school graduates as rural community health workers in a one-year training program. They will provide preventive and some curative care in hard-to-reach villages.

Community health workers should be trained as part of a healthcare team that reaches from the community to the district-level referral hospital (box 6.2). No element of the system can work in isolation. Clinics and hospitals will be

Box 6.2

Health workers to control malaria in Ethiopia

Source: Ghebreyesus and others 1966; Ghebreyesus and others 1999; Kidane and Morrow 2000.

The Tigray Region of northern Ethiopia has about 4 million people, 75 percent in sites vulnerable to malaria outbreaks, leading to high rates of malaria-induced illness and death. Health services there are poor—less than half the population lives within 10 kilometers of a health center.

The regional government introduced community-based malaria interventions for dealing promptly with malaria outbreaks. A network of 700 volunteer health workers was assigned the tasks of mobilizing communities, taking source-reduction measures, and providing clinical diagnosis and treatment. A package of interventions includes home management of cases by training of mothers and local village volunteers. District health management teams and malaria control personnel provide supervision, technical support, and free distribution of antimalarial drugs. All villages are mapped by geographic positioning systems, and HealthMapper software facilitates the surveillance of malaria and the analysis of trends.

More than a half million people receive free treatment for malaria each year through this network of volunteer health workers. Also successful is a program to recruit and train grandmothers to train neighborhood mothers in diagnosing and treating their children at home. This community-based approach led to a 40 percent reduction in deaths of children under five. It is now being implemented nationwide.

underused unless there is early recognition of conditions that require urgent care, such as serious illnesses in children and obstetrical emergencies, which often can be provided by well trained community workers. Conversely, village-level workers, who lack the skills required to provide care for serious conditions, will need to rely on functioning clinics and district hospitals.

The community worker approach is not new, but its application has tended to be piecemeal in resource-constrained settings. As part of an MDG-based poverty reduction strategy for building service delivery systems, we recommend a major scale-up of at least three types of community worker:

- Community health workers, as exemplified by China's famed "barefoot doctors."
- Community agricultural extension workers, to teach farmers about best practices in use of improved seeds, fertilizer, and small-scale water management and to mobilize communities to organize themselves to negotiate better prices for their products in local markets.
- Rural and urban community engineers, who can be trained in core tasks of infrastructure design, management, and maintenance. They would address village needs in irrigation, land reclamation, drinking water, sanitation, electricity, and vehicle and road maintenance.

Again, adequate oversight, ongoing training, and referral links to higher levels of the system should support these frontline workers.

Expanding human resources for publicly financed interventions will raise productivity and yield important macroeconomic multiplier benefits. At the same time, the public sector expansion strategy will need to be closely linked to overall employment strategies—since a major scale-up of the public labor force must be matched with the needs of the private labor market, so that private sector growth is sustained in the long term.

Infrastructure

The importance of infrastructure—including roads, ports, telecommunications networks, electricity plants and grids, public transportation, and water and sanitation networks—to achieving the Goals is emphasized throughout this report. Roads make it possible to rapidly transport women with labor complications to hospitals for emergency obstetric care and allow farmers to deliver crops to markets. Electricity grids power schools and hospitals. Water and sanitation services improve health. So scaling up interventions and coverage will be possible only if large-scale infrastructure investments are made in conjunction with the expansion of service delivery. Investments are required not only for construction, but also for operation and maintenance.

Over the past 20 years, donors have moved away from financing infrastructure for a variety of reasons, including corruption and adverse effects on communities and the environment. But there are some signs suggesting that development practice is shifting, as evidenced, for instance, by the arguments for infrastructure in the World Bank's *World Development Report 2005* (World Bank 2004d). By integrating large-scale infrastructure into their poverty reduction strategies, developing countries can increase private investment and enable scale-up of services to meet the MDGs. Learning from experience, countries will need to mitigate the social and environmental impact of such investments and ensure transparent and accountable business practices.

As discussed in chapter 5, infrastructure projects are also a learning opportunity for countries to benefit from technology transfer. By managing the relationship with foreign firms brought in to build infrastructure, countries can make sure that domestic workers and managers develop their knowledge base (box 6.3) (UN Millennium Project 2005g).

In addition to large-scale infrastructure, countries need facilities for delivering social services, such as clinics, schools, and granaries. It is difficult to teach students without a school—and impossible to save the life of a woman with obstetrical complications who needs a caesarean section without a hospital. Development partners thus need to focus on helping countries overcome these key infrastructure bottlenecks. A challenge for developing countries is determining how many facilities to rehabilitate and build. As a general rule, facilities should be built early in the 10 years remaining to 2015, since they are so vital to delivering key interventions at scale.

To prepare draft investment plans, many countries use population-to-facility ratios as guidelines for determining how many of a given facility they need to build for their population. This is a good start. But in the final strategy,

Box 6.3

Transferring infrastructure technology in Algeria

Source: UN Millennium Project 2005g.

Algeria's construction industry has been one of the "industrializing industries" since the 1970s. The government encouraged the purchase of complex and advanced systems of technology from foreign firms. Turnkey and product-in-hand contracts were used to assemble and coordinate all project operations—from conception through implementation to installation—into one package. The aim was to transfer the entire responsibility to the foreign technology supplier.

These contracts did not lead to as much technology transfer as hoped for. The turnkey contracts did not include the sourcing or training of local skills. This meant continuing reliance on external assistance for management and skilled operations—or inefficient operation by local management.

Having learned from these experiences, the Algerian government later encouraged "decomposed" or "design and installation supervised" contracts, under which infrastructure projects are more fragmented and involve more local firms than under the integrated contracts. Local firms now take charge of the phases before installation (such as exploration and planning), previously done by foreign technology suppliers under integrated contracts. With the technical assistance and supervision of foreign suppliers, local managers now carry out the projects. This new approach reduces the uncertainty in implementation and facilitates the process of learning-by-doing in local firms, thus enhancing their technological capability. It also contributes to improving managerial capabilities of local managers, because they have more opportunities to participate in implementation.

countries will obviously need to conduct a more detailed analysis of where their facilities are located, and where and how many they need to build or rehabilitate. When building more facilities, countries also need to pay attention to equity of access. For example, many developing countries have first-rate hospitals and modern schools in their capital cities, but dilapidated facilities in their rural districts. A much more equitable distribution of resources is vital to achieving the Goals. Countries thus need to create investment plans that explicitly aim to increase the percentage of the population that has access to high quality facilities, such as the percentage of the rural population with access to a functioning clinic within 10 kilometers.

Replicable and locally appropriate delivery mechanisms

The scalability of services is aided by choosing highly replicable (or algorithmic) service protocols where possible. TB treatment protocols (such as directly observed treatment, short course, or DOTS) are typically standardized, as are malaria treatment regimens and fertilizer combinations. Standardization also enables comparison of performance across regions, enhancing quality control.

Of course one size does not fit all, and any algorithms (step-by-step procedures) will need to be adapted to local conditions. But clear and simple decision algorithms will be especially important if services are to be performed by less highly trained personnel, as suggested above. While healthcare and education have increasingly adopted standardized approaches, much remains to be done to encourage similar standardized strategies in other sectors. Academia has an important role in proposing guidelines and protocols, if there is an agreed-on best way to deliver an intervention. Where evidence is less clear, academia can work with governments to help disseminate best practices that have proven effective in simplifying the delivery of core services in local settings.

Services can be delivered through the public sector, the for-profit sector, and local or international nongovernmental organizations. The UN Millennium Project stresses the fundamental responsibility of national governments to guarantee and oversee the provision of the basic services required to meet the Goals. Actual delivery can be delegated to the private sector or civil society when it is more efficient, as is possible for such infrastructure services as water, energy, or transport in urban areas (chapters 8 and 9). Regardless of who delivers the services, the government must ensure effective access to the services by rich and poor alike. In some cases this will require targeted public subsidies, even if service delivery has been contracted out to an NGO or private company.

In choosing delivery strategies, policymakers should consider not only efficiency, but also the impact on other interventions and delivery systems. Some health interventions, such as childhood vaccinations, are traditionally delivered through freestanding vertical programs, circumventing the inefficiencies of many developing country health systems. It is also possible that more sophisticated health services, such as antiretroviral therapy, could also be efficiently and rapidly scaled up by establishing dedicated treatment centers, supplied by dedicated distribution networks and funded directly by donors. Such a strategy would, however, endanger existing health services, and thus the provision of other critical interventions, by competing for limited resources in the short term, particularly trained staff. This approach would also squander an opportunity to strengthen all health services by building strong, unified systems that can sustain service delivery beyond 2015. Where possible, governments should identify synergies so that multiple interventions can be delivered with the same tools and infrastructure. More generally, the unified perspective of MDG-based planning requires taking into account the trade-offs between scaling up some services as rapidly as possible and building the systems required to meet all the Goals.

Monitoring and feedback

Improving the flow of information within the government is critical for increasing transparency, fighting corruption, and increasing the effectiveness of government. As part of their scale-up plans, countries need to develop strategies for improving data management and dissemination within and among all levels of public administration. These data will allow for monitoring progress and enable mid-course corrections. Sustained investments in modern information and communication technologies hold great promise for facilitating the dissemination of information to increase public sector transparency.

Investments in statistical services are also urgently needed. Today's ad hoc international statistical efforts are unreliable—often duplicative, inconsistent, and burdensome to national governments. Sustainable statistical capacities must be available to run population and housing censuses, conduct household surveys, set up vital statistics and health information systems, and compile indicators on food, agriculture, education, and the economy, among other areas.

Of 56 countries and areas in Africa, 19 have not conducted a population census in the last 10 years, nearly twice as many as in the previous 10. And many countries do not have a sustainable, coherent program of household surveys, or administrative systems to produce basic statistics routinely.[2] These are areas where technical assistance from development partners can be instrumental to success and where public-private partnerships can be especially fruitful. The recent Marrakech Action Plan for Statistics recommends a global framework for addressing current gaps in statistical capacity by mainstreaming strategic planning of national statistical development strategies, beginning rapid preparations for the upcoming census round in 2010, establishing an international household survey network, harmonizing donor support for survey programs, and increasing international financing for national statistical capacity building by approximately $150 million a year (World Bank Development Data Group 2004).

Monitoring should focus on measuring the impact of investments and tracking the flow of funds. Communities are ideally positioned to report on both. Community members know how often a doctor is in the clinic or how many children complete primary school. To reduce graft, district governments and local authorities should make funding flows transparent to community members. For example, publicly posted information on all budgetary outlays will allow local civil society to be a watchdog and provide advice to help direct funding to the area of greatest need.

MDG progress reports should be compiled periodically, with community participation, to share results within countries and internationally. This process is already well under way with the publication of MDG country reports by 90 countries. Once again, these reports should disaggregate results by sex, region, income, and, where relevant, ethnic group.

Another strategy is to use national and international human rights accountability mechanisms. This can complement the efforts of national economic and finance ministries to monitor progress toward the Goals. Using a rights-based approach, monitoring can measure achievement against a right, rather than a target. In other words, what did a given action or program contribute to the realization of the particular right? Evaluation often measures whether a given action contributes to reaching a target. But conceived in terms of rights, the same evaluation would measure not only those reached by a given action—but also the extent to which others are being educated about the right and are empowered to demand the right and whether the right is protected in legislation. (Human rights–based accountability mechanisms are discussed in detail in chapter 7.)

Putting communities at the center of scaling up

Supplying services to communities is only part of the equation in reaching the Goals. To have any impact, services must meet local needs and be appropriately used by communities. The best way to ensure that services are demanded and effective is to involve communities in planning and implementing their scaling up. So district and local authorities should consult their communities on the best ways to spend decentralized funds.

Information and education are essential in promoting community demand for services that may be unfamiliar or not considered a priority. Community members can be effective in providing such education—and in the implementation of programs and services, either as volunteers or as paid community workers. The message of a health program, for instance, can be stronger if delivered by a local person respected by the community. Nonspecialist community workers can perform important functions as well—for example, village traders can distribute free mosquito nets to the community. Beyond service planning and provision, communities can monitor government activity and ensure greater government accountability. To do this well they should have access to relevant information and meaningful recourse when governments do not deliver.

Cost can be an important barrier to the use of services by communities. Many countries, short of funds for delivering services, have user fees to help offset the cost of service delivery. Most studies, including a recent one from Uganda where user fees were abolished, confirm that user fees for such essential services as health and education are a significant barrier to accessing services, particularly for the poor.[3] To increase use of core services, indirect costs—say, for transportation and time away from work—may need to be covered for the poorest groups.

Promoting scale-up through long-term funding commitments and technical support

For any scale-up program for the Goals to work, funding has to be both adequate and predictable for the long term. For example, in many of the poorest countries, donors will need to support such recurrent costs as salaries (chapter 13). While donors are showing more interest in salary support, it needs to be put in place rapidly for eligible countries. In chapters 13 and 17 we discuss the required increases in ODA and improvements in the quality of aid that will be required to meet the Goals. Without sufficient funding for the next 10 years and probably beyond that, scaling up is impossible. A country cannot plan long-term investments in medical schools and water supply without guarantees that the funds will not suddenly dry up midway. Developing country governments frequently complain that planning for such long-term investments is extremely difficult because of uncertainty about a steady stream of donor funding. For their part, developing country governments need to increase long-term domestic resource mobilization and ensure budget transparency.

International momentum is growing to harmonize and align the planning and disbursement of donor aid to reduce the high "compliance costs" imposed on developing country governments by multiple sets of donor-imposed conditions and reporting requirements. Sectorwide approaches are one promising mechanism for harmonizing donor activity at the country level and for better aligning the funding with the government's sectoral priorities. Using a sectorwide approach, several donors pool their funding and direct it to the budget of the relevant ministry—rather than to donor-defined projects. Most sectorwide approaches today are in health and education, but the potential exists for expanding them to other sectors as countries create long-term plans for scaling up (chapter 13).

Technical support from bilateral donors, multilateral organizations, and NGOs is also essential for scaling up services. Many UN agencies are well positioned to offer such assistance, and some, like WHO, are increasing their assistance to countries. Bilateral donors and NGOs also offer valuable technical assistance. In the context of scaling up, this technical support must focus on sharing best practices in management and oversight—as well as on more specific areas to ensure that countries quickly build the skills to expand service delivery. Such assistance may be needed for line ministries—which are often understaffed and overtaxed and require support to create and oversee the MDG-based work plans—and for civil service reform.

Chapter 7

Governance to achieve the Millennium Development Goals

The successful scale-up of investment strategies to achieve the Millennium Development Goals requires a commitment to good governance. This includes upholding the rule of law through administrative and civil services and through legal and judicial institutions. It includes promoting human rights, particularly civil liberties and political freedom. It also includes sound economic choices, especially for macroeconomic policies and regulatory frameworks. And it includes transparent, participatory, and accountable decisionmaking processes. These critical elements of governance serve as vital complements to the scaling-up of public sector management capacity (chapter 6).

There has been great progress in quantifying and standardizing indicators of governance (box 7.1). Although the idea of "poor governance" is often still used as a euphemism for corruption in development policy circles, advances in research and measurement have helped outline the many components of governance. The result is the ability to measure the variation in governance indicators across and within countries. Some countries have high scores on an absolute scale while others, led by political reformers, score poorly not because of their leaders' actions but because of entrenched corruption, possibly as a legacy of past regimes. Still other countries are governed by corrupt rulers, while some countries fall into violent conflict, making good governance difficult if not impossible. (We return to the special issues of violent conflict in chapter 12.)

The data also show that almost every dimension of governance is highly correlated with income. This correlation signifies a two-way relationship: good governance helps achieve higher income, and higher income supports better governance.

It is broadly accepted that better governance can lead to higher economic growth as a result of more efficient divisions of labor, more productive investments, lower transaction costs, and faster implementation of social and

Box 7.1
Assessing governance: many approaches

Good governance is a challenge to quantify and measure. The attempts vary dramatically in what is measured. Here we highlight some of these attempts and their definitions of good governance.

- *Country policy and institutional assessments.* The World Bank's country policy and institutional assessment cover policy choices and institutional structures. They evaluate economic management (debt, macroeconomic and fiscal policies), structural policies (trade, financial, private sector strategies), policies for social inclusion and equity and public sector management and institutions (rule of law, financial management, efficiency of public administration, transparency, accountability, corruption).
- *Freedom House.* The Freedom in the World rankings use surveys to measure political freedoms and civil liberties. Political freedoms are measured by the right to vote, compete for public office, and elect representatives who have a decisive vote on public policies. Civil liberties include the freedom to develop opinions, institutions, and personal autonomy without interference from the state.
- *International Country Risk Guide.* The International Country Risk Guide ranks countries based on political, economic, and financial risks. Political risks include government stability, socioeconomic conditions, investment profile, corruption, conflict, quality of bureaucracy, democratic accountability, law and order, and the presence of religion and the military in government. Economic risk measurements include per capita GDP, GDP growth, inflation, and fiscal policies. Financial risk measurements include foreign debt, trade balances, official reserves, and exchange rate stability.
- *Kaufmann, Kraay, and Zoido-Lobaton and Kaufmann, Kraay, and Mastruzzi.* These data sets, produced by the Global Governance group at the World Bank Institute, rank countries based on six aspects of governance: voice and accountability, political stability, absence of violence, government effectiveness, regulatory quality, rule of law, and control of corruption.
- *Millennium Challenge Account.* The Millennium Challenge Account was announced by the U.S. government in 2002 as a new foreign aid program to assist countries that are relatively well governed. Governance is measured based on three broad categories: ruling justly, investing in people, and encouraging economic freedoms. Ruling justly is measured by scores on civil liberties, political freedoms, voice and accountability, government effectiveness, rule of law, and control of corruption. Investing in people is measured by public spending devoted to health and education, primary completion rates, and immunization rates. Encouraging economic freedoms is measured by fiscal and trade balances and the investor climate.
- *Transparency International.* Transparency International ranks countries on the basis of a Corruption Perceptions Index, a composite index that measures the degree to which corruption is perceived to exist among public officials and politicians.

economic policies.[1] But it is not often properly understood that poorer countries with low levels of human capital are less able to afford good governance, since this requires a well functioning and adequately paid civil service and judiciary, proper information technology (for registration of property or transparency in procurement), equipment and training for a reliable police force, and many other outlays for proper public administration. Moreover, richer countries generally have more literate societies, with civil society organizations and nongovernmental organizations, including the media, better able to act as watchdogs of public sector activities. Higher incomes also promote political participation and constraints on executive authority. Barro (1999), for example, has presented evidence to suggest that economic growth supports the development of democratic political institutions.

Plenty of evidence shows that human capital is a fundamental predictor of economic growth and that rising human capital in turn seems to contribute to improved institutions (Glaeser and others 2004). This is important since it suggests that external factors contributing to low human capital, such as endemic disease (malaria) that leads to high child morbidity and mortality, can have an important adverse effect on the development of government institutions. It also corroborates the findings of Sala-i-Martin, Doppelhofer, and Miller (2004), whose innovative analysis finds that human capital and geography variables were among the main predictors of economic growth in the late twentieth century.

The upshot is that while good governance can contribute to economic growth and bad economic governance can certainly impede growth, governance itself can be improved by investing in other factors (such as education and health) that support overall economic growth and human capital accumulation. This two-way causation is hugely important from the vantage point of the Millennium Development Goals. It underscores the importance of a broad-based strategy to meet the Goals, directly through good governance practices and indirectly through investments in human capital, public sector management, and infrastructure. It also underscores the point that on average a poor country is likely to have lower governance scores than a richer one, even if both governments have equally benevolent and committed political leadership. A proper assessment of a country's governance therefore requires not an absolute scale of measurement, but a measurement in relation to other countries in a similar income group.[2]

A related complication, frequently overlooked in discussions of governance, is that most available indicators—such as perceptions of corruption, government effectiveness, and risk of expropriation—are outcome indicators that only partly reflect the will and actions of policymakers.[3] For example, if one uses only outcome measures to evaluate a country's efforts in governance, a new government that is committed to ending corruption but that has inherited a system of entrenched corruption from its predecessor will be penalized for high levels of corruption. Instead of punishing such governments, the development partners should help the new leaders to root out the remaining corruption. Similarly, governance assessments cannot rely solely on absolute indicators of rule of law, civil liberties, or institutional strengths since many of those systems require real resources to be implemented. Instead, a more effective approach is to assess improvements in outcomes and to compare them with countries at similar income levels. Many government leaders in poor countries with weak governance systems are making heroic efforts at improvement, and those efforts need to be recognized and supported.

Strategies to enhance governance for the MDGs

It is the responsibility of countries themselves to strengthen their own government systems. On that, there can be no doubt. But particularly in the light of evidence suggesting an important role for human capital and other factors that contribute to governance, the international community can often support the poorest countries in strengthening both the components of governance and the elements that contribute to long-term good governance. As stated in chapter 4, we recommend that a strategy for strengthening governance be included in every developing country's long-term MDG framework. But we stress that there is no one-size-fits-all approach to undertaking highly contextualized approaches based on local needs.

To this end, we stress the need to distinguish between two broad sources of inadequate governance: bad volition and lack of capacity. The first addresses the genuinely "corrupt" governments where political power is held by larcenous leaders. The state may be run for the personal plunder of a narrow elite, a particular interest group, or an ethnic group. These are countries that consistently rank low on civil liberties, political freedoms, and human rights while rating high on corruption, with little demonstrable will to achieve broad-based poverty reduction. In these cases, the international community can play a role in humanitarian assistance and deliver aid through NGOs, but there is little hope for achieving the Goals.

At the other end of the spectrum is a second—typically overlooked—category of countries that suffer from weak governance, not because of the ill will of the leaders, but because the state lacks the resources and capacity to manage an efficient public administration. Of course most countries are somewhere between these two extremes, and it is sometimes difficult to determine levels of volition, but the distinction is critical for understanding a country's governance challenges and for shaping the appropriate responses.

The most important point is that when the limiting factor is not the volition of leaders but a lack of capacity or resources, we should view the governance challenge as an opportunity for investment in improved skills, capacities, and systems. Governance, in short, should be considered in *operational* terms, subject to investment and improvement.

Governments lacking volition

With truly rapacious government leadership, the scope for long-term development policy is limited until that leadership leaves office. Some of these governments are the outcome of deeply flawed political processes. Many such regimes in developing countries have been propped up with external support from the rich world for economic or geopolitical reasons. Indeed, memories are often short in developed countries. Public officials and the broader public tend to forget the role their countries played in supporting truly corrupt politicians and political structures, while at the same time criticizing the developing country governments of today for not having developed better institutions.

In such cases, it would be difficult for the government to produce a broadly credible MDG-based poverty reduction strategy (chapter 4), and large-scale external budget support would clearly be inappropriate. Indeed there is limited potential to achieve the Goals under such circumstances. Development strategies need to focus on humanitarian and health concerns, and aid should be channeled mainly through nongovernmental organizations. International assistance needs to be closely tied to incentives for improving governance, especially in the realms of civil liberties, political rights, voice and accountability, and anticorruption successes. Support is also needed in these countries for civil society organizations that monitor corruption, human rights abuses, government secrecy, and repression.

What about economic sanctions? Not only are they difficult to implement, but they tend to harm the population and political opposition at least as much as the government. Sanctions typically have little success in toppling bad regimes, but they can have a powerful effect in impoverishing the society.

Improving governance in poorly resourced but well intentioned governments

In low-income countries where the political will genuinely exists to meet the Goals, specific investments and policy reforms are necessary to improve governance in six areas: public administration, strengthening the rule of law, increasing transparency and accountability, promoting political and social rights, promoting sound economic policies, and supporting civil society. The remainder of this chapter focuses on these issues, all of which need to be tackled in the context of an MDG-based poverty reduction strategy.

Investments in public administration

In poor countries with well intentioned governments, the public administration should be a target of investment. The private sector long ago learned that good management requires resources; the same holds in the public sector. For example, low-income country governments with good volition but poor public administration frequently need to raise civil service pay scales to make them comparable to the salaries offered by the private sector, international agencies, and development partners. Higher pay is needed to attract and retain highly qualified public sector workers and to reduce the incentives for corruption and moonlighting. Yet impoverished countries lack adequate domestic resources to meet such challenges. Donors thus need to provide ODA to support civil service pay scales, a practice long rejected but sorely needed today. In addition to increased pay, the public sector needs to invest in ongoing worker training and capacity building, another expensive but vital activity (chapter 6).

Governments must also invest in the physical infrastructure of the public administration to improve service delivery and reduce opportunities for corruption. Some examples include:

- Communication and information infrastructure for all levels of government, including computer and telecommunications services for government offices, public hospitals, land registries, schools, and other public institutions.
- Information systems to improve the speed, reliability, and accountability of public sector transactions and systems to share information across branches of government. India, for example, is working to put all land deeds into a national database, which citizens can gain access to from anywhere in the country. This will eliminate the need for citizens to travel in order to request a copy of the deed to use as collateral in a loan.
- Modern technological capabilities for the customs bureau, to speed shipments, reduce smuggling, and control cross-border movements of illegal or dangerous goods.
- Modern technological capabilities for law enforcement, including national criminal databases, information systems to improve response times, and adequate dissemination of information to local law enforcement.
- Electronic government procurement and logistical systems, for example, to ensure reliable access to essential medicines in government clinics and hospitals.

As discussed in chapter 6, the upgrading of public administration will take center stage in the scaling up public investments and services to meet the Goals. Since all of these investments require financial resources, they should be included in the MDG-based poverty reduction strategies so that donor financing can be brought to bear in the common circumstance when domestic resources are insufficient.

Strengthening the rule of law

The rule of law, a prerequisite to sound governance, can affect the way policies are formulated and implemented. In many countries, weak institutional structures are susceptible to influence and capture by elites. Power is concentrated in a few select offices and people, and legal systems are severely overburdened, contributing to rampant corruption and mismanagement.

Setting up institutional mechanisms to respect the basic rights of all citizens and to treat them fairly is a first step to establishing the rule of law. This requires that the roles, responsibilities, and limitations of power of the different branches of government be outlined with transparent and clear accountability norms. It also requires that all three pillars of government—executive, legislative, and judicial—are well resourced and staffed to function effectively. Law enforcement is easier when the police force and the bureaucracy are well trained, adequately paid, and accountable. It is also easier when an independent judiciary has the power to consistently apply the rule of law and when the courts are well staffed and its lawyers and judges adequately paid.

Governments need to pursue an effective anticorruption strategy, by clarifying codes of conduct for public officials, by making it easier to report and track cases of corruption, and by creating more transparent procedures (box 7.2). In addition, adequate compensation to government officials is a step in enforcing strong anticorruption policies. Establishing the rule of law requires significant investments in efficient public sector management. Since low-income countries typically lack the resources to make these investments, it is difficult not just to scale up investment packages but also to create the institutional and legal framework to implement them.

Promoting accountability and transparency

Accountability requires the presence of democratic mechanisms that can prevent concentrations of power and encourage accountability in political systems. Citizens should be able to hold politicians responsible for their promises and actions through, for example, regular and fairly conducted elections in democratic forms of government and periodic reporting on electoral promises.

Implementation becomes more effective if there is a free press that can inform the public, analyze and critique government policies, monitor government performance and service delivery, and raise concerns if some parts of the population are being excluded or marginalized. Benchmarks should include general access to public information (freedom of information legislation and its effective implementation), legislative protection for the press, and specific steps to increase media freedom. For its part the press can highlight international commitments by national governments on the Goals and track progress.

For their part, parliaments are important in convening and promoting public debates on the best means of developing and implementing an MDG-based poverty reduction strategy. Parliamentarians can give voice to constituencies in remote and historically underserved areas, and they should provide a critical check on government by demanding public reviews of expenditures, by pointing out inequalities in implementation, and by making sure the policy debates on how to deliver services are linked quantitatively to the Millennium Development Goals and specific targets. The power to question government decisions publicly and to prevent the suppression of information is one of a parliament's foremost responsibilities.

In practice, accountability depends on citizens' awareness of the Goals and their corresponding rights and information about government actions. Governments need to introduce greater openness and provide full access to official data and performance indicators. They should produce timely publications of the fully audited accounts of the central bank and of the main state

Box 7.2
Actions to promote transparency

Source: Kaufmann 2004.

One of the most potent strategies to improve governance and control corruption involves transparency reforms. A government embarking on a transparency reform drive is also likely to make strides to integrate itself into the global economy and attract further FDI flows, since transparency plays an important role in investors' decisionmaking. A menu of concrete action items can be identified within a broader governance strategy. The main responsibility in implementing such actions resides with a number of key stakeholders. While the government's executive branch would generally play a key role, others—such as local governments, civil society, parliament, the private sector, and multilateral agencies—would expect to complement such efforts, and in some areas, even take the lead.

- *Empirical diagnostics of governance.* Country Governance/Anti-Corruption surveys and Public Expenditure Tracking diagnostic surveys can be conducted to assess the most vulnerable institutions and policy areas—such as procurement, customs, tax collection, public expenditure allocations to schools and clinics at the local level—and to assess progress on governance and anticorruption efforts. In natural-resource-intensive countries, a special corresponding diagnosis can be carried out. These in-depth country diagnostics have been carried out in dozens of countries, and their effectiveness is substantially enhanced where there is full transparency in publishing, disseminating, and publicly debating the implications of the results, with the objective of formulating governance action programs in a participatory fashion.
- *Access to information and freedom of the press.* Benchmarks can be set for the adoption of freedom of information acts and for publishing laws, regulations, budgets, procurement rules, incomes and assets of public officials, parliamentary voting records, and political funding contributions. Governments can also ensure timely publication of audited accounts for the central bank and major state enterprises (such as those in extractive industries). Ensuring that freedom of the press is duly protected, and that investigative journalism can be carried out and published without obstacles, is paramount.
- *Procurement assessment and action planning.* Countries can assess their procurement systems to identify priorities for reform. The role of information and communications technology can be deepened as a major protransparency tool, such as the government e-procurement system pioneered in Mexico (Compranet).
- *Public expenditure reviews (PERs).* PERs can be conducted periodically and within a governance and financial accountability framework to include a detailed assessment of all key dimensions of public expenditures (including military expenditures). Off-budget expenditures should be barred to the extent possible.
- *Public parliamentary review.* Parliamentary committees can review aid performance, with powers to question senior officials in public hearings on aid-funded projects and programs. Independent committees can also publish political and electoral financing and establish clear and enforceable rules on the use of state resources for political purposes.
- *Project-level transparency.* Transparency can be promoted by institutionalizing advance publication of all project details and rationale and convening public hearings prior to final project decisions on public investments, including those funded by multilateral agencies.
- *Civil society and private sector empowerment.* Governments and donors can implement a concerted strategy to enhance citizen, civil society, and private sector participation in transparency initiatives, with each serving a vital watchdog function by

(continued on next page)

Box 7.2
Actions to promote transparency
(continued)

publicizing information and open critiques of government action. Likewise, donor agencies and international financial institutions can foster an open and transparent environment through their own examples, for instance, by ensuring full access to their own country assistance strategy documents and the details of the government investment projects they fund. To enhance accountability of the private sector, all international organizations, including the UN system, could institute a mechanism for transparently and publicly delisting companies that have been involved in bribery in projects funded by international financial institutions (a practice at present followed only by the World Bank).

enterprises, such as those related to extractive industries. Other information for publication includes laws, regulations, budgets, procurement rules, incomes and assets of public officials and parliamentarians, and public access to parliamentary voting records and political funding contributions. There is strong evidence to show that diversions of resources decrease with greater transparency (box 7.3).

In addition to systemwide transparency, governments also need to ensure project-level accountability. This means publishing all project details and the rationale in advance, setting up public hearings and town hall meetings for public feedback before making final decisions, and adopting specific procurement and bidding safeguards to minimize manipulation.

Promoting human rights

Human rights are both a central practical objective of good governance and a normative standard agreed to by all signatories to the UN Millennium Declaration. The declaration reaffirmed the commitment of all signatory nations to respect and uphold the principles identified in the Universal Declaration of Human Rights and to fully protect social, cultural, economic, and political rights for all, including the right to development. We fully endorse this commitment and believe that a human rights framework, as outlined, for example, in the Convention on the Rights of the Child and the Convention on the Elimination of All Forms of Discrimination against Women, is an essential prerequisite to achieving all the Goals. But there has been no systematic effort to integrate development planning with a human rights framework, even though such integration has tremendous potential and relevance.

The Goals have been criticized by human rights proponents for targeting only a proportion of the population and for not referencing human rights principles, among other reasons (Alston 2004; box 8.1). To ensure the Goals are applied in a manner consistent with human rights, governments need to recognize the relevance of their human rights obligations, encourage community participation, and develop human rights–based accountability mechanisms.

Box 7.3
The power of information

Source: Reinnika and Svensson 2004a, 2004b.

In the mid-1990s the government of Uganda conducted a review of its primary education capitation grant scheme and found evidence of rampant graft and corruption. The review found that approximately 20 percent of disbursed funds were actually reaching schools, and that the median school was receiving nothing.

After some investigation, Ugandan officials discovered that most of the revenues were being captured by corrupt officials in the local agencies managing the funds. But since parents, normally active participants in school management and planning, had little knowledge or information about the capitation grant program, the large-scale embezzlement of public funds continued undetected for years.

The Ugandan government launched a new strategy to combat corruption, under which it began publishing data on monthly transfers of capitation grants to each school district in national newspapers and their local-language editions. Primary schools and district offices were also required to post notices of actual receipts of funds for everyone to see. Now, citizens could clearly compare the amount set aside for the school with the amount that schools actually received.

Equipping citizens with this information significantly improved performance of the grant program. While the median school received nothing in the mid-1990s, it received 82 percent of its entitlement in 2001. During the same period the proportion of funds lost to corruption fell from 80 percent to just 20 percent. By pursuing an inexpensive strategy of mass information, Uganda dramatically reduced corruption and improved the efficiency of its support to primary education.

Recognizing the relevance of human rights obligations

Each country should make reference in its MDG-based poverty reduction strategy to the international human rights obligations it has voluntarily undertaken. This could take the form of a human rights assessment similar to the way the World Bank conducts environmental assessments before undertaking projects. It would include:

- Acknowledging that human rights (economic, social, and cultural rights) already encompass many of the Goals, such as those for poverty, hunger, education, health, and the environment.
- Referring to international treaty obligations as well as customary law and relevant soft law standards within the national context.
- Accepting the Goals as intermediate targets that contribute to the progressive realization of basic development outcomes. The eventual development objectives can then be framed in terms of elimination of want, gender discrimination, and hunger and the promotion of health, political and social participation, and access to information for development.

National MDG-based poverty reduction strategies should be consistent with the principles of equality and nondiscrimination laid down in applicable international human rights standards. This implies that strategies are designed to reach the entire underserved population, irrespective of ethnicity, religion, regional background, or gender. It also implies taking steps to ensure that the most underprivileged and marginalized sections of society can exercise their rights. And it implies that strategies and actions do not worsen existing inequities. In particular, MDG-based strategies need to include a special focus on addressing the needs of indigenous and tribal peoples, who number about 370 million worldwide.

An outcome-based approach to the Goals, empowered within the context of human rights obligations, also encourages dealing with issues not specifically mentioned in the official targets and indicators but relevant to the outcomes to be included in the poverty reduction strategy. For example, the Goals do not refer explicitly to sexual and reproductive health, but these rights are important to achieve many of the other Goals, and are essential in themselves.[4]

Encouraging community participation based on human rights formulations

As discussed in previous chapters, governments need to affirm that broad-based and meaningful participation in decisionmaking is sought, both in design and in implementation. Such participation should always include the right to criticize official policy positions.[5] As both the *Human Development Report 2004* (UNDP 2004b) and the *World Development Report 2004* (World Bank 2003d) outline, governments need to identify mechanisms to allow groups commonly excluded from the political process to participate actively in decisionmaking processes. This is especially important in countries with rich social diversity and large indigenous and tribal populations (UN 2004c, d). Special attention needs to be paid to ensuring a balanced gender representation.

Developing human rights–based accountability mechanisms

The MDG-based strategy needs to include a commitment by the government to ensure that an appropriate legislative and legal framework will be put in place to facilitate meeting the Goals on the basis of respect for human rights (chapter 6). There is an important role for international human rights mechanisms in this, but the first line of support should be at the national level. Thus, in every country in which a national human rights institution exists, it should be given an explicit mandate to review and report on the realization of MDG targets at regular intervals. It is estimated that there are now at least 55 such institutions in existence, a dramatic increase from the 8 in 1990 (Kjaerum 2003). In countries that do not have such mechanisms, the MDG-based poverty reduction strategy could usefully recommend their creation.[6] The reporting would also need to be disaggregated to the extent possible, to take account of elements such as gender, regional disparities, and the situation of the most disadvantaged groups (which should be identified in the benchmarking process).

Promoting sound economic policies in support of the private sector

Governments need to ensure a favorable business environment for the private sector to flourish. This is the essential point stressed by two recent reports: *Unleashing Entrepreneurship: Making Business Work for the Poor* (UNDP 2004c) and the *World Development Report 2005: A Better Investments Climate for Everyone* (World Bank 2004d), both of which the UN Millennium Project strongly endorses. Building a vibrant private sector and helping the poor benefit from their entrepreneurship requires a strong foundation in the global and domestic macroeconomic environments, physical and social infrastructure, and the rule of law (figure 7.1).

The UN Millennium Project recommends, in the context of its long-term MDG framework, that each government collaborate with the local private sector to design a private sector development strategy that helps create a favorable business environment. This would include measures in seven key areas, described here.

First, the private sector needs a supportive macroeconomic framework. International and domestic macroeconomic stability minimizes uncertainty for businesses. Businesses cannot dependably buy inputs or sell their products internationally when the local currency is unstable against other currencies—or if the country is experiencing high inflation with constant readjustments of prices and a loss of confidence abroad. Nor can they operate effectively when trade barriers hinder the acquisition of inputs from abroad, preventing them from attaining international competitiveness in their own market.

Second, the private sector requires a favorable legal and regulatory environment. This includes a functioning judiciary, an effective commercial law that defines and protects contracts and property rights, and a rational public administration that limits and combats corruption. Several studies have verified that corruption raises the cost of doing business and discourages investment by raising transaction costs and uncertainty. It leads to inefficiency, misallocates talent to rent-seeking activities, increases informality, and hinders tax collection, leading in turn to higher taxes. Several measures can be taken to combat corruption—including freedom of the press, systematic auditing of

Figure 7.1
Foundations of the private sector and pillars of entrepreneurship

Source: UNDP 2004c.

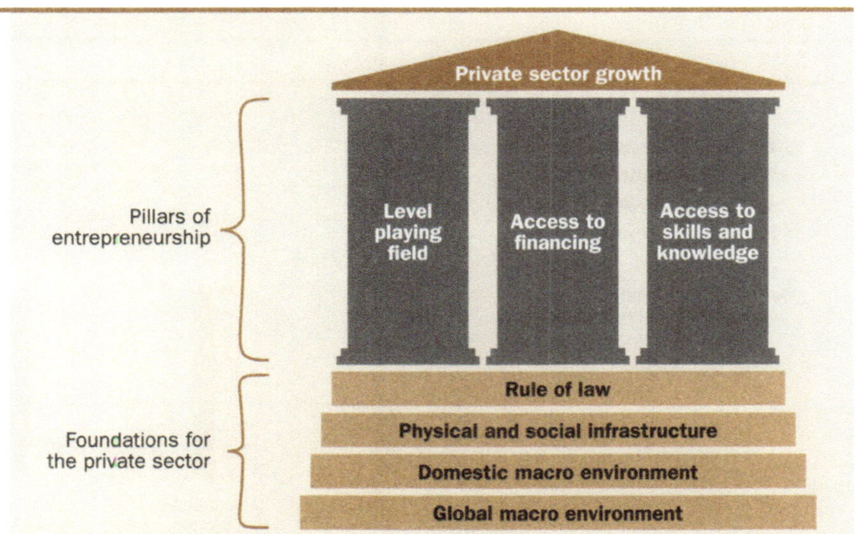

government outlays, and transparency in procurement, budget allocations, and the issuing of licenses (Gray and Kaufmann 1998).

The government can also promote more business growth by reducing the cost, duration, and number of procedures for entry and exit of firms, by improving processes to enforce contracts, and by simplifying the tax system. The World Bank's *World Development Report 2005* (World Bank 2004d) explains the range of regulation that can influence private sector performance.[7] The richest places in the world are also those where it takes the least amount of time to start new businesses, though there is a large amount of regional variation (figures 7.2 and 7.3). Ample evidence shows that policies simplifying the closing of a business (especially by enacting bankruptcy laws that maximize value, rescue viable businesses, and keep the order of claims stable) and improving processes to enforce contracts also encourage private sector activity. The World Bank has made an important contribution by measuring systematically the

Figure 7.2

Average time to start a business, by region

Startup time (days)

Note: Data are averages weighted by population.

Source: Calculated from World Bank 2004a, d.

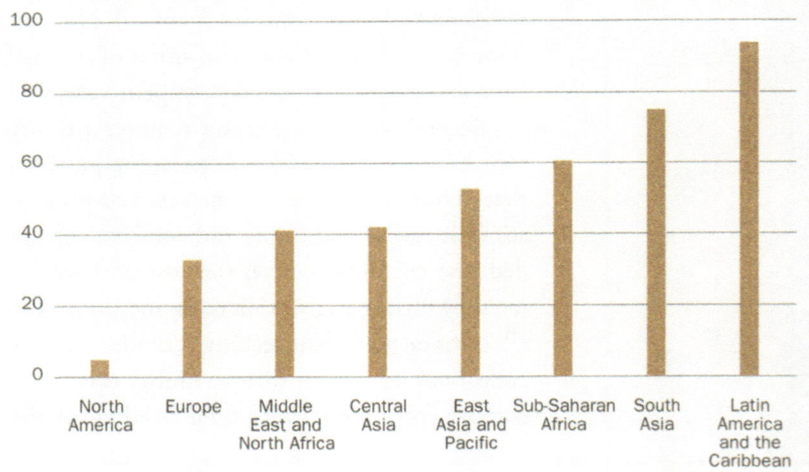

Figure 7.3

Average time to start a business, by income level

Startup time (days)

Note: Data are averages weighted by population.

Source: Calculated from World Bank 2004a, d.

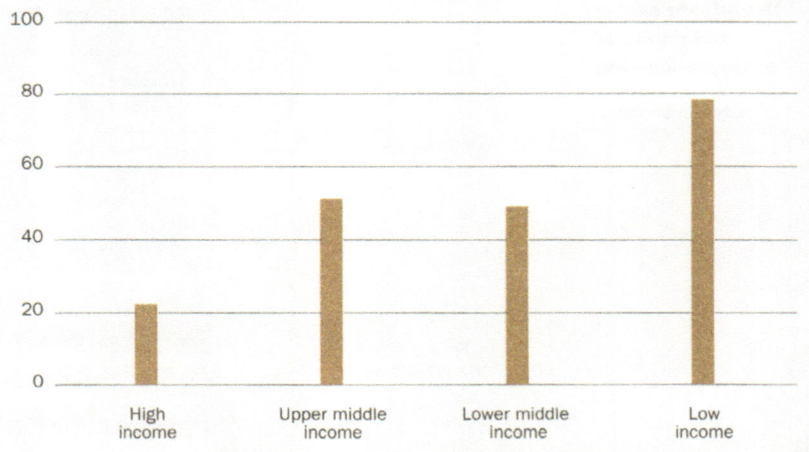

costs of doing business in various parts of the world and showing how government policies may raise or lower those costs. We strongly encourage countries to take note of these findings as a guide to lowering the costs of doing business.

All these reforms are important facilitators of private sector growth, but they are neither sufficient (the infrastructure needs explained below are equally important) nor free. Blaming the poorest countries in the world for failing to enact some of the reforms rings hollow when the reforms themselves require resources. For example, reducing the number of procedures may require sharing information across branches of government and streamlining processes. But that requires computers and complex information systems, which many governments cannot afford.

Third, a thriving private sector depends fundamentally on adequate infrastructure, human capital, and research and development. The roads, electricity, seaports, and airports financed by the public sector are critical for private sector profitability, and there are several ways of involving the private sector in the provision of these services (chapter 9). Businesses cannot operate competitively if goods cannot be transported due to badly maintained or over-congested roads, poorly managed airports, outdated seaports, or rampant criminal activity unchecked by an effective police force.

Public investments to ensure a healthy and skilled labor force are crucial for private sector productivity, since many workers in poor countries suffer frequent illness, lowering productivity and leading to high rates of worker absenteeism. The government's investments in education, through the public school system, adult education, and worker training programs, directly increase the productivity of the labor force. Through support for higher education and for research and development outlays, the government lays the groundwork for economic growth through technological advance.

Fourth, governments can actively promote business activities in science, technology, and innovation. They can foster the creation and growth of small and medium enterprises, for example, by supporting business incubators and technology parks. They can establish industry extension services and help firms establish international partnerships and linkages, and use government procurement and trade policies to support technological development (UN Millennium Project 2005g).

Fifth, governments can take specific actions to promote foreign direct investment (FDI) in the country. Many countries have difficulty attracting foreign investment, due to either their small markets or the high cost of serving as a base for exports (Charlton 2004), so a strategy for promoting foreign direct investment that will contribute to development should target specific sectors and activities. For less developed countries, a good target is commodity diversification and complementary reforms in the global trading system (chapter 14). Promoting FDI will require not just a favorable regulatory environment, but also such measures to actively attract business as special economic zones. If carefully managed, these zones allow investors to easily bring capital and technology to set up businesses, hire local labor, produce world-caliber goods, and export them through an efficient port. In many situations governments can attract more foreign companies through tax incentives, perhaps in the form of tax holidays.

Sixth, the private sector requires functioning, competitive markets for both inputs and products. Well developed financial markets, which channel resources to entrepreneurs, help reduce the cost of and increase the access to capital. In many countries, the formal banking sector needs to be strengthened through regulatory reform and increased accountability of financial institutions. Functioning labor markets and access to trade can also help ensure a competitive production environment. Although regional trade is no substitute for market access to developed countries, regional integration in free-trade areas can expand markets and contribute to productivity increases (chapter 15).

Seventh, the informal economy needs government support in several ways. The government can enable easier access to financial capital by simplifying rules for collateral, increasing flexibility for informal entrepreneurs, and providing credit at subsidized rates. The government can facilitate registrations of informal enterprises by simplifying taxation and accounting norms. And it can help small-scale entrepreneurs by providing training and skill-building opportunities, through access to vocational or other education programs.

Partnering with civil society

Civil society and governments often have an ambivalent relationship. But if governments are to implement MDG-based strategies effectively, they need a flourishing civil society—to ensure representation of diverse views and interests, to help design plans and strategies, to supplement government implementation channels, to monitor, evaluate, and review progress on the Goals, and to ensure that MDG-based strategies are sustainable beyond the short term (chapter 8). The UN Millennium Project strongly recommends that governments create the political and institutional space for civil society to operate in the following ways.

First, governments need to give civil society groups political freedom to express their views, organize, and participate in the development process. Civil society should be allowed to voice divergent views through media and other public channels in an atmosphere free from fear or threats.

Second, governments need to provide the institutional space for civil society organizations to participate in the planning and review of MDG-based strategies. This would require, for example:

- Government support for civil-society-led public dialogues and consultations before and during the design of MDG-based poverty reduction strategies.

- A formal role for civil society in the technical teams set up to draft sector strategies leading into the poverty reduction strategies.
- Formal consultations with civil society representatives to review the draft poverty reduction strategy.
- A formal role for civil society organizations in reviews and budget tracking exercises to monitor progress toward the Goals.

Third, as governments plan to scale up nationwide investment programs, civil society organizations should be seen as valuable partners in implementing plans and in local service delivery. The government can create mechanisms for learning from successful pilot projects run by local communities. It can solicit civil society recommendations for ways of implementing programs in specific contexts. It can also implement programs in collaboration with civil society.

Fourth, governments can support civil society by helping build civil society capacity—training people in technical skills and providing financial help to strengthen the ability of civil society to expand its role in development.

Fifth, as in any other sector, the government needs to lay out clear norms of accountability and transparency to make sure that civil society organizations are accountable to their constituents.

* * *

An overall message of this chapter is that governance has many dimensions. In some cases political choices are most critical. In many cases, concrete operational strategies can be implemented to improve governance outcomes. But, as is too infrequently appreciated, these strategies require investments that are often difficult for poor countries to afford.

Civil society's contribution to the Millennium Development Goals

National strategies to achieve the Millennium Development Goals require the support and involvement of civil society organizations (CSOs), who play a role quite different from the other stakeholders in development. They represent important segments of the population in a manner distinct from government as they directly reflect—and respond to—the needs of a broad range of communities.[1]

Within countries, CSOs can contribute to MDG-based poverty reduction strategies in at least four ways: publicly advocating for pressing development concerns, helping design strategies to meet each target, working with governments to implement scaled-up investment programs, and monitoring and evaluating efforts to achieve the Goals. Internationally, CSOs can also mobilize and build public awareness around the Goals, share best practices and technical expertise with governments, and deliver services directly.

CSOs have been engaged in some or all of these activities for many years. Here we highlight how their activities can be channeled to help reach the Goals. But to do so, they need political freedom, clear institutional roles, ways of partnering to implement programs, and in some cases, training and financial resources (chapter 7).

Some civil society organizations have expressed reservations about the Goals (box 8.1). Yet, most of them have spent years dedicated to the objectives that the Goals represent: promoting gender equality and fighting poverty, hunger, illiteracy, disease, and environmental degradation. Indeed, the international consensus for poverty reduction targets has come about in significant part thanks to civil society action in recent decades.

CSOs thus play a crucial role in scaling up the delivery of services to meet the Goals. In addition to their important role in representing the needs of poor people, the UN Millennium Project strongly recommends that CSOs participate in the design, implementation, and review of MDG-based poverty

Box 8.1

CSOs and the Goals: critiques and opportunities

Many civil society organizations broadly support the global partnership that the Goals encapsulate, as laid out in the UN Millennium Declaration, but remain skeptical about the Goals themselves, for several reasons. They question whether the Goals are different from previous UN goals that were not met. They have not been systematically involved in the Goal-setting process and so feel no ownership. They argue that the Goals are "top down," imposed by the international community, rather than locally developed, and that there is uncertainty about their role in achieving the Goals. They see the Goals as too narrow and unambitious, especially in comparison to the UN Millennium Declaration, leaving out critical issues of importance. For example, the Goal for gender equality falls short of the aims set in the Platform of Action at the UN Conference on Women in 1995, which governments around the world committed to. They are critical of the fact that only Goal 8, which spells out the responsibility of developed countries, lacks time-bound targets.

These are all legitimate concerns that policymakers need to address. The concerns come from a history of broken promises and systematic exclusion of CSOs from decision-making. But the criticisms focus more on the process of designing and implementing the Goals than the concept and substance of the Goals themselves. The Goals represent, at a minimum, the very objectives that CSOs have been trying to achieve for decades.

The Millennium Development Goals differ from other UN goals because, for the first time, they have been endorsed not only by the UN system and all its member states but also by other major stakeholders. They thus represent a real opportunity for global and national cooperation. They have a political momentum because of their unique link with the hopes of the new millennium. And they have already demonstrated staying power, and the ability to move major governments to recommit to goals such as 0.7 percent of GNP in official development assistance. As we argue throughout this report, achieving the Goals will require a dramatically different level of national and international effort.

Achieving the Goals will require strategies that are locally owned and developed, with full participation from all relevant constituents, including CSOs, the private sector, and other key stakeholders. Without their full involvement, the Goals cannot be implemented at the national level.

The Goals constitute a minimum set of objectives that the global community has agreed to. In several country contexts, they can provide the basis for more ambitious national objectives. Strategies to achieve them may also require a broader set of inputs than specified by the MDG targets and indicators—these could include, for example, sexual and reproductive health or energy and transport services. Achieving the Goals will require national stakeholders to agree that the Goals constitute a minimum set of objectives that can be more ambitious in different contexts. And it will require implementing the Goals in a way that focuses on reaching marginalized and underprivileged groups and regions.

If implemented in the spirit of the UN Millennium Declaration, the Goals can be a powerful framework for realizing key development outcomes. Fear of failure could become a self-fulfilling prophecy. We all have to believe in success—and to mobilize our energies and those of others to follow through.

reduction strategies in all developing countries. Given the important diversity of CSOs, one should of course not expect all CSOs to be involved in all areas. But we do recommend that, wherever possible, CSOs partner among themselves and with governments to pursue MDG-focused activities in advocacy, design, implementation, and monitoring.

Providing public advocacy for the Goals

In every country CSOs can focus public attention on the Goals and the actions under way—or not under way—to achieve them. CSOs drive broad-based mobilization and create grassroots demand that can hold leaders accountable and can help place the Goals at the heart of national debates. Strategic alliances of CSOs—with local authorities, national governments, and the international community—can raise public awareness of their government's commitments to the Goals, highlight urgent development priorities for the government's immediate action, and ensure that the needs of different groups are taken into account.

For example, the Africa Network Campaign for Education for All, a regional network of 23 national coalitions, has mobilized public opinion around the need for free, quality education for all. Its national coalition in Malawi, the Civil Society Coalition for Quality Basic Education, has advocated for including important education-related expenditures, such as teacher training and welfare expenditures, in the national PRSP.

National advocacy is also crucial in developed countries. For example, Bread for the World, a nationwide grassroots movement in the United States, has increased awareness and support for policies that reduce hunger domestically and internationally. Each year, it mobilizes about 250,000 letters to the U.S. Congress on issues of hunger and lobbies for more development assistance to poor countries. In the United Kingdom, the Make Poverty History Campaign has already galvanized remarkable support to urge the government to lead developed countries in making a major breakthrough in support for development in general and the Goals in particular.

Contributing to policy design

While many CSOs focus on advocacy, others have a key role in policy design. We endorse a formal role for CSOs in the creation of national MDG-based poverty reduction strategies (chapters 4, 6, and 7). Experience from many countries shows that the extent of openness in policy formulation can directly affect the quality of policies adopted—and the effectiveness of implementing them. CSOs can contribute to identifying priority investments, targeting priority areas and communities, helping design effective implementation strategies, setting national and local budget priorities, involving women in the design and implementation of these strategies, and ensuring that the poor and marginalized groups are central beneficiaries.

This is the approach espoused by the Country Coordinating Mechanism of the Global Fund to Fight AIDS, Tuberculosis, and Malaria, which calls for country-level partnerships, including CSOs and private sector representatives, to submit grant proposals based on priority needs. Once grants are approved, these partnerships oversee program implementation. Although not perfectly executed in all countries, the approach offers a model for how civil society can be formally included in policy design and formulation. Another important example of a CSO contribution to policy design is the work of Law and Advocacy for Women in Uganda, which is advancing policies to address female genital mutilation, women's land rights, and reproductive rights.

CSOs can also provide first-hand information about constraints "on the ground" and the full range of resources needed to implement interventions. For instance, the Tanzania Gender Network Program has been at the forefront of participatory budget analysis, bringing together officials from the Ministry of Finance to collaborate with gender experts and examine how women's needs are addressed.

If the concerns of excluded or marginalized groups are not articulated during policy debate, national MDG-based poverty reduction strategies will likely miss the very people whose needs they are designed to address. For example, indigenous groups are rarely included in planning and processes (UN Permanent Forum on Indigenous Issues 2004). Yet policies for health, education, environmental management, and infrastructure development affect them directly, so their representatives need to be part of the policy design process. Such participation need not be limited to public policy design. It can also influence private sector activity. The indigenous Secoya community in Ecuador negotiated a code of conduct with Occidental Petroleum's Exploration and Production Company in 1999 to regulate its oil exploration activities in Secoya, setting out principles of engagement based on dialogue and transparency and operational mechanisms for ensuring good environment practice.

Scaling up service delivery

The challenges in scaling up MDG-based strategies are significant (chapter 6). We recommend that civil society partner with governments to help scale up investment through four main channels: engaging local communities, building human capacity, strengthening local governance, and leading implementation and service delivery.

Local community mobilization involves immediate beneficiaries and underrepresented groups (especially women) in decisions of service delivery. Several examples of successful community engagement offer lessons for program design and implementation. The Orangi Pilot Project in Karachi, Pakistan, offers a model for a tripartite arrangement among the local community, the government, and a local CSO to provide improved sanitation services and has been replicated in settlements across Pakistan. The Hunger Project, an international NGO, has implemented a low-cost, people-centered strategy for grassroots mobilization to address hunger in rural African communities. Its "epicenter" strategy is anchored in community leadership and empowerment at the grassroots level, catalyzed by international staff and implemented by national staff working with local governments and national political leaders. Villages build a community center to house a school, a health center, a rural bank, food processing and storage, and a meeting room for adult classes in literacy, agricultural techniques, health, and nutrition. The project offers lesseons for mobilization of local resources, community leadership, and income-generating activities to build self-reliance from the start.

CSOs can train teachers, community workers, and health workers. BRAC, for example, has been training female community health workers in Bangladesh since the 1970s. By 2003 it had trained more than 30,000 health workers in almost as many villages. It provides foundation training and regular refresher training in dealing with common illnesses, such as diarrhea and dysentery, and with improving maternal health. The health workers also provide higher skilled services, such as administering DOTS to combat TB. In another example, the Association of Senegalese Women Lawyers has been training paralegal workers for 30 years to address violence against women through legal channels, winning a tremendous number of cases, including those for land claims.

Making government institutions accountable and participatory requires close collaboration between the local population and the government to set up mechanisms that enable the local community to use existing institutions more effectively. For example, the Movement for Alternatives and Youth Awareness in Karnataka, India, has strengthened institutions of local self-governance to improve school effectiveness. It has facilitated a citizens' effort to evaluate and improve school performance through existing local government structures. More than 1,000 councils have been formed in the past five years, with increased community participation and improved infrastructure for schools. Similarly, the Pamoja Trust, a Kenyan NGO, and the urban poor federation in Kenya (Muungano wa Wanvijiji) have built the capacity of local communities to self-organize, reach consensus on tenure and infrastructure upgrading decisions, and then engage with local authorities and municipalities for land and infrastructure to improve the lives of slum dwellers.

In many parts of the world, CSOs innovate to establish successful delivery models that governments can replicate on larger scale, often offering lessons for hard-to-reach areas and conflict regions. The Grameen Bank, for instance, has famously helped to provide microcredit loans for millions of Bangladeshis, most of them women, contributing to gender equality and small enterprise development at the community level (box 8.2). Other CSOs have often been pioneers in addressing issues that governments deem too sensitive to address publicly or directly, and in many countries they have emerged as a first line of defense in addressing HIV/AIDS. In Thailand, CSOs have led in providing treatment and care to marginalized populations and groups often deemed criminal, such as drug users, sex workers, and migrants who are, in turn, reluctant to deal with government officials (box 8.3). And in the central plateau of Haiti, Partners in Health has helped shatter the conventional wisdom that poor people with AIDS cannot adhere to antiretroviral treatment protocols.

Box 8.2
How CSOs help Thailand battle AIDS

Source: UNDP 2004d.

Thailand is one of the few developing countries in the world to contain the spread of HIV. Between 1992 and 2002 the HIV prevalence rate among entering army conscripts plunged from 7 percent of the population to less than 1.5 percent. Research revealed significant populationwide behavior changes, including fewer visits to commercial sex workers and more condom use during sexual intercourse. Civil society organizations played a decisive role.

A broad range of policies and actions contributed to the assault on AIDS, but civil society influenced behavior change through education and outreach. As many as 50 NGOs were working on AIDS as early as 1984, and in 1989 a national NGO Coalition on AIDS was formed to coordinate their work. And as early as 1992, the Thai government began financing their activities.

These organizations delivered participatory AIDS education, created counseling and support services, and pioneered outreach to intravenous drug users, men who have sex with men, and commercial sex workers—at-risk populations that public agencies could not reach.

Many actors deserve credit, but the coordinated efforts of civil society groups helped stem the AIDS crisis in Thailand.

Box 8.3
SPARC—partnering with local government

Source: UN Millennium Project 2005f.

In India the National Slum Dwellers Federation works with the Society for Promotion of Area Resource Centers (SPARC) and the cooperatives of women slum and pavement dwellers (Mahila Milan) to show what citizens and CSOs can do, and what governments can achieve in partnership with them. The biggest success has been to design, build, and manage community-toilet blocks where there is not enough room or funding for household provision. The program took off when local government staff saw how much better they were than contractor-built public toilets.

Working in 50 cities and with 750,000 members, the three CSOs have since been asked to work with local and national governments in redeveloping government-built tenements. They are changing policies and practices by having governments interact with communities in the provision of public services. And they have strengthened the relationship between communities and municipal authorities, setting the basis for partnership on more contentious issues of land tenure.

The Kenya Urban Poor Foundation, the Namibia Housing Action Group, and SPARC in India help build housing projects, undertake community-managed resettlements and provide and manage urban infrastructure and services at scale, showing what it takes to scale up such programs. The Rehabilitation of Arid Environments Charitable Trust (RAE) in Kenya has worked for more than 20 years on the rehabilitation of grasslands in the watershed of Lake Baringo by setting up private and communal fields protected from grazing animals by electric or live fences. The community management of land transformed severely degraded terrain into productive land within three years. PROFAMILIA, a Colombian CSO, has been promoting reproductive health services for many decades. It now provides nearly 40 percent of all family planning products in Colombia.

These diverse examples show just a few ways that civil society initiatives can complement government, ensuring that government efforts respond to the needs of local communities and implemented appropriately.

Monitoring for accountability

Civil society organizations can also monitor and benchmark progress, critical for building accountability (chapter 7). They can highlight regions where progress is slow and further efforts are required. They can engage in broad-based and transparent mapping of government spending. And they can provide real-time feedback to the general public on progress. For example, women's groups in Chile, Ecuador, and Paraguay are constructing quantitative indicators to assess how their governments are fulfilling commitments to women. With support from the Facultad Latinoamericana de Ciencias Sociales, the groups take indicators of citizen participation, economic autonomy, and women's health and reproductive rights and create a composite index that permits comparison over time, helping women hold governments accountable for progress (UNIFEM 2002).

To help monitor progress, CSOs should be given access to budget records and encouraged to disseminate them publicly. We endorse formal CSO participation in reviews of MDG-based poverty reduction strategies, and budgetary tracking and auditing exercises (see also chapter 6).

In sum, achieving the Goals within countries will require active national civil society involvement through:

- Raising awareness about the Goals, by highlighting development priorities through public dialogue and consultation.
- Designing sector strategies, by identifying priority investments and regions, effective implementation plans, and national and local budget priorities.
- Contributing to the scaling up of investments, using skills in community involvement, capacity building, strengthening governance processes, and service provision.
- Evaluating performance by participating in national MDG reviews, budget tracking, and auditing.

What international CSOs can do

Many civil society organizations extend well beyond the borders of any single country. International CSOs have a crucial role in achieving the Goals, as they mobilize cross-border support, share best practices, and contribute to direct service delivery.

Mobilizing around the Goals

CSOs across the world are creating and mobilizing global partnerships in several ways, with support from international initiatives such as the UN Millennium Campaign, which has helped facilitate civil society campaigns in more than 35 countries. Social Watch, an international network of citizens groups operating in around 60 countries, reports annually on the fulfillment of the internationally agreed commitments on poverty eradication and equality, documenting country progress toward the commitments and goals. The Latin American and Caribbean Women's Health Network has been monitoring country progress in implementing the Programme of Action of the International Conference on Population and Development (1994). Such initiatives can help hold national governments accountable for commitments to the Goals by tracking progress and highlighting relevant policy decisions in rich and poor countries alike.

Networks of international CSOs can also mobilize support across borders and build links between communities in different parts of the world. On such issues as women's rights, women's health, debt cancellation, fair trade, and the environment, such networks have shown ways of building support within and across countries (box 8.4). In 2002 Girls' Power Initiative brought young adults from all over the world to the United Nations to get policymakers to listen to their concerns and desires for greater access to reproductive and sexual health services. The World Social Forum, held each year to coincide with the World Economic Forum in Davos, brings together groups from civil society to formulate policy proposals, share experiences, and build networks for effective action. Meetings of this kind have contributed immensely to building bridges between civil society groups in different countries.

International civil society advocacy can also highlight development challenges facing poor countries—and put pressure on governments and the international community. Such advocacy is critical for all the Goals, and many successes show how it can be done. The Jubilee 2000 Campaign, and its successor, the Jubilee Debt Campaign, organized by a wide network of CSOs, put debt forgiveness for poor countries firmly in the center of international development policy discourse. It is now focusing on the eighth Millennium Development Goal on global partnerships for development. The International Gender and Trade Network has meanwhile highlighted the importance of gender issues in international trade. Oxfam International is bringing the negative effects of U.S. cotton subsidies on African farmers to the forefront of WTO negotiations. The Water Supply and Sanitation Collaborative Council, through its "W.A.S.H." initiative, is leading an international call to action for sanitation.

Sharing best practices and technical knowledge

International civil society groups can provide technical and policy assistance to developing countries and in many cases, form global repositories of technical knowledge that can help a range of countries (box 8.5). With their rich experience, they can often play a catalytic role in helping countries achieve the Goals.

Box 8.4

The women's movement and UN Conferences in the 1990s

Source: Correa 1999; UNIFEM 2000; UNFPA 2004, IWHC 2004.

Local, regional, and global women's movements actively mobilized throughout the 1990s to put gender equality and women's empowerment on the agendas of major UN conferences. Through women's advocacy efforts, the 1992 UN Conference on Environment and Development included a chapter on women's role in the environment; the 1993 UN Second World Conference on Human Rights recognized for the first time women's rights as human rights; and the 1995 World Summit on Social Development recognized the gendered implications of macroeconomic policies, especially the crippling effects of debt.

The 1994 International Conference on Population and Development in Cairo and the 1995 Fourth World Conference on Women in Beijing represent important landmarks for the global movements for women's rights and development. National, regional, and global women's organizations played important roles leading up to Cairo and Beijing, within the intergovernmental meetings and in the parallel nongovernmental forums.

At the Cairo conference, women's advocates from developed and developing countries collaborated to successfully move the overall objective of population policy away from reducing population growth to achieving women's sexual and reproductive health and rights. The Programme of Action, adopted by 179 governments, marked a new understanding among world bodies—that population and development are inextricably linked and that women's empowerment is the key to both. And, for the first time, the reproductive rights and sexual and reproductive health of women became central elements in an international agreement on population and development (IWHC 2004).

Three thousand NGOs were accredited to the UN conference at Beijing, and 30,000 people from all over the world attended the parallel NGO forum. The women's movement promoted a broad multifaceted development agenda with women's human rights at its core. Governments agreed to a Platform for Action that outlined goals and recommendations for 12 critical areas of concern (UNIFEM 2000) which further concretized sexual and reproductive rights, recognized the negative development consequences of violence against women, and renewed attention to women's economic rights.

U.S. private foundations and European governments and intergovernmental organizations enabled civil society, particularly the women's movement, to organize, network, define advocacy agendas, and participate in both the Cairo and Beijing conferences and their preparatory meetings. At each conference, there was also a Women's NGO Caucus, a democratic vehicle for consensus building that enabled women's NGOs from around the world to infuse a gender perspective into the process and outcomes of conferences. For instance, much of the language in the Cairo Programme of Action was either initiated or strongly supported by the women's organizations that participated in the preparatory process and the conference itself (Correa 1999). Similar to Cairo, the Beijing Platform for Action incorporated almost 90 percent of the NGO caucus recommendations (UNIFEM 2000).

In the years since these conferences, gender equality initiatives have proliferated. The ICPD has had effective impact at the national level across developing countries. In many countries, policy transformation has occurred, altering public actions and discussions in governments and societies (UNFPA 2004). In many countries, family planning programs have been reoriented to include a wider array of reproductive health issues (Correa 1999). Many post–Cairo and Beijing initiatives are cross-sectorial and often combine policy planning, legal reforms and community-level projects. Violence against women has gained increased visibility, and laws and national policies have been developed or improved to

Box 8.4
The women's movement and UN Conferences in the 1990s
(continued)

address violence in Latin America and elsewhere. Gender planning saw a breakthrough in countries in the Caribbean and East Asia. In West Africa, emphasis has been given to legal reform and gender and poverty programs; and in Southeast Asia and the Pacific, ratification of the Convention on the Elimination of All Forms of Discrimination against Women is cited by the women's movement as a major step forward.

Box 8.5
Grameen Trust—spreading knowledge and self-reliance

Source: Grameen Trust 2004.

From humble beginnings as a research project involving poor women artisans in rural Bangladesh, the Grameen Bank has grown into a massive success in poverty reduction. The Bank has loaned more than $4.3 billion to nearly 3.4 million borrowers in Bangladesh, with half of them crossing the poverty line. Since 1995 it has been independent of donor funding, showing that microfinance is not only socially rewarding but also financially sustainable.

Grameen has also served as a model for other organizations in Bangladesh, where microcredit programs have been scaled up to reach more than 10 million families. One of the most exciting dimensions of Grameen's experience, however, is the replicability in other contexts. Grameen Trust has mobilized knowledge from experiences in Bangladesh and other countries to support over 127 organizations in 35 countries with funds, training, and technical assistance.

Grameen Trust has provided $20 million to local partners that have reached 1.6 million families around the world with financial services. Like Grameen Bank, these local CSOs typically have a strong grassroots presence that enables a more effective outreach to poor people. The Trust's cross-country experience enables it to advocate for supportive regulatory frameworks.

Grameen experience demonstrates that community-led innovations, when scaled up effectively can reduce poverty in many contexts.

An example is WaterAid, an international NGO dedicated to the provision of safe domestic water, sanitation, and hygiene education to the world's poorest people, which uses its research and its documented good practices to influence development policies all over the world. It works with local organizations to help some of world's poorest communities set up, operate, and maintain their own water, sanitation, and hygiene projects.

Shack Dwellers International, set up in 1996 by urban federations to expand contacts with the international community, is another NGO that supports international community-exchanges, linking organizations of the urban poor in different countries. It also visits nations where federations have not yet developed or are only in early stages of development to help countries improve the lives of slum dwellers.

Leading direct service delivery

International CSOs lead service delivery in some of the most challenging places in the world, often in close collaboration with multilateral relief efforts.

Médecins Sans Frontières, for example, works in 80 countries to provide not just emergency health aid, but primary healthcare as well. It works with local teams to rehabilitate hospitals and clinics, run nutrition and sanitation programs, train local medical personnel, and treat chronic diseases. Action Aid International works with 2,000 local partners to reach almost 9 million of the poorest and most vulnerable people, helping them fight for their rights to food, shelter, gender equality, education, and healthcare and a voice in the decisions that affect their lives. CARE International, working in more than 70 countries, reaches 45 million people with emergency and humanitarian relief efforts in addition to longer term primary healthcare, education, savings and loan schemes, and agriculture programs. Such efforts can be hugely important for achieving the broad range of health MDGs.

These large international CSOs, working closely with local organizations to provide much-needed services, are often the first to reach regions hit by conflict or natural disasters, even before governments can (chapters 11 and 12). These services are especially invaluable where the government is either unable or unwilling to invest.

To recap, international civil society has an important international role for the Goals in:

- Mobilizing public opinion around the Goals and around the developed countries' commitment to Goal 8.
- Sharing best practices and technical expertise through intercountry community exchanges, direct technical support, and advice on scaling up to governments.
- Providing health, education, infrastructure services that contribute to achieving the Goals.

All told, both domestic and global civil society organizations have a crucial role to play in ensuring the Millennium Development Goals are achieved. The UN Millennium Project strongly supports the role of CSOs in achieving the MDGs, and recommends that they be supported and empowered to play central roles in each country's adoption of an MDG-based poverty reduction strategy.

Chapter 9

Contributions of the private sector

Any national strategy to achieve the Millennium Development Goals needs to include a clear framework for private sector growth. Many of the public investment strategies outlined in the previous chapters are essential for an efficient and dynamic private sector. For their own part, private enterprises can contribute directly to the Goals through core pursuits such as increasing productivity and job creation or seeking opportunities for service delivery through public-private partnerships. In all these activities, companies need to adhere to high standards of responsible corporate governance and citizenship. However, companies and their leaders can also take action to support the Goals more broadly, by contributing to MDG-based policy design, by advocating publicly for the Goals, and by pursuing various models of corporate philanthropy. In these ways, businesses can "engage as reliable and consistent partners in the development process," as called for in the Monterrey Consensus (box 9.1). In this chapter we briefly describe each of the channels through which the private sector can critically support achievement of the Goals.

Increasing productivity and creating jobs

In a market-based economy, private firms contribute to poverty reduction through many channels. They reduce income poverty when productivity rises, job opportunities increase, and competition for workers drives up wages. By producing essential goods and services in large-scale production, they can also help to keep the price of essential goods and services down, increasing the real effective incomes of poor people, a point underscored by the Commission on the Private Sector and Development (UNDP 2004c). As firms grow, they provide a larger source of tax revenues to the government, which in turn supports increased public investments.

Box 9.1

The private sector's role in development: the Monterrey Consensus

Source: UN 2002a.

The 2002 Monterrey Consensus included specific reference to the private sector as partners in poverty reduction:

"23. While Governments provide the framework for their operation, businesses, for their part, are expected to engage as reliable and consistent partners in the development process. We urge businesses to take into account not only the economic and financial but also the developmental, social, gender and environmental implications of their undertakings. In that spirit, we invite banks and other financial institutions, in developing countries as well as developed countries, to foster innovative developmental financing approaches. We welcome all efforts to encourage good corporate citizenship and note the initiative undertaken in the United Nations to promote global partnerships.

"24. We will support new public/private sector financing mechanisms, both debt and equity, for developing countries and countries with economies in transition, to benefit in particular small entrepreneurs and small and medium-size enterprises and infrastructure. Those public/private initiatives could include the development of consultation mechanisms between international and regional financial organizations and national Governments with the private sector in both source and recipient countries as a means of creating business-enabling environments."

What can the private sector do more specifically to help reduce poverty and support the Goals? In the poorest countries, agriculture forms much of the private sector so the rural productivity investments described in chapter 5—particularly those to launch a Twenty-first Century African Green Revolution—will have a direct and sizable effect in kickstarting the private sector's contributions to economic growth and rural household incomes.

Both in agriculture and in other sectors, much private sector activity in developing countries lies outside the formal economy. Of all the nonagricultural private actors in Sub-Saharan Africa and Asia, an estimated 70–80 percent are in the informal sector (UNDP 2004c). Within the informal sector, women are estimated to be about two-thirds of self-employed entrepreneurs (ILO 2003). For these informal sector entrepreneurs, significant support is often needed to climb out of poverty—but success is possible. The Grameen Bank, for example, anchors the provision of credit to poor women not in collateral but in systems of accountability, supervision, participation, and peer management (see chapter 8). Grameen estimates that 51 percent of its borrowers have crossed the poverty line through the income-generating activities made possible in part through these loans (Grameen Trust 2004). Rabobank International has also worked in Uganda and Tanzania and developed financial instruments (risk management tools such as swaps and derivatives) to help small farmers protect themselves against price fluctuations (WBCSD 2004a).

In low-income and middle-income countries with higher levels of urbanization, urban sources of income become more important. In the good scenario, manufacturing firms grow to compete internationally, providing a source of sustained economic growth. Yet in many cases, such value-added production processes are less likely to take off on a path of sustained growth until the costs of entering regional or global markets are significantly reduced through improved infrastructure and a better educated and healthy labor force. When these conditions are not met, much of the urban economy is likely to remain informal and focused in basic commodity trade and the very low-skilled service sector. To enable the entrepreneurs of the informal sector to become the pioneers of the formal sector, basic public infrastructure is needed alongside low-cost credit and regulations (as discussed in chapter 7).[1]

When major multinational corporations decide to invest in developing countries, either to enter local markets or to build production platforms for global markets, they often foster local "business ecosystems" with vertical supply chains and horizontal industry clusters. Such networking is invaluable for diffusing technologies and skills, bringing local firms into the formal economy, and increasing market opportunities for local suppliers (UNDP 2004c). Some foreign investors, when provided appropriate incentives and adequate labor standards, can also place a special emphasis on labor-intensive production technologies that create dignified employment opportunities as an instrument to poverty reduction.

At a broader level of poverty reduction, private enterprises are vital to supporting overall technological advance, the long-term driver of economic growth. Technological learning for production occurs at the level of enterprises, both public and private. This becomes more important as countries reach middle-income status and need to develop their technological base to compete internationally. Technological advances in manufacturing industries tend, by raising the productivity of workers, to increase firm growth and then increase the demand for jobs. Manufacturing enterprises also spread innovation outward to the agricultural and service sectors (UN Millennium Project 2005g).

In some important development success stories, small and medium enterprises have developed much of the local technology that laid the foundation for rapid export growth. In Taiwan (China) for example, small and medium enterprises were the engines of the economy's postwar industrial upgrading, supplying multinational corporations and foreign buyers.

As outlined in Monterrey, financial enterprises also have a special role to play in enterprise creation. Although financial markets are extremely limited in the poorest countries, they have expanded in recent years to support business development in middle-income countries, the "emerging markets." External sources of private capital, including venture capital, can promote developing country firms and entrepreneurs. In more advanced developing economies with more sophisticated financial markets, these markets often provide some capital to new and growing firms. The cost of providing such capital should be minimized by governments—not only to support physical investments but also because venture capitalists and other specialists in business development often provide important knowledge to indigenous firms. Bringing venture capital markets into developing countries can help sustain the growth of new companies, contributing to local technology diffusion, economic growth, job creation, and poverty reduction (Chocce 2003).

Service delivery through public-private partnerships

Of the many goods and services recommended for MDG scale-up programs in chapter 5, most will need to be publicly financed and managed, including, for example, those for education, research, and basic healthcare (chapter 3). But in many cases the private sector can partner with the government for service delivery. To achieve the Goals, governments need to pursue universal access to electricity, roads, and water and sanitation services. In many instances, private firms have the expertise in delivery and logistics to provide these goods and services efficiently at scale (box 9.2).

In considering public-private partnerships, it is important to separate issues of delivery from issues of finance. Even if a private firm is best placed to provide a service on efficiency grounds, that service will often need to be publicly

Box 9.2

Scaling up public services through the private sector: rural electrification in Chile

Source: Jadresic 2000.

Two billion people in developing countries do not have access to modern energy services, and most of them live in rural areas. Rural electrification is expensive, can be technically difficult, and usually fails to attract investment from the private sector. Large-scale rural electrification has been successful in many countries, using a range of public, private, or public-private partnerships. A consensus is growing that electricity and other core infrastructure services can be provided efficiently using a broad range of delivery mechanisms, with a clear regulatory framework and adequate financing.

Where governments have retained a stake in electric utilities, setting access targets is a matter of public policy. Evidence from South Africa demonstrates that such a policy can be implemented efficiently at large scale by state-owned enterprises. But where power sector reform has privatized electric utilities, rural electrification may require special approaches to involve the private sector in scaling up access.

Experience in Chile suggests that the key to expanding rural access is significant public financing targeted at encouraging private companies to develop rural infrastructure. Using domestic resources and external aid, the government of Chile in 1994 set up a special fund for rural electrification, competitively awarding subsidies to private electricity distribution companies that undertake projects. To properly target the subsidies, only projects with a positive social return but a negative economic return are considered. Participating companies design and present rural electrification projects to the government. The projects that score best on a set of objective criteria (cost-benefit analysis, lowest subsidy required, social impact) receive subsidies to cover part of their investment costs. End users pay for less than 10 percent of connection costs, and repayment can be spread out over time.

Even in a competitive environment dominated by private companies, the program increased the coverage of electricity systems in rural areas from 53 percent in 1992 to 76 percent at the end of 1999. Key success factors were well designed incentive structures combined with public financing of up to 90 percent of the capital cost. The example shows how efficient private sector delivery can be successfully combined with public financing of core infrastructure services.

financed. Enormous numbers of poor people simply cannot afford to pay even the lowest market price for the necessary inputs to a healthy and productive life. This is particularly so in rural areas dominated by subsistence agriculture, where very little output is traded and little if any cash is in circulation.

The fact that the private sector does not serve the poorest of the poor is not by itself a market failure, since markets are meant to orient their activity around high-return activities rather than around people with no money. Markets alone will not solve the short-term needs of the most impoverished communities. Instead, for the poorest and most isolated communities, public finance and subsidies will have to create the conditions for market-based economic activity to thrive.

A public-private partnership can combine the respective strengths of the private and public sectors. The private sector can leverage its advantages of greater efficiency, lower costs of distribution, and more complex delivery systems to reach new markets. The public sector can ensure universal access by providing financial support to subsidize impoverished households, thereby enabling private firms to enter large markets with guaranteed consumers. Public funding for the private provision of essential goods and services needs to cover both capital and recurrent costs to "close the revenue cycle." This is necessary to ensure that subsidized infrastructure investments do not naively project that household outlays will cover the long-term operating expenses, since in many extremely poor communities and countries, households will not be able to afford even to pay the marginal cost of basic water and energy services.

There are many instances where private firms have offered creative solutions to challenging delivery needs. In Namibia, for example, a public-private partnership between the Ministry of Health and Social Services and the United Africa Group, a private company, has allowed for efficient, timely, and regular delivery of state pensions, disability allowances, and child maintenance grants to almost all beneficiaries in rural areas. The company teams are equipped with portable automatic teller machines to disburse payments, smart cards to identify payment recipients, and small teams traveling by truck to manage the monthly delivery systems (ILO 2003).

For such arrangements to support the achievement of the Goals, it is crucial that all contracts be negotiated fairly with full transparency—to ensure that politically well connected firms do not profit unjustly (chapter 7). In cases of large infrastructure projects where economies of scale make it efficient to have a single large private contractor, the government needs to ensure that essential goods and services are universally available based on need rather than ability to pay. This may require price regulation, government subsidies, and perhaps other innovative financing mechanisms.

A promising innovation in ensuring quality service provision of basic social services by private providers is performance-based contracting—where the government contracts with private companies or NGOs to provide agreed-on services to a defined population. Payment is then based on service targets achieved and the quality of services (as confirmed through consumer feedback and inspection). This has improved the access to and quality of essential healthcare in parts of Cambodia, among other places.

Public-private partnerships can also take advantage of creative mechanisms to provide poor people with access to technologies that would otherwise be priced out of reach. Marginal cost pricing, for example, entails private firms agreeing to produce and make available a specified number of products and then to sell them to a public body for the cost of production rather than for any margin of profit. The decision of several major pharmaceutical companies to offer their antiretroviral drugs at cost rather than at patent-protected prices was an important step in this direction. So was the global recognition that low-cost generic drug manufacturers have an important role in alleviating suffering from disease in the poorest countries. In the future, tiered-pricing strategies could expand the reach of many new information technologies, biotechnologies, and other emerging breakthroughs from high-technology companies.

Responsible corporate governance and citizenship

In all their activities, private firms and their executive leadership have a unique obligation to adhere to high norms of corporate governance and citizenship. Private production processes and delivery mechanisms often have broad economic, social, and ecological consequences, which enterprises should address in a socially responsible manner (CSR Platform 2003). International treaties and instruments such as the Universal Declaration of Human Rights, the International Labour Organization (ILO) labor conventions, the 1992 United Nations Conference on Environment and Development in Rio de Janeiro, and the Copenhagen Declaration on Social Development provide guidelines for responsible business practices that help ensure private sector growth benefits society and protects the environment.

More recently, the UN Global Compact was enacted in 2000 as a UN-sponsored initiative to enable global businesses to sign up to a code of conduct that respects human rights, labor standards, and environment standards and fights corruption. The Global Compact lays out 10 basic principles for conducting business and provides useful guidelines for global corporations, in the spirit of the UN Millennium Declaration (UN 2004e). The Compact is a valuable undertaking that should promote operational guidelines and measurement tools, including targets where possible, for companies that want to contribute actively to the Goals. We endorse the UN Global Compact as a voluntary initiative and recommend that all corporations with operations in the developing world sign up for it.

For corporations with global operations, a first major priority needs to be vigilant support for human rights in their business processes, even in countries where gross violations of human rights occur. All personnel, especially security personnel, need to respect international principles of human rights (as outlined in the Universal Declaration of Human Rights) as part of the company's code of conduct. Before doing business in a country, corporations should be sure to understand how human rights may be affected by their operations.

Relevant to human rights, a healthy and decent work environment is essential for any company's work force. The International Labour Organization has promoted minimum norms for work standards, as defined in the Decent Work Initiative, that focus on steps to ensure international labor standards, provide decent and nondiscriminatory employment, offer social protection, and foster social dialogue among workers, governments, and businesses. The UN Global Compact focuses on many of these elements, such as upholding freedom of association and collective bargaining, eliminating forced and child labor, and eliminating discrimination in employment. Corporations also have an obligation to minimize work-related accidents and illness for their work force, and create safe and hygienic working conditions. We recommend that all companies adhere to these basic principles.

Another major element of sound corporate citizenship is to fight corruption. Corporations should refuse to pay bribes, especially in low-income countries where governments often struggle to strengthen the rule of law. Firms in extractive industries that operate in natural resource–rich developing countries have a special responsibility in this. The international oil companies have a particular responsibility to be transparent in their transactions with the government and local authorities, so that citizens can keep track of where the revenues are flowing. The recent "Publish What You Pay" initiative is an example of efforts by citizens in developing countries to track resource revenues and hold oil, gas, and mining companies accountable by demanding transparency in tax payments, license fees, and royalties (chapter 12).

Contributing to policy design

Many developing countries have legacies of political tension between the private sector and the public sector, often reflecting past ideological battles. As a central element of our recommended strategy for open national processes to develop MDG-based poverty reduction strategies, we recommend that the local private sector contribute to the development of these programs, alongside civil society, development partners, and the multilateral agencies. We also recommend that governments prepare a long-term private sector development strategy within the context of their MDG-based framework, consulting with the local private sector in the development of that strategy (chapter 4).

The local private sector represents a key constituency whose responsible voice should be represented alongside others in major policymaking processes. National chambers of commerce can often lead in representing the voices of private firms. Often, they are supported by the valuable work of international business organizations, such as the International Chambers of Commerce, the African Business Roundtable, the World Business Council for Sustainable Development, and the International Business Leaders Forum.

Of course, the private sector has to be balanced with civil society and other communities. Since the private sector is capable of wielding tremendous influence in decisionmaking through its financial strength, national governments have an added responsibility of ensuring transparency and balancing interests in all public discussions relating to private sector development processes.

Advocacy for the Goals

Since members of the private sector often have access to important channels of public debate, they can advocate the public investments needed to spur private economic activity, particularly in low-income countries. Domestically, private companies can have a big influence if they lobby publicly for improved infrastructure, better health services for their workers, and better education for the work force. They can help map out the sequence and scaling up of large-scale basic infrastructure over a medium-term time frame. They can also inform their home governments of the scale and type of investments needed to sustain economic growth. Major multinational firms can also lead in global advocacy efforts for the Goals, either by lobbying their local politicians that global development is a worthy undertaking, or by identifying specific practical initiatives where they might provide support to raise awareness of the Millennium Development Goals in their home countries.

Corporate philanthropy

The discussion above highlights that there are many ways in which the private sector can contribute to poverty reduction and the Goals, quite separate from direct philanthropy. Nonetheless, many companies are well positioned to make important contributions through direct giving or other philanthropic models. In aggregate, these flows can be sizable. In 2003, for example, 232 of the largest U.S. corporations alone donated $1.1 billion for international philanthropy (Muirhead 2004). Although this amount was not oriented solely to developing countries and is vastly less than what is needed to achieve the Goals, it represents considerable resources from one country and indicates an important volume of assistance that could support poverty reduction in many places. For companies seeking to increase or streamline their philanthropy, the Goals provide a framework to undertake specific activities aligned to countries' actual investment needs. Corporate philanthropic efforts can only supplement, not replace, government-led efforts. But they can often provide the trigger or "seed" funds that pave the way for increased investments.

In pursuing philanthropy, multinational corporations that operate across several countries have the ability to leverage their tremendous voice and resources to advocate and support specific development objectives. They can adopt specific Goals and base their philanthropic endeavors on these Goals in each country of operation. They can become ambassadors for those Goals and mobilize partners. The Merck Mectizan Donation Program is an example of a large-scale and long-term commitment. Since 1987 Merck has partnered with the World Bank, the World Health Organization, and other agencies to reach more than 40 million people suffering from onchocerciasis (African river blindness) in more than 30 countries, providing Mectizan free of charge and combining treatment with prevention.

Philanthropy can also be in-kind—for example, through community partnerships and education and training. Private firms can invest heavily in the development needs of their local host communities through such partnerships, signaling a long-term commitment to the welfare of the host community. Examples of such partnerships include Bayer's Integrated Crop Management Program for farmers in Guatemala and especially in Brazil, where the company trained more than 25,000 farm families in improved farming techniques. Corporations can also donate employee and management time to help public sector enterprises and governments design and implement projects. And they can contribute to developing local capacity by undertaking extensive training programs and contributing to the development of the local private sector. For example, Alcoa Aluminio in Brazil has developed an extensive training program in close collaboration with local universities that has helped build the skills of the local work force.

These are only a few examples of innovative approaches to corporate philanthropy. To systematize corporate support for the Goals, and to provide a set of benchmarks for companies to measure themselves, we encourage all Fortune 500 companies to voluntarily include a Millennium Development Goals scorecard in their annual reports. This would outline how their philanthropic and corporate governance activities have contributed to achieving the Goals, taking advantage of the quantified and time-bound nature of the Goals to measure results, plan future activities, and coordinate with other philanthropic partners.

Chapter 10

Africa's special needs

Why has progress toward the Millennium Development Goals proved so difficult in much of Sub-Saharan Africa? The standard diagnosis of Africa is that the continent is suffering from a governance crisis.[1] With highly visible examples of profoundly poor governance, as in Zimbabwe, and widespread war and violence, as in Angola, Democratic Republic of the Congo, Liberia, Sierra Leone, and Sudan, the impression of a continentwide governance crisis is understandable. But it is wrong. Many parts of Africa are well governed, especially considering the extremely low incomes of these countries, and yet even the relatively well governed countries remain mired in poverty and poverty traps. Governance is an issue, but Africa's development challenges are much deeper.

Indeed, using World Bank indicators, there is no evidence that Africa's governance, on average, is worse than elsewhere once we control for Africa's very low income (table 10.1, column 1).[2] Controlling for income is necessary in evaluating governance since good governance requires resources for wages, training, information systems, and so forth, and thus improves systematically with income levels (chapter 7).

The same findings—that Africa's governance is on par with other regions at comparable income levels—hold when we examine other measures of governance, such as the Corruption Perception Index of Transparency International (2004). Most African countries score as "good" (low corruption) or "average" after controlling for income (table 10.1, column 2). Also, many African countries have become democracies in recent years, and thus are scored as "free" or "partly free" by the well known Freedom House ranking (table 10.1, column 3).

The striking fact is that many of the better governed African countries have been unable to increase the material well-being of their populations (table 10.1, column 4). Cross-country regression results also show that, after accounting

Table 10.1

Governance in tropical Sub-Saharan Africa is no worse than elsewhere, after controlling for income

Country	Rating Based on World Bank governance indicators 2002[a]	Rating based in Transparency International index 2003[a]	Freedom House rating 2003	Household final consumption expenditure per capita 2000 (1980 = 100)
Benin	Good	—	Free	99
Burkina Faso	Good	—	Partly free	111
Ghana	Good	Average	Free	93
Madagascar	Good	Good	Partly free	64
Malawi	Good	Good	Partly free	111
Mali	Good	Good	Free	95
Mauritania	Good	Good	Partly free	105
Senegal	Good	Good	Free	100
Cameroon	Average	Average	Not free	103
Central African Republic	Average	—	Partly free	—
Chad	Average	—	Not free	—
Congo, Rep.	Average	Average	—	81
Côte d'Ivoire	Average	Average	Not free	78
Eritrea	Average	—	Not free	—
Ethiopia	Average	Good	Partly free	—
Guinea	Average	—	Not free	—
Kenya	Average	Average	Partly free	101
Mozambique	Average	Good	Partly free	79
Niger	Average	—	Partly free	—
Nigeria	Average	Average	Partly free	—
Rwanda	Average	—	Not free	84
Sierra Leone	Average	Good	Partly free	44
Tanzania	Average	Good	Partly free	—
Togo	Average	—	Not free	112
Uganda	Average	Average	Partly free	—
Zambia	Average	Good	Partly free	47
Angola	Poor	Poor	Not free	—
Burundi	Poor	—	Not free	65
Congo, Dem. Rep.	Poor	—	Not free	45
Sudan	Poor	Average	Not free	—
Zimbabwe	Poor	Average	Not free	88
Liberia	—	—	Not free	—
Somalia	—	—	Not free	—

— Not available.

Note: The table reports some common governance indicators for a 33-country sample of tropical Sub-Saharan African countries with populations of 2 million or more. Column 1 presents a ranking of African governance from Radelet (2004), who regresses a set of widely used World Bank governance indicators on GNP per capita (Kaufmann, Kraay, and Mastruzzi 2003), and ranks all countries according to the residual from the regression line, thereby standardizing the measurement of governance by the level of income. This procedure recognizes that poorer countries systematically have poorer governance measures than richer countries, since good governance itself requires resources for wages, training, information systems, and so forth.

a. Determined from the residuals of a regression of countries' governance indicators or scores on income per capita (at purchasing power parity); countries with residuals more than 1 standard deviation above or 1 standard deviation below the predicted value are categorized as "good" or "poor," respectively, and those with residuals within 1 standard deviation as "average."

Source: Reproduced from Sachs and others 2004; Kaufmann, Kraay, and Mastruzzi 2003; Radelet 2004; authors' calculations using data in Transparency International 2004; Freedom House 2003; World Bank 2004c.

for initial income in 1980 and the quality of governance, Sub-Saharan African countries grew more slowly than other developing countries, by around three percentage points a year. Africa's crisis thus requires a deeper explanation than governance alone.

Our explanation is that tropical Africa, even in well governed parts, is stuck in a poverty trap—too poor to achieve robust and high levels of economic growth, and in many places simply too poor to grow at all (chapter 3). More policy or governance reform, by itself, is not sufficient to break out of this trap. Africa's extreme poverty leads to low national saving rates, which in turn lead to low or negative economic growth rates. Low domestic saving is not offset by high inflows of private foreign capital, such as foreign direct investment, since Africa's poor infrastructure and weak human capital discourage private capital inflows. With very low domestic saving and low rates of market-based foreign capital inflows, there is little in Africa's current dynamics that promotes an escape from poverty.

The combination of Africa's low domestic saving rate and high population growth rate has led to stagnation in Africa's patterns of capital accumulation. The national income accounts indicate that tropical Sub-Saharan Africa has an average saving rate of about 11 percent, compared with 20 percent in Latin America, 18 percent in South Asia, 19 percent in the Middle East and North Africa, and 34 percent in East Asia and the Pacific (see table 3.1). The situation is even worse than it looks, however, because the national income accounts data almost surely (and substantially) overestimate Africa's true saving rate (chapter 3, figure 3.3). To a significant extent, Africa is living off of its natural capital but counting the resource depletion as income, a point stressed by Sachs and others (2004).

Africa's unique circumstances

To understand why Sub-Saharan Africa is the region with the greatest MDG investment needs, we stress five structural reasons that have made it the most vulnerable region in the world to a persistent poverty trap:

- Very high transport costs and small markets.
- Low-productivity agriculture.
- Very high disease burden.
- A history of adverse geopolitics.
- Very slow diffusion of technology from abroad.

High transport costs and small markets

To a remarkable extent, Africans live in the interior of the continent and face enormous transport costs in shipping goods from coastal ports to where they live and work. These costs are much higher than comparable costs in Asia (figure 10.1). Moreover, Sub-Saharan Africa is effectively cut off by the Sahara from high-volume overland trade with its major high-income trading partner,

Figure 10.1

Relative transport costs in a typical Sub-Saharan country are much higher than in Asia

Transport costs per ton per kilometer

Source: Reproduced from Starkey and others 2002.

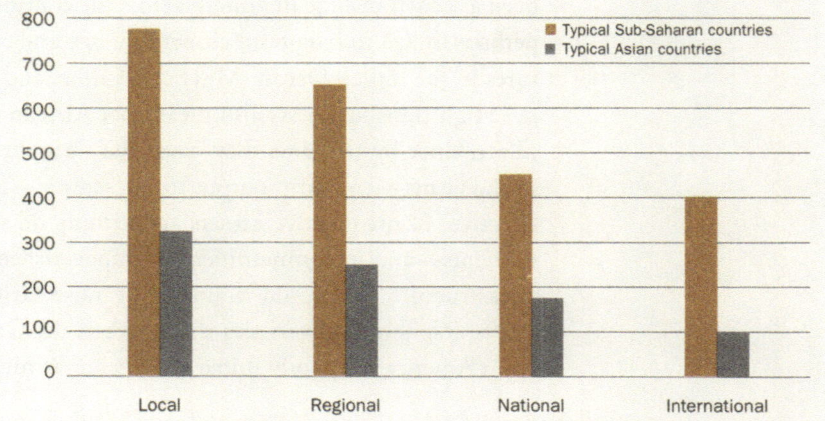

Europe, adding to the high costs of transport. Problems of isolation are compounded by small market size. Developing countries with small populations and little access to global trade tend to grow more slowly than countries with large populations or countries with small populations that have easy access to trade, such as Singapore (Sachs forthcoming).

Africans live away from the coast for several reasons. The soils are often better and rainfall more plentiful in the interior highland regions. The burden of malaria is intrinsically lower. And centuries of slave trade made it dangerous for Africans to live near the coast. This problem would not be so severe if the rivers from the interior to the coast of Africa were ocean-navigable—but they are not.

Recent evidence confirms the extremely high transport costs in Sub-Saharan Africa and their severe impact on trade. One study estimates that halving transport costs could increase the volume of transport by a factor of five (Limão and Venables 1999).[3] Before high-intensity modern trade can get started, Africa needs an extensive road system from the coast to the interior—and within the interior, where the highest population concentrations are found. These roads, however, are very expensive to build and maintain, particularly on a per capita basis in areas of low population density.

Low productivity agriculture

Africa gets no break on food productivity. Most Africans live in the subhumid or arid tropics, with few rivers to provide irrigation and a lack of the large alluvial plains, typical in much of South and East Asia, that permit cheap irrigation. As a result, Africa has the lowest share of food crops produced on irrigated land of any major region of the developing world. African agriculture also suffers from erratic rainfall, is vulnerable to high seasonal and interannual fluctuations, and is subject to high rates of evapotranspiration due to high temperatures. Indeed, of all major regions, Africa loses the highest share of its precipitation through evapotranspiration (GEMS 1995). In addition, there has been a secular decline in rainfall across the continent during the past 30 years, perhaps linked to long-term climate change and to rising sea-surface temperatures in the Indian Ocean (Mitchell, Hulme, and New 2002).

High transport costs also mean that African farmers can afford little fertilizer, since by the time they cover the transport costs of bringing fertilizer to the farms—and farm output to the market—fertilizers are no longer cost-effective. Consequently, farmers are farming on soils increasingly depleted of nutrients—and in communities too impoverished to finance roads and water infrastructure that could dramatically raise farm yields. And, as discussed below, the new seed varieties that sparked the Green Revolution in Asia and Latin America are poorly suited to African farming conditions.

Very high disease burden

Africa carries a disease burden unique in the world. In recent years the most prominent disease has been HIV/AIDS, wreaking economic and social catastrophe throughout the region. Approximately 25 million Africans were estimated to be living with HIV/AIDS in 2003, and 2.2 million died from it in the same year (UNAIDS 2004). Today roughly three-quarters of the world's annual AIDS deaths occur in Africa, with women now disproportionately affected. Sub-Saharan Africa's adult HIV prevalence in 2003 was 7.5 percent, while every other region save the Caribbean was below 1 percent (UNAIDS 2004). The spread of HIV is fueling an epidemic of TB, which takes its heaviest toll among young productive adults. In some high HIV prevalence African countries, TB infection rates have quadrupled since the mid-1980s, placing overwhelming burdens on existing TB control programs.

Africa is also home to numerous endemic tropical diseases, especially vector-borne diseases. Among these, malaria is by far the most consequential. Of the more than 1 million malaria-related deaths every year it is estimated that 90 percent occur in Sub-Saharan Africa, the great majority of them among young children (WHO and UNICEF 2003). Many casual observers make the mistake of assuming that since the United States and Europe once had malaria and got rid of it, Africa's ongoing malaria crisis is a symptom of poverty and weak institutions rather than a deep cause. In other words, many ask why should malaria have played any more of a causal (and intractable) role in Africa's development than in did in the southern United States, southern Europe, and other regions that have since eliminated the disease? The answer: disease ecology.

Africa's malaria is much less tractable than in other regions for a combination of climatic and biological reasons. The continent's temperatures, mosquito species, and humidity give Africa the highest malaria burden, as captured in a malaria stability index, a measure of the strength of transmission based solely on ecological factors (map 10.1). Unlike other parts of the world, Africa's malaria mosquitoes are almost exclusively human-biters, which

Map 10.1

Global map of malaria stability index

Note: Distribution of the actual and potential stability of malaria transmission based on regionally dominant vector mosquitoes and a 0.5° gridded temperature and precipitation data set.

Source: Kiszewski and others 2004.

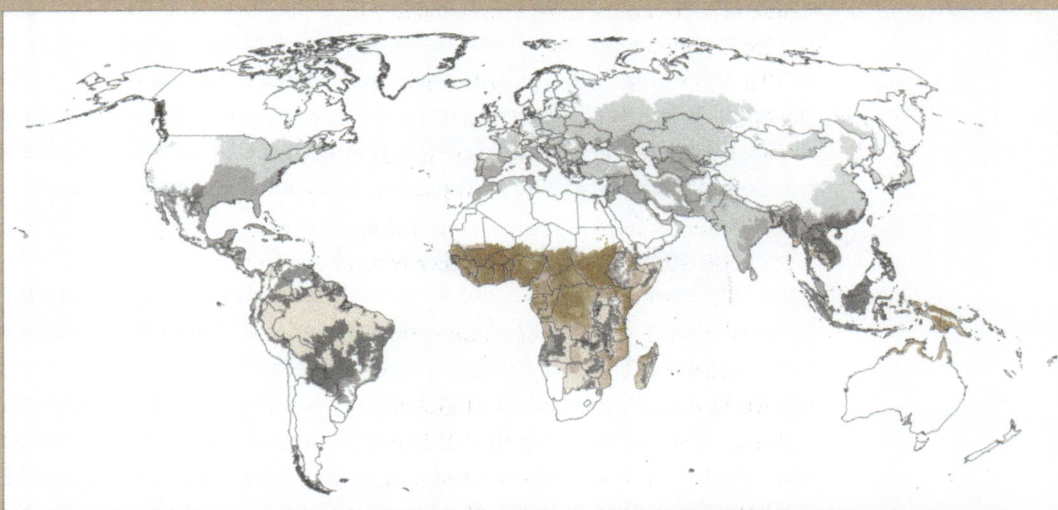

0 | 0–0.05 | 0.05–1.00 | 1–2 | 2–5 | 5–8 | 8–12 | 12–18 | 18–26 | Higher than 26

enhances the chain of human-to-human transmission. The combination of high year-round temperatures, adequate precipitation for mosquito breeding, and human-biting anopheles mosquitoes has made Africa the global epicenter of malaria from time immemorial.

Malaria contributes to a classic poverty trap. With enough investments, even Africa's malaria morbidity and mortality could be controlled, though not eliminated, with current technologies. But control would require substantially more money than Africa can afford. Thus, Africa is too poor to control malaria, while the disease reduces productivity, frustrates foreign investment, and delays or stops the demographic transition (by contributing to very high child mortality rates), helping to keep Africa poor.

A history of adverse geopolitics

On top of the structural challenges, Africa has suffered brutally at the hands of European powers for almost five centuries, and the record with Arab powers has been little better. A massive slave trade helped undermine state formation and may have depopulated Africa's coastal regions. In the nineteenth century, the slave trade was replaced by direct colonial rule and a century of exploitation by European imperial powers, who left very little behind in education, health-care, and physical infrastructure.

The starting conditions in Africa in the 1960s were far behind those of other parts of the developing world (table 10.2). Contrary to casual discourse (the common comparison of Ghana and Korea in the 1960s, for example), African countries at the time of independence had very few individuals with higher education, very few paved roads, almost no electrification of rural areas where the bulk of the population lived, and food yields far below those of other parts of the developing world. Africa had a much harder path to follow, and was much more vulnerable to getting stuck in a poverty trap. Adding to the burden, during the Cold War politics of the late twentieth century, many African countries found themselves to be battlegrounds in a global ideological struggle.

Very slow diffusion of technology from abroad

Africa has been the great laggard in technological advance, notably in agriculture and health. The uptake of technologies to prevent and treat major diseases, such as malaria, has been extremely slow. In agriculture, most of the developing world had a Green Revolution surge in crop yields in the 1970s–90s as a result of scientific breeding that produced "high-yielding varieties" combined with increased use of fertilizers and irrigation. Africa's uptake of high-yielding varieties was the lowest in the developing world for very clear reasons.

The Green Revolution high yielding varieties, designed for conditions in Latin America and Asia, did not easily transfer to the agronomic and economic

Table 10.2

Comparative indicators across developing regions around 1965

Note: Data are averages weighted by population.

Source: World Bank 2004c; average schooling data from Barro and Lee 2000.

	Life expectancy at birth (years) 1965	Under-five mortality rate (per 1,000 live births) 1970	Literacy rate (percent of adults) 1970	Cereal yield (kg per hectare) 1965	Average schooling (years in total population) 1965
East Asia and the Pacific	53	124	54	1,764	4.0
Latin America	58	123	73	1,250	3.0
Middle East and North Africa	50	201	25	881	0.8
South Asia	47	205	33	1,268	1.4
Sub-Saharan Africa	42	239	24	801	0.9

conditions of Africa's rainfed, fertilizer-scarce, subhumid, and arid tropics. The absence of a Green Revolution had a clear impact. Sub-Saharan Africa has the lowest cereal yield per hectare of any major region and the slowest gain in yields in the last two decades (table 10.3). Indeed, it is the only major region with a (slight) decline in food production per capita during 1980–2000.

MDG-focused investments for Africa

The structural impediments outlined above are real. But in countries where governments are committed, they can be overcome if addressed through an intensive investment program that directly confronts high transport costs, low agricultural productivity, high disease burden, the colonial legacy of weak infrastructure and poor educational attainment, and the like. Ending the poverty trap in Africa and meeting the Millennium Development Goals will require a comprehensive strategy for public investment in conjunction with improved governance. This section lays out priorities for such an investment strategy, drawing on many of the key interventions in chapters 5, 6, and 7.

Before outlining some specific investments, we begin with an overarching point. When African countries design their own packages, they need to give careful consideration to gender equality—specifically, to improving the social and economic status of Africa's women. To a great extent, women are the farmers, caregivers, and child-raisers of Africa. They carry a triple burden. They care for children, the elderly, and the sick. They spend long hours gathering water and fuelwood and processing and producing food. And they work on farms and family enterprises for little or no income. Many of the UN Millennium Project's recommended investments—such as improved access to

Table 10.3

Agricultural technology and productivity, by developing region

	Share of area planted to modern varieties (percent)[a]				Contribution of crop genetic improvement to yield growth[a, b]	Cereal yield (kg per hectare)[c]	Average annual growth in cereal yield (percent)[c]	Average annual growth in food production per capita (percent)[d]
	1970	1980	1990	1998	1960–98	2000	1980–2000	1980–2000
Asia	13	43	63	82	0.88	3,662	2.3	2.30
Latin America	8	23	39	52	0.66	2,809	1.9	0.90
Middle East and North Africa	4	13	29	58	0.69	2,660	1.2	1.00
Sub-Saharan Africa[e]	1	4	13	27	0.28	1,112	0.7	–0.01

a. From Evenson and Gollin 2003.

b. Measured as a share of increase in productivity.

c. From World Bank 2003a.

d. From FAO 2003b.

e. "Sub-Saharan Africa" refers to all countries in columns 1–5 and in columns 6–8 refers to 33 countries in "Tropical Sub-Saharan Africa" as defined in Sachs and others 2004.

Source: Adapted from Sachs and others 2004.

water supply, modern cooking fuels, enhanced transport services, and better soil nutrients—will have a special benefit for women.

Rural development

The first investment area focuses on raising rural productivity, since three-quarters of Africa's poor live in rural areas (chapter 2). In particular, the investments in farm productivity will increase rural incomes and reduce chronic hunger, predominantly caused by insufficient agricultural productivity. A Twenty-first Century African Green Revolution is needed to help launch an environmentally sound doubling or more of agricultural productivity. Additional interventions in roads, transport services, electricity, cooking fuels, water supply, and sanitation all provide a basis for higher productive efficiency.

Urban development

Throughout Sub-Saharan Africa the large cities have not yet generated internationally competitive manufacturing or service-based industries. In conjunction with abject rural poverty, which fuels rural-urban migration, the lack of urban jobs has led to extremely high levels of urban poverty that are rising across much of the continent. An MDG-based urban strategy needs to focus on urban infrastructure and services (electricity, transport, water, sanitation, waste disposal, and so forth) and slum upgrading. Successful urban development and viable export industries across Africa are contingent on improving access to rich countries' markets, particularly for apparel and light manufacturing, and the flexibility to use targeted industrial policies as needed. Since urban populations are growing very rapidly across the continent, countries must develop investment strategies to provide alternatives to slum formation.

Health

Investments are needed to address Africa's extraordinary disease burden, widespread micronutrient deficiencies, and extremely high fertility rates by focusing on health, nutrition, and family planning. This package includes health system–based interventions to improve child health and maternal health; prevent the transmission of and provide treatment for HIV/AIDS, TB, and malaria; improve nutrition; and provide reproductive health services. Halting the AIDS and TB epidemics is of enormous importance.

For HIV/AIDS, scaling up prevention initiatives, improving testing and counseling services, and increasing public awareness are important first steps to containing the disease. Treating people already infected with the disease is equally important. These measures will require concerted financial and technical support from developed countries. The World Health Organization's 3 by 5 initiative is a promising start, aiming to get 3 million of the world's AIDS patients on antiretroviral treatment by 2005. But such interventions will need to be scaled up much more in the next decade to reach the 25 million Africans currently infected with HIV (UNAIDS 2004).

As another important element of a health system, we recommend an entire package of sexual and reproductive health services, including family planning to enable a significant voluntary reduction in Africa's very high total fertility rates and population growth rates. Access to reproductive health services and contraception, girls' education, women's empowerment, and off-farm employment opportunities for young women can lead to a dramatic reduction in the total fertility rate in just a few years.

Education

Today, only 57 percent of children in Sub-Saharan Africa have access to primary education (UNDP 2003d). Secondary education enrollment is much more variable, ranging from 6 percent in Niger to 43 percent in Zimbabwe (World Bank 2004c). MDG-based strategies in Africa should aim for universal completion of primary education and increased access to secondary and tertiary education. In designing this package of interventions, particular attention needs to be paid to increasing girls' completion rates through additional demand-side interventions, such as incentive payments to poor households to encourage them to keep their daughters in school.

We also argue that secondary school enrollments need to be increased, since the returns to education now and especially in the future will depend on secondary education. Large numbers of secondary school graduates will be required to deliver the other MDG intervention packages (secondary school graduates will become community health workers and agricultural extension workers). Targeted adult literacy programs designed to raise educational attainments among the adult population will complete the investments in human capital.

Human resources

To achieve the Goals in Africa, significant investments in human resource development are needed urgently, since health, education, agricultural extension, and other critical social services cannot function without cadres of properly trained staff. HIV/AIDS, years of public sector wage ceilings and hiring freezes, outward migration, and poor working conditions have stripped Africa of the human resources needed to deliver needed interventions. The Joint Learning Initiative (2004) estimates that Africa now faces a shortage of a million health workers. Qualified teachers and other service providers are also in short supply. To build Africa's capacity to deliver the services and interventions to achieve the Goals, major coordinated investments in preservice training (such as degree and certification programs) will be needed to build a qualified work force of service delivery staff. These will need to be complemented by in-service training for existing staff, adequate salaries, and human resource management systems.

Given the need to reach rural and often remote areas, we put great stress on scaling up the training of vast numbers of community workers in health, agriculture, and infrastructure, with programs of one-year training. These community workers will play a vital role in enabling villages to make the basic MDG investments in health, education, water and sanitation, electricity, irrigation, soil nutrient replenishment, and other areas of vital need. This process of scaled-up community-based training should start right away in 2005.

Gender equality

As indicated above, any MDG-based investment program for Africa should pay particular attention to promoting gender equality, both as a goal in itself and as a crucial input to achieving all the other Goals. This includes ensuring full access to reproductive health rights and services, as well as guaranteeing equal property rights and access to work, backed by affirmative action to increase political representation. Of particular concern in many parts of Sub-Saharan Africa are persistently high levels of violence against women and girls, which need to be confronted with public awareness, legislative and administrative changes, and strong enforcement.

Science, technology, and innovation

An essential priority for African economic development is to mobilize science and technology. Tropical Sub-Saharan Africa produces roughly a twentieth of the average patents per capita in the rest of the developing world (U.S. PTO 2001). And it has only 18 scientists and engineers per million population compared with 69 in South Asia, 76 in the Middle East, 273 in Latin America, and 903 in East Asia (World Bank 2004c). We stress the need for increased investments in science, higher education, and research and development targeted at Africa's specific ecological challenges (food, disease, nutrition, construction, energy).

Regional integration priorities

Regional integration is essential for African economic growth. With much of Africa landlocked (15 Sub-Saharan countries), the interior countries have little chance to develop unless they have ready access to the coast with efficient low-cost infrastructure. And from a global perspective, individual African countries are very small markets.

Regional integration will raise the interest of potential foreign investors by increasing the scope of the market that accompanies an operating presence in Africa. It is also important in achieving scale economies in infrastructure networks, such as electricity grids, large-scale electricity generation, road transport, railroads, and telecommunications—and in eliciting increased R&D on problems specific to Africa's ecology but extending beyond any single country (public health, energy systems, agriculture). Regional programs, such as those advanced by the New Partnership for Africa's Development, thus require greatly increased support (chapter 15).

Public sector management priorities

Although governance in Africa is not systematically worse than that in other countries after controlling for income, many of the government systems are still weak on an absolute scale and require significant investments in public administration (chapter 17). Information management systems and investments in the training of public sector managers will undoubtedly be crucial. Addressing this issue should be closely linked to reversing and treating the AIDS pandemic, which is taking the lives of hundreds of thousands of civil servants throughout the continent.

Chapter 11

Investment priorities for reaching the Millennium Development Goals in other regions

Development is not a one-size-fits-all process, so national MDG-based strategies will need to be adapted to specific regional and national situations. This chapter describes regional investment priorities outside Sub-Saharan Africa. We begin by identifying interventions from chapter 5 that are especially relevant in Asia, home to two-thirds of the world's poorest people: 271 million in East Asia, 430 million in South Asia and the rest in Central Asia (Chen and Ravallion 2004). Although human development indicators are improving overall, the region is marked by a high degree of variation in social and economic conditions. East Asia has been one of the fastest growing regions in the world for the last half century and has made major strides in reducing poverty, hunger, illiteracy, and disease. But pockets of the region, especially in Lao People's Democratic Republic (PDR), Cambodia, and parts of Indonesia, remain trapped in poverty. China embodies the contrasts starkly—the western and northern rural provinces lag far behind coastal regions, which have grown explosively. In South Asia, India is home to more than 360 million people below the poverty line (Chen and Ravallion 2004). It has seen rapid development in its southern states, but serious challenges remain in the north. Most of the rest of South Asia shares a similar experience of growth mixed with variable progress on social development indicators. The Central Asian states face extreme geographic isolation, ecological stress, and rapidly deteriorating development indicators.

Countries in Latin America and in the Middle East and North Africa face their own challenges in meeting the Goals. These two regions together accounted for only an estimated 57 million of the world's poorest people in 2001, or 5.2 percent of the total (Chen and Ravallion 2004). But both regions face serious development challenges such as economic stagnation, environmental stress, rural isolation, and deeply entrenched social inequalities. The transition countries of Central and Eastern Europe also face unique challenges, many of which derive from the collapse of the Soviet Union. For countries with special needs, including the Least Developed Countries, landlocked countries, small island developing countries, and countries at high and regular risk of natural disasters, specific interventions will be required.

East Asia

China

China's economic performance in the last two decades has been nothing short of spectacular. Real per capita economic growth rates have averaged 8.2 percent a year. Output has quadrupled. The incidence of rural poverty declined from 30 percent in 1990 to 11 percent in 2002 (Woo and others 2004). China is on track to meet many of the Goals, including the targets for poverty, hunger, primary enrollment, and health.

Despite these achievements, China still has 102 million people below the poverty line, and income growth is unlikely to maintain the pace of the past two decades. Rural income growth in particular declined from 5.7 percent in the 1990s to nearly 4 percent during 1997–2002, leading to more disparities between the coastal, more urbanized provinces—such as Guandong, Jiangsu, and the metropolises of Beijing and Shanghai—and the interior, more rural southwest, central, and northwest provinces—such as Henan, Shanxi, and Gansu (map 11.1) (Woo and others 2004).

In addition, China faces challenges in meeting the Goals for gender equality, HIV/AIDS, access to clean drinking water, and environmental sustainability. The sex ratio at birth in China increased from 111 males per 100 females in 1989 to 118 in 2000, reflecting the strong preference for boys (Woo and others 2004). The public health system is under severe stress, with rural health insurance coverage falling from 90 percent to less than 10 percent between the 1970s and 1998 (Bogg and others 1996; Liu, Rao, and Hsiao 2003). As a result, the vast majority of people in rural areas must pay out of pocket for all health services, which can result in financial catastrophe for those with serious illness and has been found to be a major contributor to rural poverty in China's villages (Liu, Rao, and Hsiao 2003). HIV prevalence rates have risen sharply; the number of people living with the virus is estimated to be 1 million, but estimates suggest it could increase to 10 million by 2010 without an immediate effective response (UNAIDS 2004). With China's rapid urban growth, urban poverty has risen from less than 1 percent in 1984 to 3.4 percent in 2000. Environmental degradation has been another serious consequence of economic growth. More than 90 percent of China's grasslands are degraded, more than 75 percent of the water in rivers in urban areas is unfit for human contact, and 6 of the world's 10 most polluted cities are in China (Woo and others 2004).

An MDG-based investment strategy for China would need to include four key elements:

Map 11.1
Per capita incomes in provinces of mainland China
2002 US$

Source: National Bureau of Statistics of China 2003.

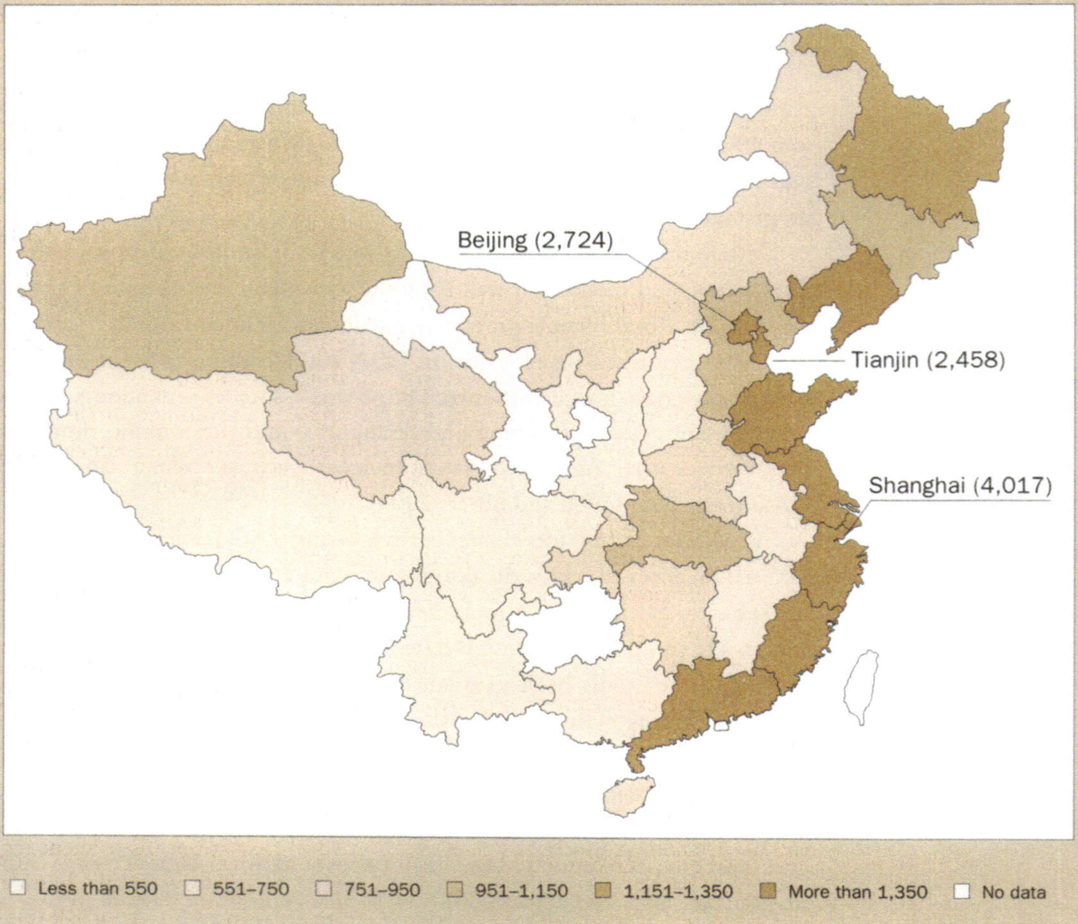

- *Rural investment strategies.* Rural areas require better roads and transport services, energy, sanitation, and water supply. Investments should also seek to increase agricultural productivity through increased research on improved seeds, biotechnology for agriculture, improved local livestock, better extension services, and support for agroprocessing technologies (as outlined in chapter 5). A central challenge across China is to contain rising water consumption by agriculture and industry through improved production techniques and better water management. Some of these interventions are already embedded in government programs such as the Outline for Poverty Alleviation and Development in China's Rural Areas and the Western Development Strategy, but they need to be supported and expanded.

- *Health and education.* Resources are required to provide universal access to basic education and primary health services. The primary healthcare system requires support, particularly in rural areas, with special attention to reducing the out-of-pocket healthcare costs for access to services through increased government funding and rural cooperative health insurance. China needs to mount an urgent HIV prevention effort and to focus on the health needs of girls and women.
- *Urban development.* Continuing migration from rural areas is likely to increase pressure on China's cities, which in 2001 already had more than 175 million people living in slum-like conditions (UNDESA 2004). Current projections estimate that the urban population could increase from 36 percent to roughly 57 percent of the total population by 2025 (United Nations Population Division 2003a). Managing this increase will require significant investments in slum upgrading, low-cost public housing, such basic services as health and education, infrastructure, and environmental management for waste disposal, recycling, and wastewater treatment.
- *Environmental management.* Critical investments to combat pollution and environmental degradation include research, new technologies, regulatory capacity building, water treatment plants, mechanisms to monitor pollutants from private industries, and active conservation programs for degraded ecosystems. Investments in clean energy are particularly important given China's rapidly increasing consumption of fossil fuels.

Other East Asian countries

The countries of East Asia have, as a group, moved closer toward achieving the Goals, but progress has been uneven within the region, within countries, and across the Goals. Some countries have already come close to achieving most of the Goals and have even committed themselves to more ambitious, MDG-plus targets, but others remain significantly off track for meeting the original Goals.

Thailand and Cambodia illustrate some of the disparities. In Thailand poverty fell from 27.2 percent in 1990 to 9.8 percent in 2002, the proportion of underweight children dropped from 18.6 percent to 8.5 percent between 1990 and 2000, and progress has been significant on the other Goals (United Nations Country Team and government of Thailand 2004). By contrast, despite strong government commitment to the Millennium Development Goals, Cambodia is off track for most of the Goals. The under-five mortality rate actually increased between 1990 and 2000, TB prevalence is rising, and rapid deforestation continues (UNDP 2001). Although conditions vary by country, Lao PDR, Indonesia, Myanmar, and the Democratic Republic of Korea are also struggling to meet the Goals, particularly those related to poverty, health, and gender equality.

Regional challenges include achieving targets for health, gender equality, and environmental sustainability, as many countries remain off track. For example, while the prevalence of HIV has fallen slightly in Cambodia and Thailand, it has risen in other countries, including Myanmar, Nepal, and Viet Nam (UNAIDS 2004). In many countries in the region, health systems are undergoing a transition, with more participation from the private sector in financing and service delivery. While this has in many cases improved quality, it threatens to reduce access for the poor who cannot afford user fees or copayments. In terms of gender equality, a majority of countries in the region have sizable male-to-female enrollment disparities at all levels of education. Meanwhile, 9 of 11 countries in Southeast Asia experienced declines in forest cover of more than 1 percent of land area between 1990 and 2000 (UN ESCAP 2003).

Investment priorities differ between strong performers and countries still struggling to make progress. For countries that are struggling to meet the Goals, such as Cambodia, Indonesia, Lao PDR, and Myanmar, the most urgently needed investments include:

- *Health and education.* Stagnant or worsening health and education indicators in Southeast Asia reflect financial and human resource constraints in social service delivery. Salaries in the social sectors are low, increasing absenteeism, impeding performance, and encouraging the charging of informal fees that pose a barrier to access, especially for the poor. Strategies to ensure that the poor have access to health services in the face of increasing privatization are essential. In the education sector, greater infrastructure investments are needed, as are investments in teachers' salaries and training, other recurrent costs, and demand-side interventions.
- *Rural investment strategies.* While urban poverty is rising rapidly, extreme poverty and hunger in Southeast Asia continue to be concentrated in rural areas. These regions require investments in basic infrastructure in addition to productivity-enhancing investments, such as improved agricultural inputs and agricultural extension services.

Investments needed in all the countries of the region, including such strong performers as Malaysia, Thailand, and Viet Nam, include:

- *Environmental management.* Deforestation and biodiversity loss remain critical in Southeast Asia, as do industrial waste and pollution. In combination they lead to an increasing degradation of terrestrial, marine, and coastal ecosystems. Investments in monitoring, regulatory capacity, and enforcement are required to prevent further degradation.
- *Science and technology.* While parts of East Asia have built technology-based industries, the region as a whole needs to invest in higher education

and infrastructure for science and technology development to develop higher value industry and provide employment opportunities for youth.

- *Public management.* Many of East Asia's economies remain highly vulnerable to external changes in the world economy. Strengthening the ability of domestic institutions to respond to adverse economic shocks will lay the foundation for sustained progress toward the Goals.

South Asia

India

India has seen strong economic performance over the past decade. Per capita incomes have grown by almost 4 percent a year, fueled largely by strong agricultural growth, a rapidly expanding services sector, and an increase in export-based and other manufacturing activities (World Bank 2004c). Rapid growth has led to significant declines in poverty rates. The World Bank estimates that the percentage of people below the poverty line has dropped sharply to 35 percent in 2001, and national estimates show levels falling from 37.5 percent in 1990 to 26 percent in 2000 (Chen and Ravallion 2004; Bajpai, Sachs, and Volavka 2004). Remaining challenges include high rates of undernutrition, large numbers of children out of school, poor health indicators, and wide disparities in social and economic indicators, particularly for women, girls, and members of low-caste and tribal populations.

Regional disparities are wide. The northern states are among the weakest economic performers in the country, while the southern and coastal states are generally enjoying rapid economic growth. Since the waning of the growth spurt induced in Punjab and Haryana by the Green Revolution, rapid growth in India since the early 1980s has been driven largely by the big coastal urban centers and information technology–based cities, such as Bangalore and Hyderabad in the south.

India faces several challenges in meeting the Goals. In 2001 the government spent just $4 per capita on health, while out-of-pocket private spending was $24 per capita. Not only is the total expenditure likely insufficient to deliver a quality package of essential services, the high level of out-of-pocket spending is a major financial burden and a barrier to access to health services for the population, especially the poor (Bajpai, Sachs, and Volavka 2004; WHO 2001). While India has a well developed primary health system in theory, in practice access to services is compromised by high rates of health worker absenteeism and often inadequate supplies and poor infrastructure. This leads the majority of the population to turn to largely unregulated private providers.

India today has 4.6 million people with HIV, the second highest national number of HIV-positive people in the world, after South Africa (UNAIDS 2004). The infection is concentrated in groups at high risk of infection, such as injecting drug users and sex workers and their clients. But it could spread to the general population without significantly expanded HIV prevention efforts and a concerted focus on battling stigma. India is also one of 22 TB high-burden countries that together account for 80 percent of global TB infections. Despite a strong effort to provide DOTS, only 31 percent of cases are detected in DOTS programs, so most people with TB do not have access to the most effective treatment (WHO 2004).

Discrimination against girls and women remains pervasive, reflected in the systematic underallocation of food and education to women and girls within households. Literacy rates among girls, though increasing, are much lower than among boys (map 11.2). Women have very limited access to sexual and reproductive health services, limiting their ability to make decisions about their own fertility. Minority groups (such as India's scheduled castes and scheduled tribes) do not have equitable access to public services or infrastructure, restricting their ability to participate in the economy. Although most of the population

Map 11.2

Literacy rates for girls, though increasing, are much lower than for boys

Ratio of female to male literacy rates in India, 1991 and 2001

Source: Vanneman and Barnes 2000; ML Infomap 2003.

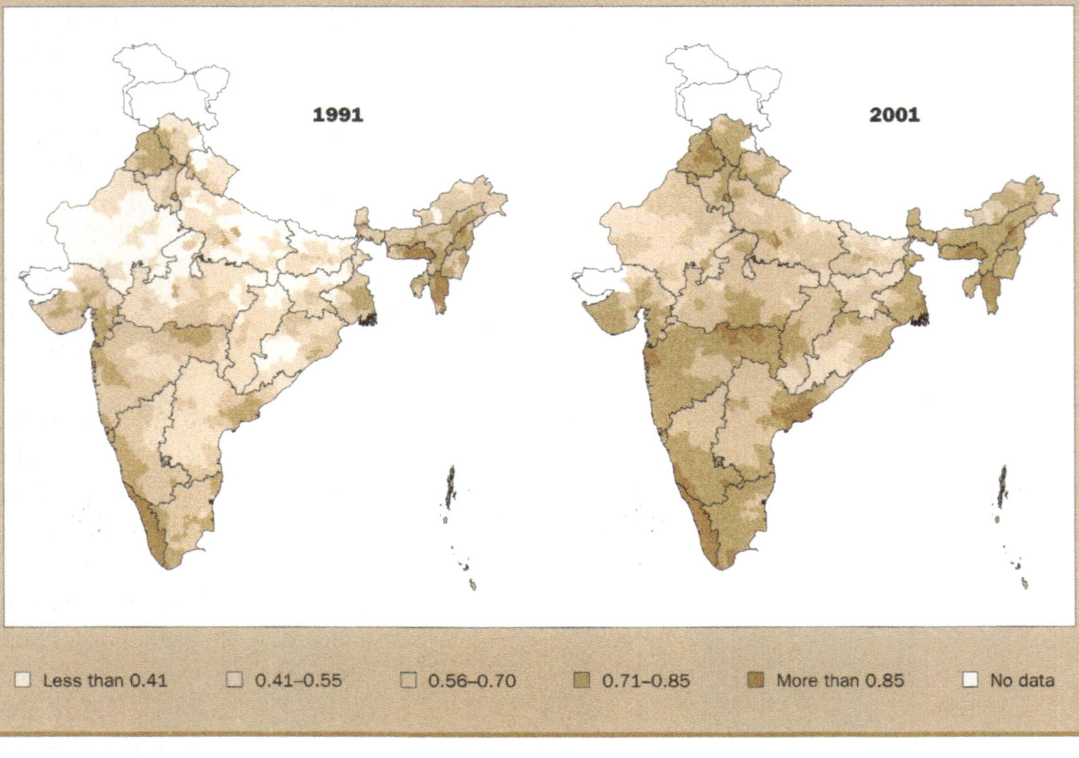

now lives within one kilometer of a primary school, student completion rates remain very low due to systemic weaknesses, including teacher absenteeism, poor facilities, and low quality of instruction.

In a country where more than 60 percent of the labor force works in the agricultural sector, the scale of the rural poverty challenge remains high (World Bank 2004c). Despite the strong agricultural performance in recent years, agriculture remains dependent on the rains. Given the many constraints, India's most pressing challenge is a massive scaling up of public investments in the rural areas of the worst performing states, but concerted action is also required to achieve the other Goals. An MDG-based investment strategy for India needs to include the following elements:

- *Health.* Basic health infrastructure and services, especially sexual and reproductive health services, require scaling up, particularly in rural areas. Investments in health need to rise to strengthen human resource capacity, provide needed supplies, and build the infrastructure required to deliver essential health services. In addition, there must be a sharper focus on working with private providers on priority health programs, such as expanding DOTS, since those providers deliver most of India's health services but they are subject to little regulation or oversight to ensure quality services. India needs to make a large investment in HIV prevention to forestall further transmission to the general population—and to provide treatment to those requiring it.
- *Education.* Given the challenges in scaling up the quantity and quality of education services, a full complement of investments will be necessary, including demand-side interventions, improvements in systems of management and accountability, infrastructure, teacher training, and funding for other recurrent costs. Gender-parity targets will probably require extensive support through special subsidies targeted at girls.
- *Rural investment strategies.* India's economy remains highly agrarian, with comparatively poor infrastructure. Many areas require significant improvements in roads, transport, electricity, other energy, water and sanitation facilities, environmental management, and agricultural research and development. Falling water tables, a growing concern, should be addressed through harvesting rainwater, naturally recharging aquifers, and managing demand.
- *Urban management.* A growing number of India's poor live in urban slums. Using the range of policy reforms and interventions described in chapter 5, cities need to upgrade slums and strengthen urban infrastructure to promote the development of labor-intensive industries and services. Well organized civil society organizations throughout the country's cities can play a critical role in slum upgrading and urban management.
- *Public management.* Improvements in service delivery and accountability are needed to ensure that the investments reach intended recipients.

This requires increasing investments in information systems and modern management systems to track disbursements and to remove bureaucratic bottlenecks. It also requires the promotion of transparency in government procedures, and encouragement for local communities to participate in implementing public investments.

- *Targeting underprivileged populations.* Resources need to reach the least privileged groups, including women and girls in low-income families, the so-called low caste and outcaste communities, and tribal populations. Systematic targeted investments and educational campaigns are needed to provide these communities with basic social services, skills, and income-generating opportunities to enable them to benefit from the country's economic growth.

Other South Asian countries

For the other South Asian countries, the challenges of poverty remain great, even though the region as a whole has seen moderate economic growth, a diminishing disease burden, and rising food yields. Sri Lanka's development outcomes stand out for their success. It has achieved universal primary school completion, relatively low under-five mortality rates (19 deaths per 1,000 live births), low maternal mortality ratios (92 deaths per 100,000 live births in 2000), and relatively high access to water and sanitation (World Bank 2004c; UNDP 2004b; WHO and UNICEF 2004). Bangladesh has seen mixed success, having made enormous strides in education, infant mortality, and gender equality, but still has a high maternal mortality rate and very high levels of poverty and hunger. Rural areas in Bangladesh suffer from poor water quality and arsenic contamination. It is estimated that 46–57 million people in Bangladesh may be exposed to arsenic poisoning (PRB 2002). Since 1990 Pakistan has seen persistently high poverty and slow economic growth accompanied by modest declines in undernourishment and significant reductions in child mortality. Afghanistan meanwhile faces daunting development challenges in nearly every sector, together with postconflict reconstruction. Many other countries in the region, including Nepal and Sri Lanka, are pursuing the Goals in the context of political instability and insecurity.

The links between environment, infrastructure, and poverty are strong. Nearly all South Asian countries depend on agriculture for a large proportion of GDP, and most of them face problems of water scarcity, soil erosion, and salinization. The scale of urban problems is also immense, with more than 262 million people living in slums in South and Central Asia (UN-HABITAT 2003).

In general, MDG-based investment strategies in the region will need to include the following:

- *Health and education.* Basic education, nutrition, and health services (especially reproductive health) need to be extended to the majority of South Asia's population, especially to people in remote rural areas. Women and girls need to be specifically targeted. As in India, demand-side interventions, infrastructure, and support for recurrent costs will be important.

- *Rural investment strategies.* Special investments are needed to address rural poverty and isolation. The full range of rural infrastructure investments is important, including roads, transport services, water, sanitation, and energy. A particular challenge is improving rural water supplies in Bangladesh to reduce the devastating impact of arsenic poisoning.
- *Slum upgrading.* With large shares of the urban population in the region continuing to live in informal settlements or slums, countries need to accelerate citywide slum upgrading programs in all cities.
- *Public management.* Many countries would benefit from greater transparency, better systems of management to track disbursements and the use of development assistance, and more community mobilization to monitor local implementation of social programs.
- *Environmental management.* Better environmental management is required where the extent of poverty is often the most severe and the vulnerability to natural hazards extreme, requiring investments in disaster prevention and management mechanisms (for example, in Nepal, Sri Lanka, and coastal Bangladesh). With climate change, these investments will take on an added importance over coming years, as discussed later in this chapter.
- *Conflict resolution and peace-building.* Social conflict afflicts nearly every country in the region, hampering development outcomes. Investments in conflict resolution and postconflict reconstruction are needed in the countries that suffer from ethnic, sectarian, or other divisive tensions (chapter 12).

Central Asia

Central Asian economies that were part of the Soviet Union have experienced a sharp deterioration in poverty and other social indicators. Under Soviet rule, massive subsidies supported inefficient and resource-intensive industries in poorer republics. Independence brought this support abruptly to an end. Although most economies of the Commonwealth of Independent States (CIS) have recovered from the deep recessions of the early 1990s and are now growing reliably, per capita GDP remains well below pre-independence levels. In Tajikistan, GDP per capita is less than 20 percent of the level before 1991 (UNDP 2003c). While human development indicators remain high relative to income, they are likely to decline without significant and well managed investments in social services and infrastructure. Indeed, significant investment will be needed just to replace the aging and often obsolete capital stock inherited from Soviet central planning.

Many of Central Asia's major challenges also stem from extreme geographical isolation. The region is landlocked, far from seaports and major world markets, and its larger countries are water-scarce. While Kazakhstan hopes its oil wealth will become a centerpiece of economic recovery, most other Central Asian countries lack major natural resources and must depend on agriculture, manufacturing, and services. Perceptions of corruption and weak governance throughout the region remain high, depressing both foreign and domestic investment (Gray, Hellman, and Ryterman 2004).

As a result of economic regress and stagnation, social spending has declined significantly, and health indicators mirror this collapse. In Kyrgyzstan health financing per capita in 2001 was only 32.5 percent of the 1990 level. Not surprisingly, the number of TB patients increased more than threefold in the same period (UNDP 2003b). Kazakhstan and Tajikistan are both significantly off track to meet the child and maternal health Goals. All countries in the region face infectious disease problems, including mounting HIV transmission rates (UNDP 2003c; Government of Kazakhstan and UN Country Team 2002). Environmental degradation is also a major problem, particularly desertification and inadequate water supply.

To halt declines and achieve the Goals, major investment priorities in the Central Asian countries should include:

- *Urban and rural investment strategies.* Basic infrastructure systems, such as those for energy, transport, and water and sanitation, require expansion and support in both rural and urban areas. In Tajikistan, for example, only 23 percent of the population has access to sewage facilities (UNDP 2003c). In the region's mountainous and rugged terrain, these investments will be expensive, technically challenging, and likely to need external support.
- *Cross-border cooperation for infrastructure.* Regional transport and energy infrastructure is needed to help the landlocked countries of Central Asia overcome the barriers imposed by geographical isolation. Investment priorities are clearly defined in the Almaty Programme of Action for landlocked developing states (discussed further below) (UN 2003).
- *Cross-border cooperation for integration and government cooperation.* Given the geography of Central Asia, regional cooperation is essential to coordinate environmental and water resource management, develop transportation and communication networks, and deal with other transboundary issues (chapter 15). While Central Asia already has a plethora of institutions with these aims in mind, their impact has been limited. These institutions must be adequately resourced and encouraged to stimulate economic integration, transparent cross-border regulation, and better regional governance.
- *Public management.* Central planning during the Soviet era left little viable administrative capacity in its wake, and widespread corruption has filled the vacuum. Investments in human resources, incentives, institutional structures, and information systems are needed to rebuild the capacity to govern.

- *Health and education.* Declining health indicators reflect overstretched and poorly resourced health service delivery systems. These systems also suffer from the Soviet legacy of inattention to primary care. In many cases, primary care systems must be built almost entirely from the ground up. On the positive side, there are many trained health workers, though salaries must be dramatically increased to ensure retention and the quality of care. Limited financing in Kyrgyzstan, for example, meant that the government was able to finance only 18 percent of its National Program for HIV/AIDS for 1996–2000 (UNDP 2003b). While education indicators reflect high enrollments in some countries, investments are needed in modernizing and improving the curricula and the quality of education.

Latin America

Among developing regions Latin America has the highest per capita income, but significant disparities and pockets of severe poverty exist both within and between countries. The region is highly urbanized, with a majority of the extreme poor living in urban slums (see table 2.3). Much of the urban population is either unemployed or employed in the informal sector, lacking social safety nets and access to essential services.

In many Latin American countries, severe economic inequalities are linked to deep-rooted social divisions. The poorest communities remain the indigenous and the black populations, who form the majority of the rural and urban poor in Latin America's pockets of extreme poverty, such as northeast Brazil, the Central American highlands, and the Andean region. These regions' extreme levels of poverty are captured by the high infant mortality rates (map 11.3).

The inequalities have been compounded by Latin America's generally weak and unstable economic growth over the past decade. All too often, economic crises and natural disasters have wiped away years of hard-won gains. Structurally, Latin America remains largely a primary commodity exporter, leaving economies subject to the volatility of primary commodity markets and the long-term decline of prices for such exports. There has been a shift toward manufactured exports, which nearly quintupled from $43 billion in 1990 to $200 billion in 2000 (UNCTAD 2003). Still, the region's manufactured exports were less than half its merchandise exports in 2000, much lower than the 80 percent typical of high-income countries (World Bank 2004c). And nearly all that export growth is accounted for by Mexico's NAFTA-driven industrialization and a few key industries in Brazil.

Subregions of Latin America have specific economic development features and concerns. On the positive side, NAFTA has spurred growth in Mexico, particularly in the north, and Chile has enjoyed consistent economic gains. And though Argentina and Brazil suffered from significant financial crises in the 1990s, they are rapidly beginning to recover.

Map 11.3

High infant mortality rates point to pockets of extreme poverty

Deaths per 1,000 live births

Source: CIESIN 2005a.

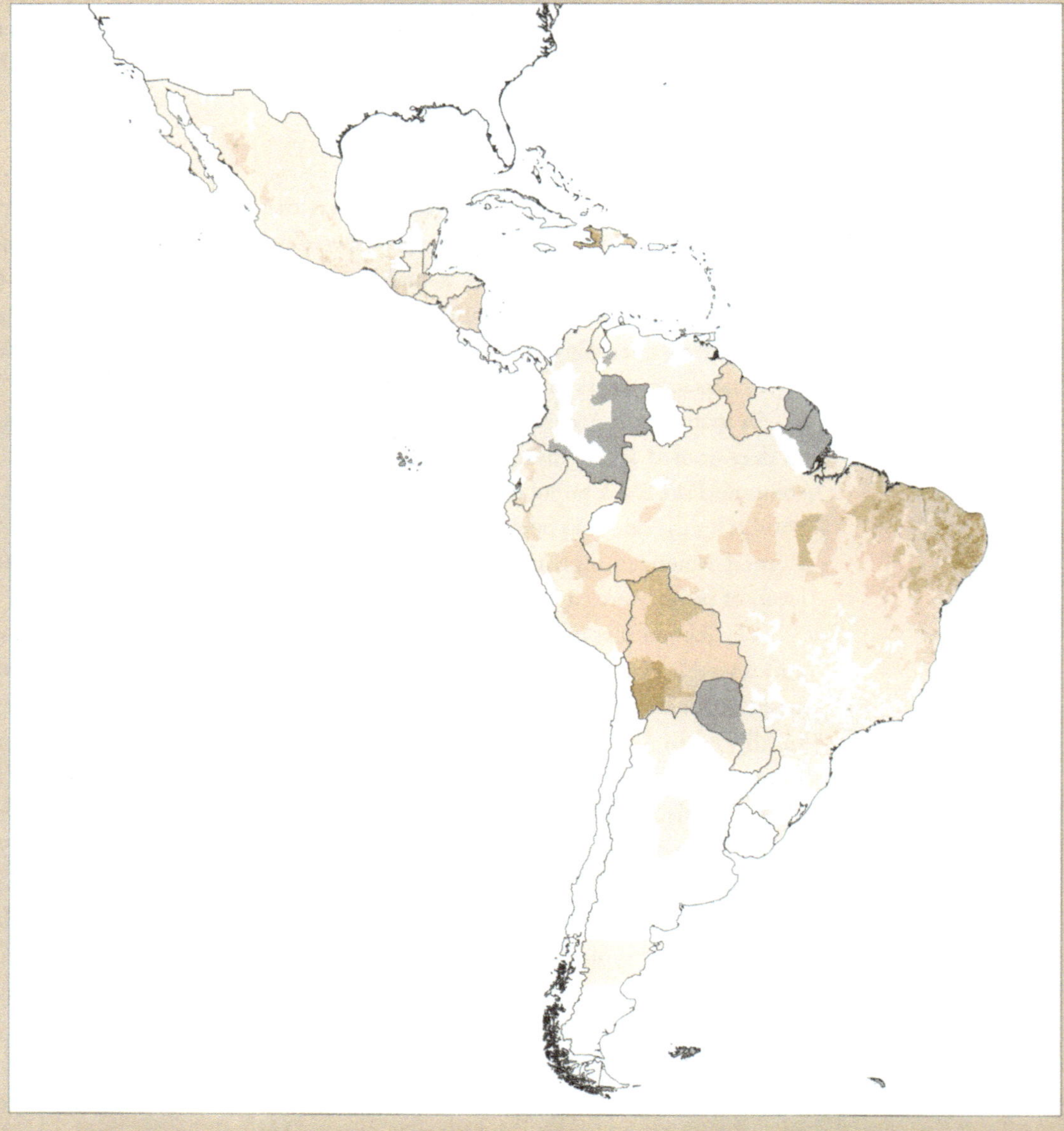

Fewer than 20 | 20–40 | 40–60 | 60–80 | 80–100 | More than 100 | No data

Less fortunate in recent years, the Central American countries are small and disaster-prone. Indeed, among all developing regions, Central America stands out for its susceptibility to natural hazards, hampering progress toward the Goals (table 11.1). Countries in Central America also have large and frequently excluded indigenous populations, and they are typically recovering from decades of political instability. For example, recent estimates suggest that the proportion of people in extreme poverty exceeds 50 percent in Nicaragua and 30 percent in El Salvador, and these trends are stagnant rather than improving (UNDESA 2004).

In the high-altitude Andean countries rural poverty—especially among the indigenous populations—is relatively high, in large part due to geographic isolation, high transport costs, and limited infrastructure. In Peru, for example, the population of the extremely poor has increased from 9 percent in 1994 to 18 percent in 2000 (UNDESA 2004). In Bolivia, nearly 55 percent of the population lacks access to improved sanitation (WHO and UNICEF 2004). UNDP projections suggest that several Central American and Andean countries are unlikely to meet the Goals on current trajectories.

In health, Latin America has made significant progress in reducing child mortality and infectious diseases, but it still faces a major challenge in reducing maternal mortality, despite high rates of attended deliveries. Though good overall, health status varies among social groups and geographic areas, indicating seriously unequal access to clean water, sanitation, and good health services. Approximately 1.6 million people live with HIV in Latin America (UNAIDS 2004). Brazil, with the largest HIV-positive population in the region, has made a major effort at both prevention and treatment and has dramatically reduced death rates from AIDS. In education there is a reverse gender gap, with more women than men enrolled in postprimary education. This gender gap both

Table 11.1

Mortality risks due to natural hazards, by region

	Drought index	Earthquake index	Landslide index	Floods index	Cyclones index	Volcanoes index	Average
Central Asia	3.94	1.76	0.66	2.19	0.00	0.00	1.43
East Asia and the Pacific	3.40	0.90	0.44	5.95	2.65	0.23	2.26
Europe	1.19	0.42	0.21	3.40	0.08	0.05	0.89
Latin America and the Caribbean	3.52	1.17	0.92	6.31	0.54	0.24	2.12
Central America and the Caribbean	2.60	3.17	2.29	7.58	2.45	0.95	3.17
Middle East and North Africa	3.89	0.71	0.19	2.50	0.00	0.00	1.22
North America	2.02	0.39	0.05	4.77	2.35	0.00	1.60
South Asia	6.54	0.34	0.17	6.30	0.51	0.00	2.31
Sub-Saharan Africa	3.96	0.17	0.09	3.22	0.33	0.01	1.30

Note: The index for each hazard is created based on the population distribution, the severity of each hazard, and the hazard mortality rate for 1981–2000. See Dilley and others (2005) for further explanation. The index is based on authors' calculations, by multiplying the country's percentage of population exposed to different severities of each hazard to create a 0–10 index, where a higher number indicates more severe hazard exposure or a larger percentage of the population exposed.

Source: Calculated from Dilley and others 2005.

reflects and reinforces a lack of economic opportunities for the young, as youth unemployment rates rise in the region.

One of the reasons for Latin America's slow economic growth, in contrast with fast-growing Asian economies, is that there has been little concerted effort at enhancing the region's technological and scientific capacities. Research and development expenditure as a percentage of GDP was significantly lower than that in East Asia and the Pacific in 1999 (0.52 percent, compared with 1.09 percent) (World Bank 2004c). In another key indicator East Asia had 17.9 patents per million population in 2000, compared with only 0.6 for Latin America (U.S. PTO 2001).

In the context of these challenges, national MDG-based strategies in Latin America will need to include:

- *Targeting marginalized regions and populations.* Integrating excluded indigenous groups and disadvantaged regions will require targeted investments in health, education, energy, water, and sanitation to enable them to participate more fully in the economy and society (chapter 5).
- *Urban investment strategies.* Latin America has some of the highest numbers for slum dwellers as a share of the population. Slum upgrading and the regularization of informal settlements must be a central priority. In addition to improving the security of tenure, cities must expand social services and urban infrastructure into informal settlements, with a focus on affordable mass transit systems.
- *Science and technology.* Though primary completion rates are generally high, especially in middle-income countries, the region should dedicate significantly more resources toward becoming a technological innovator. Investments in research and higher education are needed to shift the region from primary commodity–based growth to knowledge- and innovation-driven growth that is more broad-based and sustainable. Such a push would entail a major increase in spending on research and technology, up to around 2 percent of GDP (as in leading Asian innovators), partly through increased public support for laboratories and universities and partly through incentives for private research and development.
- *Cross-border cooperation.* Countries need to expand regional infrastructure for transport and energy, enhance economic cooperation, and strengthen regional political collaboration (chapter 15). While many MDG investment needs can be supported domestically, regional cooperation will bring enhanced financial and nonfinancial resources to bear on these important issues. Cross-border cooperation will be especially important for meeting the Goals and achieving sustainable economic growth in landlocked Bolivia and Paraguay.
- *Rural investment strategies.* Better regional transport links are needed to lower costs in rural landlocked and mountainous regions. Investments

in basic infrastructure such as water and sanitation need to be prioritized and are likely to have cross-sectoral benefits, such as better health outcomes.

- *Environmental management.* Latin American countries need to strengthen regulatory mechanisms and policy coordination to improve environmental management—with a particular focus on deforestation and biodiversity loss, which are accelerating across the region. Urban areas require major investments in water treatment, solid waste disposal, and measures to curb air pollution.
- *Investments in health.* Improvements in the quality of care will be needed to reduce maternal mortality in the region. For several countries, the healthcare infrastructure will have to expand to address the needs of geographically isolated populations. Expanding prevention and treatment of HIV/AIDS among socially excluded populations will also be essential.

Middle East and North Africa

The oil-rich states of the Middle East and North Africa have reasonably good infrastructure and social services. Meanwhile, the oil-poor states have much weaker infrastructure and social services. Poverty, high in several countries, increased significantly in Algeria, Djibouti, and Morocco in the second half of the 1990s (UNDP 2003a). Despite the presence of a few strong performers, overall economic performance in the region has been stagnant.

Health indicators reflect the stark differences in income. Maternal mortality ratios are well under 50 per 100,000 live births in the oil-rich states, but they exceed 500 in Djibouti, Mauritania, and Yemen, and are closer to 1,500 in Somalia (UNDP 2003a). Gender inequality and poor access to emergency obstetric care are the main contributors to the high maternal mortality. HIV surveillance is not strong in much of this region, but prevalence rates appear to be approximately 0.2 percent (UNAIDS 2004).

Gender inequality has sharply limited regional progress toward the Goals. The region has a total fertility rate of 3 children per woman, the third highest in the world after Sub-Saharan Africa and South Asia, contributing to an exploding youth population (World Bank 2004c). Despite recent progress in closing the gender gap, female participation rates in the economy remain very low.

Chronic underinvestment in higher education, science, and technology have limited the ability to sustain a robust and internationally competitive private sector and generate broader economic diversification. Highly bureaucratic management systems, clientelist political relationships, and social unrest also plague many countries in the region (Yousef 2004). At least eight countries in the region experience serious water scarcity. In nine countries rural access to safe water is less than 70 percent (UNDP 2003a).

Solutions to all these problems will require strong regional approaches as well as national programs of action:

- *Gender equality.* Priority interventions in gender inequality include guaranteeing women equitable access to education, income-generating opportunities, political rights, and sexual and reproductive health services, including family planning.
- *Health systems.* Investments in emergency obstetric care and skilled birth attendance will be required, along with efforts to improve gender outcomes, to bring down the extremely high maternal mortality ratios in several countries. Urgent investments are needed to strengthen health systems, beginning with village clinics and district referral hospitals.
- *Education.* New investments should focus on vocational and post-primary education to help young people acquire skills that can ease their entry into the labor force. Such programs need to be accompanied by interventions to expand private sector opportunities to absorb the growing numbers of job seekers.
- *Rural investment strategies.* Access to water remains a serious issue in both rural and urban areas. Given this region's generally arid climate, investments will be necessary to ensure sustainable management of water resources and to combat desertification. Particularly needed are research and extension services to address the needs of dryland agriculture.
- *Science and technology.* Investments in higher education and in research and development are critical for building a knowledge base that will allow countries to compete in higher value global markets. A particular focus must be on strengthening research and teaching capacities in engineering and the natural sciences.

Transition and CIS countries of Central and Eastern Europe

The transition and CIS countries of Central and Eastern Europe tend to receive insufficient attention from international efforts to achieve the Millennium Development Goals, largely because of the impression that decades of communism solved their development problems. It is true that effective reforms, favorable geography, and strong support by the European Union have helped the central European countries to become development success stories. They have moved from the collapse of state socialism to membership in the European Union in just over a decade.

But many European countries that were part of the Soviet Union have seen a deterioration in development indicators regress in the 1990s. Poverty and long-term unemployment, income inequality, declining access to public services, and the spread of diseases such as HIV and TB present greater challenges at the beginning of the new millennium than they did at the beginning of transition. For example, Moldova's GDP fell sharply after the collapse of the

Soviet Union, and despite strong growth since 2000, its GDP was less than 40 percent of its 1991 level (World Bank 2004c).

Health indicators also worsened significantly in many of the transition countries, in part due to worsening poverty and in part to crumbling post-Soviet health systems. HIV infection rates have climbed in nearly every transition country, with the highest adult prevalence rates in Ukraine at 1.4 precent, followed by the Russian Federation and Estonia, both at 1.1 percent (UNAIDS 2004). The main transmission route for HIV in the region is injecting drug use, which has increased tremendously since the demise of the Soviet Union. More than 80 percent of people infected with HIV are under 30 (UNAIDS 2004). TB incidence rates have risen as well. In Bulgaria, for example, incidence rates rose from 25 cases per 100,000 in 1990 to 48 in 2000 (United Nations Country Team and government of Bulgaria 2003).

Progress toward the Goals has been hampered in several countries by corruption and weakly governed public institutions, deteriorating service delivery systems, and inequitable internal distributions of resources. School curricula are often of poor quality and ill suited to current social and economic realities. Social and economic inequalities are high. In Romania the rural rate of severe poverty is at least twice the urban rate. In Albania the poverty headcount in rural areas is 50 percent higher than in urban areas (United Nations Country Team and government of Romania 2003; UNDP Albania 2004). Social exclusion of minority groups, such as the Roma, means that they are more likely to be poor and to lack access to education and health services.

Meanwhile, environmental issues remain largely unresolved, with countries experiencing industrial pollution, poor rural sanitation, insufficient regulation, and patchy oversight and enforcement. Armenia, for example, has identified the lack of a coordinated national strategy for sustainable development policy and regulation as an obstacle to achieving the environment targets of the Goals (UN 2001).

Several investment priorities follow from these observations:

- *Public management.* Governance capacity needs to be increased through the development of human resources, information systems, and appropriate institutional structures to implement and oversee public programs and interventions. Examples include improving financial control systems and overhauling the administration of educational systems to increase efficiency and adapt curricula to the demands of market-oriented economies and an informed civil society.
- *Health and education.* Given rapidly deteriorating health indicators, reinforcing health service delivery systems is an urgent priority, responding to existing institutional constraints. Reversing HIV transmission trends, with a focus on effective prevention strategies for injecting drug users, should be a priority given the rising threat of an epidemic.

Education outcomes have declined in the absence of increased investments to improve, or in many cases maintain, quality.
- *Environmental management.* Better water, sanitation, and waste management infrastructure and pollution abatement systems are needed, as is technical and financial support for monitoring, regulation, and enforcement.
- *Targeting excluded groups.* Investments to improve the social and economic status of marginalized groups and the rural poor are necessary to alleviate persistent poverty.

Special challenges

In addition to regions, some categories of countries experience specific challenges and constraints in pursuing the Millennium Development Goals. Here we highlight the special needs of four such groups: Least Developed Countries, landlocked developing countries, small island developing states, and countries vulnerable to natural disasters.

Least Developed Countries

This report outlines the investments for all developing countries to reach the Millennium Development Goals, but the needs of the Least Developed Countries lie at the very heart of the analysis. Least Developed Countries have per capita incomes of less than $750, poor health and education outcomes, and economies vulnerable to shocks (UN OHRLLS 2004). In the Least Developed Countries of Sub-Saharan Africa, nearly 50 percent of the population lives on less than $1 a day, and in Asia 30 percent. They face challenges on nearly every dimension of poverty. For example, the percentage of people in Least Developed Countries with less than the minimum daily caloric intake increased from 37 percent in 1990 to 38 percent in 2000.[1]

Least Developed Countries require special and sustained assistance because they cannot meet their basic needs from domestic resources—regardless of the quality of their policies or governance. The Brussels Programme of Action for the Least Developed Countries suggests that they require greatly increased official development assistance, since private capital flows will not finance needed public investments. The Programme outlines several priority areas for cooperation including human and institutional resource development, removing supply-side constraints and enhancing productive capacity, protecting the environment, and attaining food security and reducing malnutrition. Progress on the Programme is fully consistent with achieving the Millennium Development Goals, so we urge all governments to support it.

Landlocked developing countries

The 30 landlocked developing countries face four kinds of dependencies. They rely on a neighboring country's infrastructure to access world markets, on sound cross-border political relations, on peace and stability in neighboring countries, and on neighbors' administrative practices (Faye and others 2004). Many landlocked developing countries are further disadvantaged in other ways, with rugged terrain or high disease burdens that make investments costly and technically challenging. Many of the solutions will lie in regional integration strategies to simplify bureaucratic procedures, invest in regional infrastructure, and resolve political tensions. The UN General Assembly recently endorsed the Almaty Programme of Action for landlocked developing countries to address their special challenges. MDG-based strategies in landlocked countries should focus on the following recommendations, many of which are included in the initial roadmap for implementing the Almaty Programme and are discussed further in chapter 15:

- *Cross-border infrastructure.* Major investments in road and rail infrastructure are needed to reduce transportation costs to and from national borders. Donor assistance will need to cover both capital and maintenance expenditures. Regional investments in electricity generation, joint transmission networks, and communication technologies are also necessary to develop integrated regional trade links.
- *Cross-border government cooperation.* Cross-border agreements are needed to develop standardized border procedures, provide guarantees for access to transit routes, and set transit fees and other border costs. These arrangements can be coordinated through regional and subregional organizations, such as the Common Market for Eastern and Southern Africa (COMESA).
- *Trade facilitation.* Developed countries should emphasize special and differential treatment for landlocked developing countries in the trade facilitation negotiations at the World Trade Organization (chapter 14).
- *Private sector development.* Landlocked countries can benefit from industries that are less reliant on transport costs and from products that have high value-to-weight ratios. This may require investments that support a shift in private industrial activity away from primary products to service and manufactured export industries.

Small island developing states

Small island developing states face specific challenges of their own (UN 2002c). Because of their size and geographic exposure, the economy and ecology of the islands are closely interlinked. Nature-based tourism, production of primary commodities, export agriculture, mineral extraction, or some combination of these activities form the base of many of their economies. They are threatened by global climate change, which is projected to lead to rising sea levels, increased salinization, coral bleaching, and increased incidence of vector-borne diseases (IPCC 2001a). The islands are also highly susceptible to natural disasters, including hurricanes and cyclones, which will likely become more frequent and severe as a result of climate change (see table 11.1). And they face deep structural epidemic challenges. Their distance from global markets creates high transaction costs, and small local populations hinder specialization and economies of scale, making development even more challenging. Moreover, many Caribbean islands suffer from high external debt burdens.

We urge the implementation of the Programme for Action for Sustainable Development of Small Island Developing States, adopted in Barbados in 1994. The Programme was the first intergovernmental policy initiative to integrate the small islands into the world economy and address their specific development problems, setting out actions and measures in 14 priority areas at the national, regional, and international levels.[2] A decade after the 1994 Barbados conference, the Programme has yet to be implemented because of the lack of external resources. Drawing on the Barbados Programme for Action, an MDG-based investment strategy will need to address the specific challenges of small island developing states, but also include strategies to meet the universal challenges of disease, hunger, and poverty. In particular, it would need to include the following elements:

- *Science and technology.* New technologies, including information and communications technology, can help small island developing states overcome barriers of scale and isolation and integrate with global markets. Focusing on technological development will also contribute to service-based industries (tourism, financial services), enabling a move away from primary commodity dependence and providing more productive employment for young people.
- *Urban and rural investment strategies.* Investments in airports, seaports, roads, clean water, and sanitation are crucial to encourage economic growth, particularly in the tourism industry. In addition investments in sustainable energy infrastructure, including renewable energy systems, are needed to serve energy needs while minimizing the environmental impact on fragile island ecosystems.
- *Climate change interventions.* Small island developing states require investments to respond and adapt to global climate change. Interventions could include sea walls, antistorm systems to predict and prepare for hurricanes, and infrastructure and housing to cope with resettlement. However, success in responding to climate change depends in large part on industrial countries taking the lead in stabilizing greenhouse gas emissions.
- *Environmental management.* Interventions are needed to preserve and manage fragile island ecosystems, assess and monitor fish stocks and coral reefs, and maintain traditional knowledge and bioresources. In many countries, water services, sanitation, and waste disposal systems are in short supply, particularly in rural regions and outer islands.

- *Strengthening management systems for natural disasters.* As discussed at the end of this section several types of investments are required to improve the detection and prevention of natural disasters, as well as postdisaster management.
- *Regional goods.* Small island developing states are particularly reliant on improved regional goods, such as economic cooperation or regional infrastructure (chapter 15).

In addition to these recommendations that apply across small island developing states, the major regional clusters of island countries have their own specific needs. In the Caribbean these include:

- *Health services.* At 2.3 percent and rising, HIV prevalence in the Caribbean is second only to Sub-Saharan Africa (UNAIDS 2004). Maternal mortality remains high in many countries, reaching 680 deaths per 100,000 births in Haiti, and disease and malnutrition remain severe problems in some (UNDP 2004a). Priorities include improved training of health personnel and assistance in meeting high treatment costs for communicable diseases.
- *Education.* Given the Caribbean dependence on external trade, the erosion of trade preferences for bananas and the final implementation of the Agreement on Textiles and Clothing will require adjustment to develop new industries to replace declining ones. Job training and skills development will be particularly important to improve human capital across the Caribbean and to support workers displaced by global market adjustments.

Many of the Pacific small island developing states are widely dispersed archipelagos that are far from world markets. In several of these countries, national averages mask striking developmental disparities between inner and outer islands, and between rural and urban areas. The countries' special MDG investment needs include:

- *Health investments.* Though many countries feature strong health indicators, others face damaging resource shortages. The government of the Solomon Islands, for example, has been unable to pay doctors and other health workers regularly, causing declines in health service delivery (Asian Development Bank 2003).
- *Education and science and technology.* Many Pacific countries have had difficulty generating productive employment opportunities for young people and consequently have seen increases in the number of women turning to prostitution. Investments in vocational training, advanced education, and technology development will help these islands diversify and expand the range of available opportunities.

Countries vulnerable to natural hazards

Many developing countries experience frequent natural hazards, including drought, floods, cyclones, earthquakes, and landslides. The countries most vulnerable lie in the Caribbean, Central America, Oceania, southern and eastern Africa, and Southeast Asia (map 11.4). Many are vulnerable to several types of disaster, often several times a year. As indicated above, many small island developing states have especially high disaster risks.

A large number of the highest-risk countries are in the low-income category, with droughts and floods particularly prominent among their risks.[3] In Sub-Saharan Africa, disaster risks related to drought are particularly high. Meanwhile, flood-related risks are especially high in Asia, Central America, and Andean and Southeastern South America. In the tropics and subtropics, heavy rainfall events that lead to flooding can also be accompanied by outbreaks of infectious disease, including malaria. Earthquakes, prevalent at tectonic plate boundaries around the Pacific Rim and across Central Asia, are particularly destructive to low-income countries, where infrastructure is seldom built to appropriate seismic safety standards or to deal with related risks like tsunamis.

Natural disasters cause enormous damage. They leave large proportions of the population at risk of losing their livelihoods, their homes, and often their lives. But smaller, more frequent disasters also have significant impacts,

Map 11.4

Areas at highest natural disaster–related mortality risk

Mortality decile, all causes

Source: Dilley and others 2005.

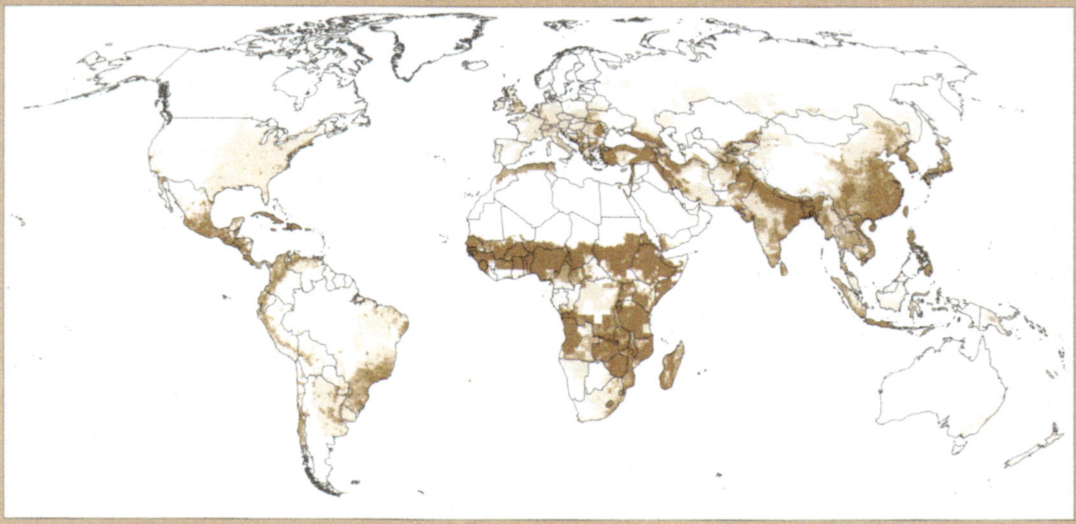

especially in developing countries with poor infrastructure and weak response capacities. Population groups that are typically most at risk include smallholder farmers, rural landless poor, fishers, and the urban poor. Apart from the direct economic loss, vulnerability to natural disasters also contributes to economic volatility, which in turn contributes to higher risks and higher costs of investments. Evidence suggests that in some cases natural disasters and the ensuing environmental stress, such as drought-induced crop failures, can fuel conflict (chapter 3).

It is important to stress that the furies of nature systematically claim the lives of the poor in much greater numbers than they do the rich. When disasters strike regions that include both developed and developing countries, as the devastating 2004 hurricanes in the Caribbean and Southeast United States did, it is in poor countries where most lives are lost. People in low-income countries are four times as likely to die in a natural disaster as people in high-income countries, and in developing countries disasters cause annual economic losses of 2–15 percent of GDP (Kreimer, Arnold, and Carlin 2003).

What the rich world suffers as hardships, the poor world often suffers as mass death. The rich, unlike the poor, can afford to live in fortified structures away from floodplains, riverbanks, and hillsides. They have early warning systems, such as seismic monitors, weather forecasts, and disease surveillance systems. And they have cars and trucks that enable them to leave on short notice when natural disaster threatens. Moreover, rich countries, unlike poor ones, can quickly mobilize food, drinking water, backup power generators, and doctors and emergency medical supplies in the aftermath of a disaster.

In recent decades, population and economic growth have been higher on average in coastal regions that are highly exposed to cyclones and floods. Losses per event will rise in the future, particularly in rapidly growing urban areas, unless systematic efforts are made to reduce vulnerability. Climate change is further expected to affect rainfall and temperature patterns, altering the frequency, severity, and location of climactic hazard events. Climate changes may result in rising sea levels, which would contribute to storm surges and flooding and enhance the hydrologic cycle, altering patterns of extreme events such as drought (IPCC 2001a).

For countries at risk, strategies to reduce losses associated with natural disasters need to be mainstreamed into MDG-based poverty reduction strategies and fall under four broad categories: infrastructure investments, investment in safety nets, early warning systems, and emergency and contingency plans.

First, in places at risk of natural disasters, investments in infrastructure are necessary to minimize the damage from natural disasters. These can include earthquake-proof buildings, emergency shelters, and protective embankments. Such preemptive planned investments can minimize not just the loss in lives and incomes, but also the disruption of strategies for achieving the Millennium Development Goals. They are particularly important in rapidly growing cities since uncontrolled urban growth increases vulnerability to disasters. For this reason urban planning must include systematic risk assessments, design appropriate land use plans, and set construction standards.

Second, ongoing investments in social safety nets are necessary to safeguard lives during times of crisis. Governments need to make these investments during noncrisis periods in order to establish an institutional presence capable of effectively delivering services when shocks or disasters occur. During drought, for instance, employment guarantees, microfinance schemes targeted at the hungry, and the protection of livestock-based livelihoods can all play a role in protecting communities exposed to crisis.

Third, governments should invest in building and strengthening national and local early warning systems to monitor conditions and provide advance warning of potential disasters. Few natural disasters can be prevented, but their impacts can be mitigated by advance planning, advance notice, and a comprehensive response. Surveillance systems can include field monitoring, remote sensing, and meteorological forecasting. Properly designed early warning systems can provide a critical window of opportunity to act before a crisis strikes. Public information campaigns are important for raising awareness of the risks of natural disasters and adequate responses. Early warning capabilities for managing climate hazards have been greatly enhanced in recent years through seasonal-to-interannual climate forecasting.

Fourth, precrisis emergency and contingency plans need to be drawn up so that early warning systems can yield an early and effective response. Plans should include strategies for evacuation, emergency safety zones, insurance schemes, and the prelocation and financing of humanitarian resources for rapid distribution. As part of contingency plans, governments must establish mechanisms for delivering emergency services after a disaster has occurred, especially immediate healthcare to prevent the outbreak of disease amongst displaced populations. Developed countries should establish a far more systematic financial mechanism for disaster response, including contingent credit investments for individual countries. Once triggered, responses need to be more rapid, and disbursement decisions need to be made more flexible than they are at present. For example, the UN's Immediate Response Account has recently been funded at only $35 million, which would cover only two weeks' worth of food in a large operation. A level of $300 million would be more appropriate to facilitate rapid response at the outset of a food crisis (UN Millennium Project 2005d).

Chapter 12

Strategies for countries affected by conflict

The UN Millennium Declaration rightly emphasizes the critical role of peace, security, and disarmament as fundamental for human well being and eradicating poverty in all its forms (UN 2000). Many of the poorest people in the world live in fragile states where ethnic or geopolitical tensions and vulnerability to conflict or regular natural disasters undermine efforts to achieve the Millennium Development Goals. We define "fragile states" broadly as countries facing natural or manmade stress that threatens their ability to function effectively and, in extreme cases, their existence as viable states. Countries at regular risk of natural disasters are discussed in chapter 11. In this chapter we discuss conflict situations.

Conflict can take many forms. It can be latent or explicit; it can range from isolated violence to full-scale civil war; it can cross borders and result in large movement of populations. As chapter 3 discusses, outbreaks of conflict have a strong negative correlation with per capita incomes. On average, poor countries—even those not in conflict—risk conflict in the future. If a low-income country has a 3 percent risk of an outbreak in any given year, the cumulative effect is a more than one in four chance of a major conflict during a 10-year period.[1] It should thus not be altogether surprising that of the 34 poor countries farthest from reaching the Goals, 22 are in or emerging from conflict. Without effective strategies to forestall conflict, a significant number of national MDG-based strategies will likely be thrown off course by violent conflict in the course of implementation. We therefore recommend that any international or national strategy to achieve the Goals include a focus on conflict and conflict prevention.

For many fragile countries, if not most, the Millennium Development Goals can be powerful in promoting long-term stability by offering a coherent long-term development vision that is currently lacking. The Goals represent prospects for decent education, healthcare, access to basic infrastructure, and freedom from hunger and want. Given the positive relationship between human capital growth and institution building (chapter 7), extreme risks of conflict are not a reason to abandon development initiatives. If anything, investing in development is a key step toward averting conflict.[2]

To ensure that development policy accounts for the risks of conflict and responds appropriately, several practical steps need to be undertaken, at both the national and international levels. The Report of the Secretary-General's High-Level Panel on Threats, Challenges, and Change (UN 2004a) outlines the key elements of an international response to the threat of conflict, especially for international and internationalized conflicts. We endorse the Panel's recommendations and focus here on country-level actions that will respectively help prevent internal conflicts, end conflict, and support the transition to peace in countries emerging from conflict.

Investing in conflict prevention

To prevent conflict, MDG-based scale-up programs need to narrow rather than widen existing ethnic, regional, or communal divides. In practice, countries must ensure the equitable provision of services and infrastructure to all groups in society, including people in disadvantaged regions, minorities, and those with special needs, such as refugees. Major projects in countries at risk should conduct regular peace and conflict impact assessments to ensure that they do not exacerbate existing tensions and increase the risk of conflict.

Specific actions can minimize the chances of an outbreak of violent conflict. First, conflict early warning systems are essential in regions vulnerable to increased violence. Elements of individual warning systems will differ by political context, but there are some guiding principles for their development and use. They should emphasize the continual collection of behavioral indicators (not just structural indicators) to reveal whether the implementation of MDG-based strategies is contributing to violent conflict.[3] They should also link with regional and international institutions watching out for early signs of conflict.

To be useful, conflict early warning systems need to be well integrated with response strategies. Within countries, their findings should be integrated with the work of governments, particularly ministries of finance and planning. At the international level, the systems need to be well coordinated through the UN secretariat.

Second, conflict prevention should be supported through the careful design of MDG-based poverty reduction strategies. This entails a particular focus on marginalized groups, migrants, and displaced persons. If growth-enhancing policies are likely to produce or worsen severe regional or ethnic inequalities, compensatory investments should be made in disadvantaged areas, with steps to facilitate migration to areas of faster growth, encouraging the return of remittances. We urge countries to undertake disaggregated reporting on how different regions or groups within a country are progressing toward the Goals in order to shed light on patterns of inequality across groups and help mainstream conflict prevention.

Third, decentralized and participatory decisionmaking structures, with direct involvement of marginalized communities, can help reduce the risks of conflict by providing political space to address grievances and aspirations. Civil society organizations (CSOs), particularly human rights groups and leaders of indigenous people, can be pivotal in preventing conflict (chapter 8). They can present early warnings of crises, serve as conduits to understanding the root causes, and act as intermediaries, facilitating links and dialogues between contending groups and affected communities. In Brazil, for example, CSOs representing Afro-Brazilians have focused government attention on racial inequality and contributed to defusing potential conflict. In India, the efforts of civil society organizations have prevented riots from taking place in many cities (Varshney 2002).

Fourth, conflict prevention can be supported by increasing transparency in the flow of public funds, and accountability for how revenues are spent or distributed by government. This is particularly important in countries where volumes of aid need rapid scale-up. For countries dependent on extractive sectors, participation in the Extractive Industries Transparency Initiative can help. Already such countries as Nigeria and Sierra Leone have agreed to take part in this noteworthy initiative.

Multinational firms have a special responsibility to promote transparency when engaged in low-income countries (chapter 9). Mechanisms such as the UN Global Compact and the UN "Norms on the Responsibilities of Transnational Corporations and Other Business Enterprises with Regard to Human Rights" need to be strengthened to this end. The Publish What You Pay campaign has advocated that all payments to overseas governments should be published as a precondition for such privileges as listing in major stock exchanges. Multilateral initiatives such as the Kimberley Process to regulate conflict commodities also need support and expansion. Any corporations receiving payments to deliver MDG-related goods or services in high-risk countries should receive them only on compliance with such basic norms. The criminal responsibility of corporate employees should be invoked where international corporations are complicit in conflicts and human rights violations (Alston 2004; Clapham 2001).

Fifth, MDG-based poverty reduction strategies should include specific investments aimed directly at enhancing peace and security. Measures are needed to strengthen state capacity, including investments in justice and security. Training the police and military to provide effective, accountable, and rights-based public services is itself a priority for development planning (UNDP 2002). Proper regulation of weapons is also important. Integrating these programs typically requires better coordination between development ministries and defense departments both domestically and in donor countries (OECD/DAC 2004c).

Investing in countries in conflict

For countries in conflict, large-scale government budget support is often inappropriate, and assistance strategies need to be targeted to the local situation. Conflicts can affect societies in different ways, based on their scope (for example, local in northern Uganda, national in the Democratic Republic of the Congo, or regional in Sierra Leone and Guinea), their nature (genocides, secessionist wars, insurgencies aimed at altering government policies, bids to capture the state, or quasi-criminal conflicts aimed at securing control over lucrative commodities), and their duration (short-term, intermittent, or long-term). In each case, the strategies for humanitarian and development work will differ significantly. The UN Millennium Project has not attempted to outline the specific needs and strategies in all these diverse situations. Instead, we highlight some shared implications for national governments and the international community.

The first priority for countries in conflict is to bring the conflict to a close. Efforts to achieve the Goals cannot be separated from conflict management. Humanitarian assistance is essential for survival—to enable people to meet the basic needs of food, clothing, healthcare, reproductive health services and care, and security from sexual violence.

The immediate implication of an outbreak of violent conflict is generally a diversion of government finances to military efforts and, at a minimum, a disruption of basic services and infrastructure. In these circumstances, cutting external assistance to governments can have devastating consequences for human development—so any such decisions should be taken with extreme care. Where possible, donor agencies should aim to provide ongoing MDG-based financial and technical assistance to maintain or restore basic infrastructure and the provision of social services, delivered in a way that reaches refugees and people in conflict zones without worsening the conflict. Since the Goals address many of the same needs as humanitarian relief—security from want, hunger, illiteracy, and disease—wherever possible they should guide humanitarian activities over the medium term, as described in the principles of the Good Humanitarian Donorship Initiative established in 2003.

In some cases of civil war, rebellion, and insurgency, the state is no longer a credible actor, lacking the authority to implement either humanitarian programs or development strategies. In such cases, external support for governments can worsen the conflict. Some long-term development assistance may still be feasible, but will then need to be channeled through multilateral agencies or NGOs, trusted and seen as impartial by all sides at war, but not providing new targets for warring groups. This would include financing for schools and teachers and providing primary healthcare and access to water and basic sanitation facilities.

Another typical first consequence of an outbreak of violence is the movement of people—internal migration or cross-border movement. Of the 175 million international migrants in 2000 nearly 16 million (9 percent) were refugees. In addition, conflict-related internal displacements are estimated to have affected 25 million people in more than 47 countries in 2002 alone (Commission on Human Security 2003). Associated development and humanitarian challenges include organizing the settlement of these people, providing security for migrant groups, ensuring access to basic services of education and health, and creating opportunities for long-term productive activities and income generation. In many cases, national governments simply do not have the capacity to provide for large numbers of refugees and internally displaced people—but this does not mean that the needs of these groups can be ignored. International support is essential to help governments identify ways of addressing the economic and political needs of such groups.

A major additional issue in most conflicts is the need to support women and girls, who are often at the epicenter of conflict's harmful effects. Sexual violence as a strategic weapon of war is widespread, as seen in recent conflict situations in Bosnia, East Timor, Rwanda, Sierra Leone, and Sudan. The consequences include higher HIV/AIDS prevalence, the spread of other sexually transmitted infections, increased trafficking of women and girls, and long-term psychological damage. While male relatives are in combat, women are often the sole providers for their families. They are also active combatants in many armies and rebel groups. Humanitarian and relief work needs to provide basic health services for women and girls, especially reproductive health services and care, and ensure security from sexual violence.

Investing in countries emerging from conflict

Statistically, countries emerging from conflict show a 44 percent tendency to relapse into conflict within the first five years (World Bank 2003b). Such post-conflict countries require policy and investment measures, by both national governments and the international system, to avoid sliding back into violence. Some countries have adequate institutional structures to manage large-scale budget support. Others that do not have the same institutional strengths will require support to build public administration capacity and project support for key investments.

Large investments are needed to rehabilitate wartorn areas, refugee populations, internally displaced people and former combat troops and to rebuild basic infrastructure such as roads, electricity, schools, and hospitals. Health systems especially need rapid investments. Multiple studies now show that the health impacts of conflict continue long after fighting stops. In Sierra Leone, for example, infant mortality rates in the Kenema district were reported as high as 303 deaths per 1,000 live births, even after fighting had abated (International Rescue Committee and Ministry of Health and Sanitation, Sierra Leone 2001).[4]

Countries emerging from conflict also need much larger investments in rebuilding the general administrative capacity of the government. Several countries emerging from conflict, including Mozambique, highlight the critical role of investments in sectors that lay the foundation for rapid recovery and much faster economic growth. What is needed in the early stages of peacebuilding is the development of long-term MDG-based frameworks that focus on income-generating opportunities, healthcare services, primary and higher education, and access to basic infrastructure such as housing and electricity. Humanitarian and development assistance should be seen as complementary and sometimes sequential investments, which together can help populations during and after conflicts.

Since health needs loom especially large in conflict and postconflict situations, the WHO has taken the lead in proposing a sound health-based strategy for fragile countries, especially those emerging from crisis (box 12.1). Postconflict health services should include HIV/AIDS treatment and care, treatment of sexually transmitted infections, and social and psychological support services for victims of trauma (UN 2002d). In the immediate aftermath of conflict, early and sustained support of this nature is a priority for successful peacebuilding.

A postconflict society needs to focus on ways of building peace and security to prevent a relapse into conflict. Here again we support the recommendations

Box 12.1

Responding to health needs in fragile settings

Source: Nabarro, Colombo, and Griekspoor 2004.

Although some reversals and declines are inevitable when dealing with the complex problems of fragile states, a coordinated and concrete approach focused on sustaining health will lay the groundwork for eventual recovery—and ultimately the achievement of the Millennium Development Goals.

A realistic strategy should be built on the following principles:

- *Focus on essentials.* Interventions must focus primarily on providing the essentials for life, including access to water, sanitation, hygiene, food and nutrition, shelter, security, public health and disease control, and sexual and reproductive health information and services.
- *Integrate health into stabilization efforts.* Health concerns need to be priorities in the overall process of stabilizing fragile states to guarantee that health service delivery continues securely and reliably even in the face of political and economic changes.
- *Deliver predictable and coordinated support.* In practice, maintaining health services delivery requires stable and predictable financing. Financing should be channeled through a single comprehensive and results-based planning instrument that uses concrete milestones and an inclusive system for managing and coordinating implementation.
- *Repair now, reform later.* In periods of instability or rapid flux, maintaining and repairing existing health systems is more important than initiating major reforms, which may disrupt service delivery systems precisely when they are most urgently needed.
- *Develop managerial capacity.* Substantial assistance will be needed to develop capacity for effective resource management and stewardship.

of the Secretary-General's High-Level Panel on Threats, Challenges, and Change, which offer a range of international responses for peacebuilding, such as greater coordination through the United Nations, a standing fund for peacebuilding with at least $250 million to finance the recurrent expenditures of nascent governments, and critical agency programs in the areas of rehabilitation and reintegration.[5]

At the country level, an important element of peacebuilding is a successful strategy for demobilizing, disarming, and reintegrating combat troops. The United Nations Department of Peacekeeping Operations is developing such a strategy to register people, monitor and control decommissioned weapons, and disarm soldiers and train them for alternative careers. Such efforts need more systematic inclusion of women ex-combatants (UN 2002d) and better coordination with development planning in postconflict contexts.

Much greater international commitment is needed to collect and destroy weapons in the aftermath of conflict. Too often, collected weapons later come back into circulation. Individual countries need to tighten up the civilian regulations governing military-style arms. Exporting countries should adopt broad and internationally consistent mandatory codes of conduct to better regulate the sale of arms, reduce the likelihood of resale, and allow more transparent arms sales monitoring.

Finally, to prevent countries from relapsing into conflict governments must also of course address the legitimate grievances of their people. Many grievances, though not all, are tied to social or economic development outcomes. But addressing them requires political structures that engage different groups, allowing them to voice their concerns while also meeting their needs. The civil society function of voicing the public interest is often a critical starting point for social transformation and should be supported where necessary. This leadership role of CSOs—representing women, indigenous people, ethnic and racial minorities, and ex-combatants—has been exemplary in resolving conflict and bringing about reconciliation in many parts of the world, for example, in Ecuador, the Mano River countries, and the Philippines. But CSOs cannot succeed in isolation. A sustainable conflict prevention strategy requires the combined commitment of national and local governments, the international community, and civil society to resolve long-standing conflicts and prevent their resurgence.

3

Recommendations for the international system to support country-level processes

Fixing the aid system

Many national strategies will require significant international support. But the international system is ill equipped to provide it because of a shortage of supportive rules, effective institutional arrangements, and above all resolve to translate commitments to action. Here we diagnose why the development system is not yet up to the task of the Millennium Development Goals (MDGs), and how it needs to scale up its financial and technical support. That system has the potential to help countries achieve the Goals, but it needs a significantly more focused approach to do so.

The 10 central problems with the aid system today

1. Lack of MDG-based aid processes

Although the system has identified the Goals as its common aspiration, it lacks a coherent MDG-based approach to reducing poverty. For example, the Bretton Woods institutions can do much more to help countries design and implement MDG-based poverty reduction strategies.[1] IMF program design has paid almost no systematic attention to the Goals when considering a country's budget or macroeconomic framework. In the vast number of country programs supported by the IMF since the adoption of the Goals, there has been almost no discussion about whether the strategies are consistent with achieving them.

In its country-level advisory work the UN Millennium Project has found that multilateral and bilateral institutions have not encouraged the countries to take the Millennium Development Goals seriously as operational objectives. Many documents referred to the Goals either in passing or as a lofty ambition, but no country has been supported to plan seriously around the Goals. Many low-income countries have already designed plans to scale up their sector strategies, but due to budget constraints could not implement them. In other cases, countries are advised not even to consider such scaled-up plans. Some experiences in operationalizing the Millennium Development Goals at the country level are discussed in box 13.1. Fortunately, the Bretton Woods institutions are now showing more interest in basing the country programs that they support on the Millennium Development Goals, and it is important for them to follow through on that expressed interest.

2. Development partners do not approach country-level needs systematically

Since development partnership is not driven by a coherent set of operational targets, there are no clear criteria for evaluating the types or amounts of development assistance required by individual countries. These are often difficult questions because there is no established framework, for instance, for differentiating support to countries with corrupt governments as opposed to those that are weak but willing (chapter 7). Two governments with the same governance "scores" but entirely different underlying governance dynamics are unlikely to receive appropriately different forms of support. There is also a deep paradox in development assistance when "governance" is frequently discussed as the most important condition for official development assistance, but the countries that tend to receive the greatest support are those emerging from conflict, where systems of governance have typically broken down completely.

3. Most development processes are stuck in the short run

Development is a long-term process, but the key processes for international partnership are short term in their orientation. Most important for low-income countries, Poverty Reduction Strategy Papers are typically three-year strategies, tending to take many constraints as given rather than identifying ways to overcome them over time. In many cases the actual planning cycles are even shorter, dictated by the annual meetings of Consultative Groups. Without any predictability of resource flows, developing countries are neither encouraged nor able to make real long-term strategies based on long-term assistance. Too often, they are then chastised for not thinking or behaving with a long-term vision.

4. Technical support is inadequate for MDG scale-up

Most low-income countries require technical support from the international system to put forward scaled-up investment plans to achieve the Goals. Yet the international agencies that are the global repositories of sector-specific knowledge—such as FAO or IFAD for agriculture, UNICEF for child health, UNIDO for industrial development, or WHO for health systems and disease control—are usually asked instead to focus on small pilot projects. In general, the technical UN agencies on the ground are not prepared to help countries scale up national programs.

Box 13.1
Selected pilot country experiences in operationalizing the Millennium Development Goals

As part of its effort to develop practical recommendations that will be most useful to developing countries and to the UN system, in 2004 the UN Millennium Project began supporting UN Country Teams and governments in a selected number of developing countries to identify the best ways to integrate MDG targets and time horizons into ongoing policy processes, in line with the recommendations of chapter 4. The "pilot countries"—Cambodia, Dominican Republic, Ethiopia, Ghana, Kenya, Senegal, Tajikistan, and Yemen—were identified in consultation with the UN Development Group and represent a geographic cross-section of countries currently off track to achieve the MDGs, but where it was generally recognized that the MDGs are indeed achievable if the international community provides adequate support.

Many of these countries are producing MDG-based strategies in 2005 that will provide a basis for fast tracking en route to the MDGs (chapter 16). Here we describe early experiences in Ghana, Kenya, and Tajikistan. Progress in Ethiopia is discussed in box 13.5.

Ghana

The MDG needs assessment had a gradual start in Ghana in 2004 due to concerns that development partners might not support an outcome-based, resource-unconstrained analysis for the Millennium Development Goals in Ghana. Another challenge was creating mechanisms for coordinating among the government's own ministries, departments, and agencies on the analysis. However, responding to government initiative, both the UN Country Team and technical experts from donor agencies played an instrumental role in the needs assessment by providing data in the health, environment, education, and water and sanitation sectors.

Overcoming these early challenges, a team of analysts working closely with the government, the UN Country Team, and other partners has prepared a draft needs assessment, which was reviewed by all partners including civil society in November. The needs assessment will be expanded and strengthened over the coming months in order to inform the next version of the Ghana Poverty Reduction Strategy scheduled to be completed in 2005.

Kenya

Kenya has repeatedly affirmed its desire to integrate the Goals into its national development processes. Recently, the government launched the Economic Recovery Strategy for Wealth and Employment Creation as its main national policy framework. Kenya has decided to undertake an MDG-based needs assessment to form a more detailed picture of what resources would be required to achieve the Goals. This assessment will borrow from and build on all existing and ongoing work, including the costing conducted within the sectorwide approaches framework and the Public Expenditure Review. But it will focus on the specific investments needed to meet the Millennium Development Goals and will establish the cross-sectoral linkages and tradeoffs between competing investment requirements.

Under the leadership of the Ministry of Planning, government-appointed sector working groups, comprising a broad range of stakeholders, have been charged with the primary responsibility for conducting the needs assessment, which is in the early stages. Key development partners have engaged with the working groups. The donor agencies and the UN Country Team have also been integral in providing data and technical support to the various sectors. To ensure the needs assessment is relevant to lives of ordinary Kenyans, civil society involvement will continue to be crucial in both the research and the policy dialogue.

(continued on next page)

Box 13.1
Selected pilot country experiences in operationalizing the Millennium Development Goals
(continued)

Tajikistan

The Tajik government's political commitment to achieving the Goals provides the country with an opportunity for achieving equitable and sustained development. The government will base its next Poverty Reduction Strategy Paper on a comprehensive MDG-based needs assessment that estimates the investments and financing needed to achieve the Goals.

As an early step in the process, the UN Country Team, in conjunction with the UN Millennium Project, developed an initial needs assessment, focusing on three core sectors: education, health, and water and sanitation (see chapter 17). In early 2005 several remaining Goals—poverty reduction, gender equality, food security, and improved international cooperation—will be integrated into the analysis. Meanwhile, the UN Country Team will continue to work with government-appointed working groups (which include the government, donors, UN agencies, and civil society) to refine the first three analyses. Early results will be shared with a broad set of stakeholders, including civil society organizations. The draft needs assessment document is scheduled to be shared for comments by early 2005, and the final report released by the government by mid-2005.

5. Multilateral agencies are not coordinating their support

Multilateral organizations frequently compete for donor government funding to implement small projects, instead of supporting country-scale plans and budgets. The various UN agencies, programs, and funds have begun to coordinate their efforts through the structure of the UN Development Group at headquarters and the UN Country Teams at country level, but this is still often more a forum for dialogue than real coordination. Moreover, the UN agencies are frequently not well linked to the local activities of the Bretton Woods institutions and regional development banks, which tend to have the most access in advising a government since they provide the greatest resources.

6. Development assistance is not set to meet the Goals

As the IMF Managing Director has recently written, it is the developed world that has the greatest responsibility for ensuring the achievement of the Goals (box 13.2). Public investments cannot be scaled up without greatly increased official development assistance. This is particularly important in low-income countries where assistance levels are generally set more by donor preferences than by developing country needs. Although long-term sustainability and capacity building in the poorest countries require support for recurrent costs—such as salaries and maintenance—donors have historically refused to support them, thus preventing any hope of *true* sustainability. Similarly, even though worker shortages are often the major bottleneck for countries trying to deliver basic social services, donors do not systematically invest in long-term preservice training of health, education, and other key workers (as described in chapter 6). Aid flows are also not growing as fast as promised. Since even the much-heralded Monterrey commitments have not fully materialized, developing countries wonder whether developed countries are genuinely committed to the Goals.

Box 13.2
What advanced economies must do for the Goals

Source: de Rato y Figaredo 2004.

In a recent opinion piece published throughout Africa, IMF Managing Director Rodrigo de Rato y Figadero described how developed countries bear the greatest responsibility for supporting developing countries to achieve the Millennium Development Goals.

"If we are to achieve the Millennium Goals, the heaviest responsibility inevitably must fall on the advanced economies, which have a dual task. First, they must meet their commitment to provide higher levels of aid, whenever possible on grant terms. Current aid flows are insufficient, unpredictable, and often uncoordinated among donors. Better coordination and multiyear commitments are keys to making development assistance more effective.

"Second, the developed countries must improve access to their markets for developing country exports and dismantle trade-distorting subsidies. The framework agreements reached at the World Trade Organization last July are welcome, and place the Doha Round back on track. This needs to be followed by determined progress to maintain the momentum and achieve the goals of the Doha development agenda. In doing so, both rich and poor countries carry responsibilities in promoting the fuller integration of developing countries into the global trading system."

7. Debt relief is not aligned with the Goals

The targets for debt relief are based on arbitrary indicators (debt-to-export ratios) rather than MDG-based needs. Many heavily indebted poor countries (HIPCs) retain excessive debt owed to official creditors (such as the Bretton Woods institutions) even after relief. Many middle-income countries are in a similar situation and receive little or no debt relief.

8. Development finance is of very poor quality

The quality of bilateral aid is often very low. It is too often:

- Highly unpredictable.
- Targeted at technical assistance and emergency aid rather than investments, long-term capacity, and institutional support.
- Tied to contractors from donor countries (table 13.1).
- Driven by separate donor objectives rather than coordinated to support a national plan.
- Overly directed to poorly governed countries for geopolitical reasons.
- Not evaluated or documented systematically for results.

Low-quality assistance has fostered the serious misperception that aid does not work and has thereby threatened long-term public support for development assistance. Aid works, as discussed in chapter 3, and promotes economic growth as well as advances in specific sectors, when it is directed to real investments on the ground in countries with reasonable governance. The problem is not aid—it is how and when aid has been delivered, to which countries, and in what amounts. For low-income countries, after one subtracts the money counted as official development assistance that is actually interest payments on debt, technical cooperation payments to consultants from developed countries, food aid (emergency and nonemergency), and debt forgiveness, only 24 percent of

Table 13.1
Tying of donor assistance, 2002
Percent of net disbursements

Country	Untied	Partially tied	Tied
Ireland*	100	0	0
United Kingdom*	100	0	0
Norway	99	0	1
Switzerland	95	0	5
France	92	5	3
Netherlands	89	3	8
Germany	87	0	13
Japan	83	8	9
Finland	82	0	18
Denmark	82	0	18
Sweden	79	9	12
New Zealand	76	8	16
Austria	69	0	31
Canada	61	0	39
Spain*	60	0	40
Australia	57	0	43
Portugal*	33	0	67
Greece*	14	0	86
Belgium	—	—	—
Italy	—	—	—
Luxembourg	—	—	—
United States	—	—	—

— Not available.
* Gross disbursements.

Source: OECD/DAC 2004a.

bilateral aid can actually finance real MDG-based development investments on the ground (table 13.2). The corresponding ratio is only 9 percent for middle-income countries. (We separate ODA for direct MDG support from ODA for MDG capacity building, that is, training, as discussed more in chapter 17.) The proportion for multilateral aid is significantly better—at 54 percent for low-income countries and 52 percent for middle-income countries, although still well short of ideal. Another 21 percent of bilateral flows supports capacity building in low-income countries through technical cooperation, but this is mainly for training qualified professionals rather than training new professionals.

9. Major MDG priorities are systematically overlooked

Development programs routinely overlook needed investments in regional integration, environmental management, technological upgrading, efforts to promote gender equality, and even such core investments as roads, electricity, adequate shelter, disease control, soil nutrients, and sexual and reproductive health.

10. Policy incoherence is pervasive

Many developed countries have identified incoherence as a core problem in their policies. For instance, a government might provide aid to support agriculture

Table 13.2

Estimated official development assistance for direct MDG support and MDG capacity building, 2002

Billions of 2002 US$

	Low-income countries			Middle-income countries		
	From bilateral sources	From multilateral sources	Total	From bilateral sources	From multilateral sources	Total
Estimated ODA for direct MDG support						
Grants	16.7	4.2	20.9	12.2	2.8	15.0
Gross loans	3.3	9.8	13.2	4.5	2.8	7.3
Principal repayments	–1.9	–2.8	–4.7	–4.2	–1.1	–5.3
Net ODA	**18.1**	**11.3**	**29.4**	**12.6**	**4.4**	**17.0**
Interest payments	–0.8	–0.9	–1.7	–1.7	–0.5	–2.2
Technical cooperation	–5.4	–0.8	–6.2	–6.2	–0.5	–6.8
Development food aid	–0.8	–0.2	–1.0	–0.3	0.0	–0.3
Emergency aid	–1.9	–0.9	–2.8	–0.5	–0.3	–0.8
Debt forgiveness grants	–3.2	–0.3	–3.5	–2.0	0.0	–2.0
Support through nongovernmental organizations[a]	–0.9	0.0	–0.9	–0.8	0.0	–0.8
Estimated maximum support through government budgets	**5.2**	**8.1**	**13.4**	**1.1**	**3.1**	**4.2**
25 percent of non-MDG support through government budgets[b]	–1.3	–2.0	–3.3	–0.3	–0.8	–1.0
40 percent of support through NGOs for MDG investments[c]	0.4	0.0	0.4	0.3	0.0	0.3
Total ODA for direct MDG support	**4.3**	**6.1**	**10.4**	**1.2**	**2.3**	**3.5**
Share of net ODA (percent)	**24**	**54**	**35**	**9**	**52**	**20**
Estimated ODA for MDG capacity building						
Technical cooperation for capacity building in support of MDGs[d]	3.2	0.5	3.7	3.7	0.3	4.1
MDG capacity building through NGOs[e]	0.5	0.0	0.5	0.5	0.0	0.5
Total ODA for MDG capacity building	**3.8**	**0.5**	**4.3**	**4.2**	**0.3**	**4.5**
Share of net ODA (percent)	**21**	**4**	**14**	**33**	**7**	**27**

Note: Numbers in table may not sum to totals because of rounding.

a. Based on DAC estimates.

b. Assumes that 75 percent of "maximum flows through government budgets" directly finances MDG investment needs.

c. Assumes that 40 percent of official development assistance provided through NGOs directly supports MDG investments.

d. Assumes that 60 percent of technical cooperation contributes to MDG capacity building (OECD–DAC assumption).

e. Assumes that 60 percent of official development assistance provided through NGOs contributes to MDG capacity building.

Source: OECD/DAC 2004d; authors' calculations.

in a food-exporting country while also applying market access barriers to the same agricultural exports. Similarly, a finance ministry might collect debt payments that negate the benefits of aid being disbursed by the development ministry. Incongruous policies highlight the need for a clear set of measurable objectives to align developed country policies.

Key measures to improve aid delivery

Each of these problems is significant. But each is also solvable through committed and specific actions on the part of development partners (table 13.3). Here are 10 key "to do's" for donors and multilateral agencies.

1. Confirm the Goals as concrete operational targets for countries

The deepest challenge to be overcome in development policy is the lack of a coherent MDG-based approach to reducing poverty. This is particularly crucial

Table 13.3

Recommendations for reforming development partnership

Shortcoming	Recommendation
Purpose and process	
1. Aid processes are not MDG-based	Development partners should affirm the MDGs as the operational objective of the development system, with country-level MDG-based poverty reduction strategies as the anchoring process for development support, based on needs.
2. Support strategies are inadequately differentiated by country need	Development partners should differentiate support by country needs, whether for budget support, emergency assistance, or simply technical support.
3. Development is a long-term process, but short-run processes dominate	Development partners should support countries to put forward 3- to 5-year MDG-based poverty reduction strategies that are anchored in a 10-year needs-based framework through 2015. In conflict countries, a shorter time frame may be more appropriate.
Technical support	
4. Technical support is not adequate for scaling up to the MDGs	Technical support should focus on supporting governments to develop and implement nationally owned MDG-based poverty reduction strategies.
5. Multilateral agencies are inadequately coordinated	The UN Resident Coordinator system should be dramatically strengthened to coordinate agencies' technical contributions to the MDG-based poverty reduction strategies.
Development finance	
6. Development finance is not needs-based or set to meet the MDGs	ODA should be set by the MDG financing gaps outlined in the MDG-based poverty reduction strategies. For many well governed low-income countries, this will imply a substantial increase in ODA and funding of recurrent costs.
7. Debt relief is not aligned with the MDGs	"Debt sustainability," particularly Paris Club debt, should be evaluated as the debt level consistent with countries achieving the MDGs. This will imply a dramatic acceleration of debt relief for many heavily indebted countries.
8. Development finance is of very poor quality	For well governed countries, a much larger share of ODA should take the form of budget support. For all developing countries, donor agencies should also follow through on their commitments outlined in the Rome harmonization agenda.
System issues	
9. Major MDG priorities are systematically overlooked	Within the needs-based approach to development assistance, development partners should increase attention to issues like long-term scientific capacities, environmental conservation, regional integration, sexual and reproductive health, and cross-border infrastructure.
10. Donor countries show a persistent lack of coherence in their policies	Donors should evaluate their development, finance, foreign, and trade policies for coherence with respect to supporting the MDGs. Donors should subject themselves to at least the same standards of transparency as they expect of developing countries, with independent technical reviews.

for low-income countries. It is important to reiterate that many of these countries already have scaled-up sector strategies. But they are typically never implemented or even adopted due to budget constraints. Indeed, without a guarantee that their MDG-based strategies would receive the assistance needed to implement them, government leaders and international officials in each of the countries we are working with expressed private concerns about the risk of even pursuing the Goals. They fear that it would be politically dangerous to build expectations that the international community would provide the resources it has promised to help achieve the Goals. In many cases, even the most committed and sincere representatives of the international agencies—people whose organizations are vocally advocating for the Goals—need to be convinced that it is worthwhile for a country actually to draft an MDG-based strategy.

Even when targets are set today, they typically are very narrow in scope, with incremental progress targeted in one dimension, and with stagnation or retrogression in others that might be equally if not more important. For instance, many low-income countries have been lauded by the international community for balancing budgets and lowering inflation, while health and education outcomes have stagnated or worsened. Thus it is that many African countries too poor to invest in AIDS treatment and prevention have been congratulated for successes in macroeconomic stabilization—while life expectancies have turned steeply downward.

In some instances the international community, not willing to provide additional budget support, still requests reductions in health sector budgets in the midst of health crises, in order to attain macroeconomic stability. Meanwhile, in the majority of countries with Poverty Reduction Strategy Papers (PRSPs), the targets are generally not linked to the actual budget targets in the medium-term expenditure framework (MTEF), so the public strategy has no direct link to the actual public investment program.

Because of the still casual attitude toward operationalizing the Goals, the international financial institutions and bilateral donors have not yet clarified any kind of common methodological approach on how to work back from the Goals to set investment plans. As a result, even when targets are set in alignment with the Goals, there is no common professional reference point for evaluating whether a country's investment plans are being properly set to actually meet the Goals (see box 4.2). Since mid-2003 the UN Millennium Project has collaborated closely with the chief economists of the IMF and World Bank in pursuit of such a common approach, making tremendous progress. But much technical work remains to be done.

The lack of serious MDG-orientation in the development system has suppressed expectations for many years. Development practitioners today see little hope for the developed world to follow through on the Millennium Development Goals promise when it has not followed through on assistance promises before, such as the long-standing target for 0.7 percent of GNP for official development assistance. The same experienced professionals working either for donor agencies or for specialized technical agencies have stressed to us the need not to be "unrealistic" in believing that the resources to achieve the Goals will be forthcoming. For the same reason, they insist that the countries should therefore be "realistic" and not even aim for the Goals. All told, in low-income countries where increased assistance is integral to achieving the Goals, the lack of follow-through on promises made by the developed world has meant that the notion of taking the Goals seriously remains highly unorthodox among development practitioners.

The only solution can be for multilateral and bilateral development agencies and other relevant international institutions to make explicit their support for MDG-based poverty reduction strategies. The international financial institutions have a particularly important agenda-setting role (box 13.3). To achieve the Goals, every institution needs to start with the MDG targets and work back from them, asking what needs to be done by 2015 and what constraints need to be overcome. The goal for every development-focused institution should be to act as if its legitimacy and value depended on its contributing to countries achieving the Goals.

2. Differentiate donor support according to country-level needs

Donor governments need to distinguish among countries so that aid is focused where it will make a difference, and so that donors do not shortchange the countries that need the most help by focusing on those with greater geopolitical attention. To be sure, countries like Afghanistan and Iraq, for example, have urgent needs that require donor support but these should not divert attention

Box 13.3

Governing for the Goals at the international financial institutions

The international financial institutions—including the IMF, the World Bank, and the regional development banks—have played a crucial role in converting the Goals from broad reference points to serious policy objectives. As the world's major development finance institutions, they wield enormous influence in setting the agenda for the Goals and development more broadly.

Given the lack of MDG orientation in a majority of their core country programs, the governing bodies of the international financial institutions have a crucial role in ensuring that the Goals are met. The respective executive boards set policy and approve finances for program assistance, so it is important that they set policy guidelines for a reporting system fully aligned with the Goals. The executive boards should ask, for instance, that every program presented for approval include an evaluation of Millennium Development Goals progress in the relevant country—and for an assessment of how the program under consideration links to achieving the Goals. In instances where a country is not on track to achieve the Goals and the program will not be sufficient to support achieving them, the boards should ask for full information on why the country is off track and what the key constraints are. If there is an MDG financing gap (as defined in chapter 17) that even the new program cannot fill, this shortfall should be made explicit in the program proposal.

and financial resources from other worthy countries. Debt cancellation for Iraq, for example, without similar debt cancellation for Nigeria is unjustified on grounds of equity, merit, and need.

Different types of support will be needed for middle-income countries, well governed poverty trap countries, and poorly governed poverty trap countries (box 13.4). Special attention should also go to conflict countries and developing countries with special needs, such as landlocked states, small island economies, and countries vulnerable to natural disasters.

3. Support 10-year frameworks to anchor 3- to 5-year strategies

To address long-term development needs systematically, countries should be supported to produce an MDG needs assessment through 2015 and a corresponding 10-year policy framework. This framework should then guide the more detailed and shorter term MDG-based poverty reduction strategy (chapter 4). This would enable a major scale-up of capacity and force donors to think through their genuine long-term commitments to development in any country. Support will need to be predictable and long term to meet the needs of scaling up for the Goals. For example, training large numbers of nurses or building medical schools will take several years and concrete financial support. By making explicit their commitment to the longer term, the development partners could promote long horizon planning at the country level.

4. Coordinate technical support around the Goals

Developing multisector 10-year scale-up plans will require a great deal of technical work. In many instances developing countries will require technical support to take on this challenge, and in most cases the international specialized agencies can play an important advisory role. We recommend that the international agencies take advantage of their repositories of highly skilled technical experts across a range of development topics to support countries in the development of their MDG-based public investment programs. Pilot projects testing new ideas and programs could still continue, but they should be seen as test cases of ideas that help refine the UN agencies' core mission of providing substantive technical support toward meeting the Goals. The multilateral and bilateral agencies should organize their technical efforts around supporting countries to develop and implement MDG-based poverty reduction strategies.

We recommend that agency specialists in all sectors—such as health, gender, education, and agriculture—be trained to complement their sector-specific knowledge with basic skills to support country-level budgeting processes.[2] With such combined skills, sector specialists can help governments translate their sector plans into a budget framework for the poverty reduction strategy and assist in assessing the capital and recurrent costs of sector programs to reach the Goals. The Bretton Woods institutions and regional development banks are likewise repositories of highly skilled economists and other specialists who also need to

Box 13.4

Differentiating development support by country needs

Middle-income countries

Most middle-income countries can finance the Goals largely through their own resources, nonconcessional flows (market-based loans from the World Bank and regional development banks), and private capital flows. Donor efforts should be directed at helping these countries to eliminate the remaining "pockets of poverty." Some middle-income countries also need further debt cancellation, especially on debts owed to creditor governments (Paris Club debt). The successful conclusion of the Doha Development Agenda of multilateral trade negotiations, with increased access to rich world markets, will bring benefits to middle-income countries. Many middle-income countries, such as Brazil, China, and Malaysia, already are donor countries. We recommend that they and other successful poverty-reducing countries, such as India, step up their donor efforts, including financial contributions and technical training for low-income country partners.

Well governed poverty trap countries

For well governed countries caught in a poverty trap, even a significant increase in domestic resource mobilization will not be enough to achieve the Millennium Development Goals. Substantial co-financing through official development assistance is required, especially for Least Developed Countries, to scale up the needed investments in infrastructure, human capital, and public administration. The key for well governed poverty trap countries is to base aid on a true MDG-needs assessment, and then to ensure that aid is not the binding constraint to scaling up. These countries should be fast-tracked in 2005.

Poorly governed poverty trap countries: lack of volition

For countries like Belarus, Myanmar, the Democratic People's Republic of Korea, and Zimbabwe, where the problem is the will of the political leadership, there is little case for large-scale aid. Aid should be directed to humanitarian efforts or through NGOs that can ensure delivery of services on the ground. Any aid directed through the government should be conditional on significant improvements in human rights and economic policies.

Poorly governed poverty trap countries: weak public administration

When the volition exists in government leadership but public administration is poor because of a lack of sound public management, one key step is to invest in public-sector capacity. This will also raise the "absorptive capacity" for aid in later years. Donors should view the poor public administration as an investment opportunity, not a barrier to achieving the Goals. Early efforts should be directed at building the government's analytical and administrative capacity at national, regional, and local levels—and building the technical expertise at the grassroots level in health, education, agriculture, and infrastructure. We expect that these countries will significantly outperform current expectations. In many countries international expectations are low but the country's potential is very high if timely donor support and debt cancellation are brought to bear, and phased in over time.

Conflict countries

Countries in conflict, just out of conflict, or falling into conflict present urgent special cases for the international community. Rapid responses are essential. A delay in well targeted aid can mean the difference between a consolidated peace process and a resurgence of conflict. Aid should be targeted at ending the violence and restoring basic services, directed in a manner to ease tensions among competing groups. Carrots (offers of an expanding aid effort) generally are much more powerful than sticks (international sanctions) in crisis countries, for example, Haiti and Sudan. Yet sticks are more typically applied, with few lasting results.

Box 13.4
Differentiating development support by country needs
(continued)

Geopolitical priorities

Countries with geopolitical priority (such as Afghanistan and Iraq) have urgent needs, to be sure, but may take up a disproportionate share of donor funding and public attention. If the major donors are to devote substantial efforts to these countries, they must ensure that the efforts do not divert attention and financial resources from other worthy countries. Debt cancellation for Iraq, for example, without similar debt cancellation for Nigeria would be unjustified on grounds of equity, merit, and relative need.

Countries with special needs

Developing states with special needs include:

- Small island states (isolation, small markets, natural hazards).
- Landlocked states (isolation and high transport costs).
- Mountain states (isolation and high transport costs).
- States vulnerable to natural disasters.

The geographically isolated states require special investments in transport and communications—and geopolitical help to support regional cooperation and regional integration. Hazards are rising in frequency, intensity, and impact, and traditional ad hoc responses are too slow and underfinanced. Donors should establish special emergency funds for natural hazards (droughts, floods, pests, disease) and steady funds for long-term improvements to cope with disasters.

provide more technical support. The World Bank, for instance, will need to dedicate its considerable expertise to the MDG-based public investments, with special emphasis on the problems of scaling them up.

This bolder substantive focus will require greater country-level coordination of the UN system. The UN Development Group should guide resident UN Country Teams in their MDG support, and the UN Development Assistance Framework (UNDAF) should identify the specific ways in which the Country Team will support the government to achieve the Goals. The UNDAF should be produced as a follow-up to the poverty reduction strategy to identify ways for members of the UN Country Team to support the national strategy. A sound MDG-based poverty reduction strategy crafted with full UN support would obviate the need for the Country Team's Common Country Assessment, simplifying the workload of the UN Country Team, aligning the UN system's local priorities with those of the government, and directing agencies' thinking to the proper (MDG-based) scale of activity. We also recommend the establishment of multi-agency, cross-sectoral regional technical centers to support governments and UN Country Teams in developing, financing, and implementing MDG-based poverty reduction strategies.

5. Strengthen the UN Development Group and the UN Resident Coordinator

As the senior UN representative on the ground, the UN Resident Coordinator's office needs dramatic strengthening, both to coordinate UN organizations through the UN Country Team and to manage a core technical staff to support the host government in developing and implementing the MDG-based poverty reduction strategy. The UN requires a technical support unit in each country, with 8–10 technical staff members, to focus on all of the priority intervention areas within the poverty reduction strategy (chapter 5). UN agencies would be encouraged to second staff members to this office for 12–18 months around the poverty reduction strategy preparation process. The United Nations Development Group at the headquarters level should support the strengthening of the Resident Coordinator position.

The local representatives of the international financial institutions should work closely with the UN Country Team in support of the host country poverty reduction programs, following the impressive model of such collaboration in Ethiopia as has arisen during the UN Millennium Project's pilot advisory work there (box 13.5). Since implementing the MDG-based poverty reduction strategies will require significant additional grant assistance in many countries, the IMF will need to provide technical support to develop a macroeconomic framework to manage the resources. The needs-based financial assessments for infrastructure and social investments should feed into the medium-term expenditure framework that the IMF helps the government establish. Moving ahead, we recommend that these questions be fed into the IMF's core tasks by asking its staff to focus on a new question: how to identify the MDGs expenditure framework that can be implemented given sufficient resources?

6. Set ODA levels according to proper needs assessment

ODA levels should be guided by the MDG needs assessment, rather than being picked for political reasons or on the basis of incremental budgeting, as is now the case. By partnering with local research institutes in a series of countries, the UN Millennium Project has undertaken the first ever bottom-up needs assessments of the country-level investments required to achieve the Goals (see chapter 17) (UN Millennium Project 2004b). Although these first estimates need to be refined through the real country-level processes we are recommending in this report, the results show that the total cost of investments in low-income countries is on the order of \$70–\$80 per capita per year in 2006, increasing to \$120–\$160 per capita per year in 2015 (more detailed discussion is included in chapter 17). Middle-income countries will generally be able to afford these investments on their own. But the low-income countries, even after they initiate a major increase in their resource mobilization, will require \$40–\$50 per capita in external finance in 2006, rising to \$70–\$100 in 2015. As Landau (2004) has argued, to ensure the sustainability of programs, development assistance should also cover recurrent costs (such as public sector salaries, operations and maintenance) in addition to capital costs.

Since only a small share of current official development assistance actually supports MDG-related investments (see table 13.2), we recommend, as a practical measure, that the OECD's Development Assistance Committee create

Box 13.5
MDG-based planning in Ethiopia

Source: UNDP Ethiopia 2004.

Ethiopia has made a credible and ambitious commitment to achieving the Millennium Development Goals by placing them at the center of its national development strategy. In 2003 Prime Minister Meles Zenawi reaffirmed Ethiopia's commitment to meeting the Goals and established an institutional framework for integrating MDG targets into the next revision of the Sustainable Development Poverty Reduction Program, due in 2005.

Under this framework the Ministry of Finance and Economic Development, designated as the lead ministry to ensure this integration, has chaired a task force to supervise MDG-based needs assessments and integrate the results into national development planning. The task force includes the heads of the technical teams in each of the relevant line ministries and representatives from the UN Country Team and the World Bank. Line ministries conducted needs assessments for relevant sectors, with technical assistance and support from the World Bank, the UN Country Team, and the UN Millennium Project. The process has been characterized by local ownership by the government and close cooperation between the UN system and the World Bank. It could be improved even more by giving a greater role for other important stakeholders, such as bilateral donors and civil society organizations, which have expertise on needs and strategies in different parts of the country.

Close working relationships between the Ethiopian government and its international partners have been a key success factor in the MDG-based planning process. The government has engaged in ongoing technical discussions with the UN Country Team, the UN Millennium Project, and the World Bank to refine its needs assessment models and overall policy framework. The UN Country Team decided to use the government's needs assessment as its next Common Country Assessment and as the basis for the UN Development Assistance Framework. Meanwhile, the World Bank demonstrated strong interest and involvement in ensuring that the Sustainable Development Poverty Reduction Program is aligned with the Goals. Close coordination between the government, the UN, the World Bank, and other stakeholders has created a common vision around the Goals.

To build on this foundation for cooperation, the Ethiopian government seeks to integrate the MDG-based needs assessments into an ambitious 10-year strategic vision and the 5-year program to achieve the Goals. Success will depend on achieving a broad-based consensus among stakeholders on the results of the needs assessment. But already the MDG-based planning process has challenged policymakers in Ethiopia to move away from the traditional approach of planning based on hard budget constraints and to ask a fundamentally different question: "What does Ethiopia need to achieve the Millennium Development Goals?" With broad support from the international community, this could be a first major step toward Ethiopia actually achieving the Goals.

a new indicator for tracking these real investments. This would be a subcategory of aid flows, reported annually by developed and developing countries, to evaluate the level of assistance that should be counted as directed toward the specific practical investments needed to achieve the Goals in each country.

7. Deepen and extend debt relief and provide grants rather than loans

"Debt sustainability" should be redefined as "the level of debt consistent with achieving the Millennium Development Goals," arriving in 2015 without a new debt overhang. For many heavily indebted poor countries this will require

100 percent debt cancellation. For many heavily indebted middle-income countries this will require more debt relief than has been on offer. For some poor countries left off the HIPC list, such as Nigeria, meeting the Goals will require significant debt cancellation. A corollary for low-income countries is that current and future ODA should be grants rather than loans.

8. Simplify and harmonize bilateral aid practices to support country programs

To empower national ownership of MDG-based strategies and to limit the transaction costs of providing financial support, bilateral donors should use simplifying coordination mechanisms—including sectorwide approaches, direct budget support, and multilateral financing, such as that through the International Development Association (IDA). They should also follow through urgently on the actions they set for themselves in the Rome harmonization agenda. Multilateral official development assistance is typically of much higher quality than bilateral assistance (see table 13.2), but the bilateral agencies have the opportunity to increase the quality of their aid dramatically this year, notably at the March 2005 High-Level Development Assistance Committee meeting in Paris.

Although the coordination of aid efforts represents a complicated agenda that the UN Millennium Project has not focused on extensively, we recommend that any program for harmonization and simplification consider the comparative advantages of different donor sources.

Multilateral financial institutions. The multilateral financial institutions, such as the IDA of the World Bank and the European Development Fund, are the largest providers of predictable, large-scale, multiyear funding needed to support the Goals in low-income countries. The IDA in particular needs to play a core role in the scale-up of public investments in the MDG-based poverty reduction strategies. And for the poorest countries, the IDA needs to shift to an all-grants facility rather than its current structure of 20 percent grants and the rest concessional lending.

Regional development banks. The regional development banks, with their particular expertise in regional topics, should focus on issues unique to their region, such as addressing ethnic inequalities in Latin America or managing the environment in Asia. They should also take the lead in supporting cross-border initiatives, such as regional transport corridors and energy programs, and in implementing regional policy initiatives, such as the harmonization of customs procedures.

Global multilateral funds. Specialized multilateral funds—such as the Vaccine Fund operated in conjunction with the Global Alliance on Vaccines and

Immunizations, the Global Environmental Facility, and the Global Fund to Fight AIDS, Tuberculosis, and Malaria—are effective for pooling donor resources when strong targeted programs are needed in the recipient countries. These multilateral funds can have a high public profile and can generate policy attention in the recipient countries. Care should be taken, however, to ensure that the vertical programs supported by multilateral funds are themselves properly integrated into the broader MDG-based poverty reduction strategy. For example, the control of AIDS, TB, and malaria should be part of the overall development of functioning health systems, rather than a stand-alone set of programs.

UN specialized agencies, funds, and programs. Several UN agencies, funds, and programs have mandates to carry out large-scale operational and advisory work within recipient countries. Such activities are not only meritorious, but vital to humanitarian relief and long-term development. UN bodies need increased budgetary support, and a seat at the table of country-level donor groups, as well as on the UN Country Team. The key requirement, however, is to harmonize the work of the UN organizations with the MDG-based poverty reduction strategies, so that the UN technical strengths are geared to support scaling up at the country level.

Bilateral development agencies. Bilateral agencies provide roughly two-thirds of all ODA. Most of this is grant based, although significantly less currently goes to direct MDG-related investments than does multilateral ODA (see table 13.2). Further, individual bilateral agencies work at a smaller scale and tend to impose higher transaction costs than multilateral funders (because the recipient country must deal with up to two dozen bilateral donor agencies). Nonetheless, bilateral agencies are frequently important for broadening industrialized countries' public support for development. Operationally, they can make major contributions to the MDGs by providing budget support to developing countries. They can also importantly provide:

- Technical assistance and technology transfer, drawing on the scientific and technical expertise within their own countries.
- Support for NGOs from the agency's country and in the host country.
- Support for student exchanges, scholarships, and fellowships and other people-to-people exchanges in sports, culture, and the like to increase international public understanding.
- Support for innovative projects, to test new concepts of delivering aid or to introduce new technologies on a trial basis.

In recent years, bilateral agencies have increasingly harmonized their work through sectorwide approaches, in which donors agree to give joint support to the scaling up of a particular sector. These represent a major improvement on past approaches that would see a dozen or more disparate bilateral projects operating in parallel.

Middle-income countries as donors and advisors. Many middle-income countries have much to offer low-income countries, including institutional knowledge of how to manage scale-ups, well established centers for learning and intellectual exchange, networks of experts that can be leveraged to support the development of MDG-based poverty reduction strategies, and in some instances financial resources that can be allocated to assistance for low-income countries. The successful industrializers of Asia, for instance, have significant expertise based on their own successes and failures. And their civil servants and project managers could be usefully deployed to advise public sector managers in low-income countries on how to implement projects and programs at scale. As an example of political leadership for the Goals, President Lula of Brazil has committed his government to supporting a global antihunger program, using his country's experience to focus world leaders on fighting hunger worldwide.

Private foundations. Private foundations can support global science, innovative projects, and civil society organizations that are trying new ways of service delivery. The Bill and Melinda Gates Foundation, for example, has been a world leader in advancing scientific research for the poorest countries, as has the Rockefeller Foundation. Other foundations, such as Ford, Hewlett, MacArthur, and Packard, have all identified key issues left off the mainstream intergovernmental agenda and supported groups advancing them (chapter 15).

9. Focus on overlooked priorities and neglected public goods

Development partners should support developing countries in promoting neglected public goods, including long-term scientific capacities, environmental conservation, regional integration, and cross-border infrastructure, much of which is discussed in chapter 15. Other key priorities such as maternal health, gender equality, and preservice training are discussed in chapters 5 and 6.

10. Measure policy coherence against the Millennium Development Goals

Donors should evaluate their development, foreign, and financial policies with specific reference to the Goals. Donors should subject themselves to at least the same standards of transparency and coherence as they expect of developing country governments. Some countries have made progress by initiating their self-evaluating Goal 8 reports, but independent technical groups should publish evaluations of donor policy impacts and donor coherence, with data made publicly available to permit reanalysis. The independent evaluations would also apply sound methodologies to distinguish between donor impacts and other reasons for development outcomes.

Chapter 14

A global breakthrough in trade

International trade can be a powerful driver of economic growth and poverty reduction. It is not, however, a magic bullet for achieving development. The slogan "trade, not aid" is misguided, particularly in the poorest countries. Trade reforms are complementary to other development policies, especially scaled-up investments in infrastructure and human capital, macroeconomic stability, and institutional development.

As outlined in the Monterrey Consensus, an MDG-based international trade policy should focus on two overarching issues:

- Improving market access and terms of trade for the poorest countries.
- Improving supply-side competitiveness for low-income countries' exports through increased investments in infrastructure (roads, electricity, ports) and trade facilitation.

The second of these is often overlooked when discussing trade reforms in the context of the poorest countries. For many middle-income countries, basic infrastructure and productive capacities are in place so that market access issues pose the greatest challenge to increasing trade. Yet for many low-income countries, increased market access will provide only limited direct benefits, since there is often little agricultural surplus available to trade, a weak to nonexistent manufacturing base for export, and insufficient infrastructure to achieve competitiveness in nontraditional exports.

The poorest countries should pursue open trade and negotiate vigorously for greater market access in the high-income markets. But they should also emphasize, and receive, additional aid to overcome their supply-side production barriers. Overcoming the supply-side limitations will require significant investments in agricultural productivity (rural electricity, irrigation, soil nutrients, transport and storage facilities; see chapter 5), other key infrastructure linked to trade (roads, ports, airports, telecommunications, electricity), and human capital (health, education, training).

The context of trade negotiations

Achieving more open and fair markets for the promotion of development is the mission of the multilateral trading system, which has evolved progressively since the end of World War II and delivered impressive results for many countries. Throughout most of its existence, however, the trading system has mainly served the interests of developed countries. Developing countries, sometimes by their own decision and other times by explicit exclusion dictated by richer countries, have not been influential in the system's design. Moreover, most of today's multilateral rules have emulated to a great extent the policies, the practices, and most important, the laws and regulations of only a few developed countries.

The system is thus unbalanced against the interests of developing countries. Balancing it will give developing countries greater economic growth potential, a major stake in developing multilateral trade rules and disciplines and in pursuing trade liberalization, and more effective capacity to expand trade and defeat poverty. That goal was the motivation underpinning the Doha Development Agenda Round of trade negotiations launched in November 2001, at least according to the rhetoric.

But this sense of purpose was short lived. With key deadlines missed and progress practically nil on every issue contained in the Doha Development Agenda, the WTO Ministerial of September 2003 in Cancun collapsed amid acrimony. There is no single explanation, but the developed countries' failure to lead by example was notable. WTO members have since made a courageous effort to revive the Round, but a lot more effort will be required. The 2004 Doha Work Programme framework, while necessary to prevent the collapse of the Round, is far from sufficient to sustain it.

The real work remains to be done, and a sense of urgency is required if the Round is to be completed by 2006 (UN Millennium Project 2005l). If this narrow window of opportunity is missed, it is hard to see how the Round can be completed in time to contribute to the achievement of the Millennium Development Goals by 2015. All members of the WTO must identify the core priorities of a real development round and make concrete political and financial commitments to achieving them.

A real development round is achievable but will require high-level political leadership—from both developed and developing countries—as part of a coherent policy approach to meeting the Goals. The year 2005 offers a rare opportunity to harness the momentum of the Goals to seek a major political consensus among heads of state, particularly in the lead-up to the September 2005 UN summit, to shape the multilateral trading system for the future. This grand vision would keep focused the eyes of negotiators at the Sixth WTO Ministerial Conference in Hong Kong (China) in December 2005.

Based on the work of the UN Millennium Project's Task Force on Trade, we recommend that leaders agree on two key guidelines for the future path of the trading system (UN Millennium Project 2005l). First, in a conveniently distant long term, such as by 2025, the multilateral trading system should deliver the total removal of barriers to all merchandise trade, a substantial and extensive liberalization of trade in services, and the universal enforcement of the principles of reciprocity and nondiscrimination in a way that supports attainment of the Millennium Development Goals. This target is ambitious but not impossible, with political leadership and appropriate support for adjustment. There is also a base to build on: APEC economies have already committed to free trade by 2010 for developed members and 2020 for developing members. Second, the most useful WTO would be focused on trade and be relieved of other global economic governance tasks, which can be better accomplished by other international instruments or entities.

Key areas under negotiation

In the Doha Round, the most important negotiation topics are likely to remain agriculture, nonagricultural market access, and services. Other important issues include contingent protection, standards, preferences, trade facilitation, intellectual property rights, and special and differential treatment.

Agriculture—the biggest and costliest aberration

The biggest and costliest aberration of the trading system is found in agriculture. Farm producers in rich countries receive support in excess of $250 billion, pushing their farmgate prices almost one-third higher than world prices (UN Millennium Project 2005l). Consumers in those countries pay for that protection through higher taxes and higher food prices. This is their choice. But it must be stressed that by so doing they also impose a heavy burden on other agricultural producers, particularly in developing countries. Agricultural protection in both developed and developing countries is most assuredly a contributor to poverty in a large number of poor countries.

That rich countries should lead farm liberalization is beyond question. They should deliver substantial liberalization under all three pillars of the agricultural negotiations. They should shift their farm policies to income support—helping the poor and small farmers in rich countries to adjust to more open farm markets. Export subsidies should be totally and definitively eliminated, as agreed in the Doha Development Agenda framework of August 2004. These should be removed by 2010. This will send a powerful signal to developing countries, which will follow suit with their own deeper market opening without the danger of export subsidies greatly distorting trade and competition. All countries should decouple all support payments to farmers by 2010 and cap all domestic support measures at 10 percent of the value of agricultural production (on a byproduct basis) by 2010 and at 5 percent by 2015.

Negotiations on farm trade liberalization should also broaden their focus beyond elimination of export subsidies to stress reductions in tariffs—themselves a powerful discipline on export subsidies—and reduction in domestic support. By 2015 no bound farm tariff should exceed 5 percent for OECD countries. Market access negotiations must address both the unacceptably high tariff peaks that remain in agriculture and tariff escalation, which continues to frustrate developing country efforts to move up the value chain. All nontariff barriers, including tariff rate quotas, should be removed by 2010.

Economic growth of the poorest countries depends crucially on a more dynamic agricultural sector. The fragility of these countries, however, suggests that, as a result of the Doha Round, they should reduce only their bound tariffs—since most of their applied tariffs are moderate—and their applied tariff peaks, which cost their poor consumers dearly without bringing public revenue. Additional complications for the poor countries that may be hurt by this liberalization, particularly the low-income food-importing countries that will pay higher import prices, should be dealt with by a substantial increase in international aid. The increased aid would serve partly to cover increased food import costs and partly to stimulate a new Twenty-first Century Green Revolution in food-insecure regions, especially in Sub-Saharan Africa. By 2015 no bound farm tariff should exceed 15 percent for the poorest countries and 10 percent for other developing countries.

Meanwhile the Green Box of minimally trade distorting subsidies should be maintained for the poorest countries—with clarifications or such marginal additions as support for diversification, transportation subsidies for farm products, consumption subsidies for domestic food aid, and public assistance for establishing farm cooperatives or institutions promoting marketing and quality control.

Nonagricultural market access—all countries should liberalize

Trade barriers in nonagricultural products, though not as severe as in farm products, continue to be significant and particularly detrimental to developing countries. For example, developing countries' exports to developed countries face tariffs that are, on average, four times higher than those faced by the exports of other developed countries. Developing countries' exports suffer from mega tariffs, tariff peaks, tariff escalation, and quotas imposed by rich countries on goods of great export potential. Developed countries should bind all tariffs on nonagricultural merchandise at zero by 2015, the target date for achieving the Millennium Development Goals. A midterm goal could be for no tariff higher than 5 percent by 2010.

Over the last few decades, developing countries have undertaken an unprecedented level of trade liberalization, both autonomously and in the context of multilateral and regional negotiations. They still suffer, however, from their own protection, which reduces not only their competitiveness in world markets but also the enormous opportunities of increased trade among themselves. Developed countries bear a special responsibility to liberalize in the Doha Round, but developing countries should also do so because they are important markets for each other, including the poorest countries. While still less than full reciprocity, the poorest countries should nonetheless bind their tariffs at uniform and moderate rates in their own development interests. Ideally, developing countries should all be at zero tariffs by 2025. As soon as possible, these countries should bind all their tariffs in coherence with their applied rates. The poorest countries should also aim to bind all tariffs at a uniform and maximum rate. Adjustment costs should be economically and socially sustainable in developing countries—for example, by phasing in tariff reductions and providing international technical and financial assistance.

The Uruguay Round Agreement on Textiles and Clothing was supposed to phase out quotas progressively by January 1, 2005. But phaseouts have been heavily backloaded, with more than 50 percent of quotas—covering the most commercially valuable products—still to be removed (as of the end of 2004). Backloading has robbed developing countries of one of the major gains expected from the Uruguay Round and given rise to legitimate doubts about the willingness of the major importers to honor the agreement. It has also undermined any chance of gradual and orderly adjustment in the sector. The abrupt removal of the remaining quotas on January 1, 2005, will create adjustment problems for importers and exporters alike and is unleashing powerful protectionist forces in high-income countries. These must be effectively contained—for example, by restraining the proliferation of contingency protection measures. The correct answer lies not in pursuing protectionism by other means, but in providing adjustment support to small suppliers highly dependent on this sector through trade and development measures.

This has led some to call for an extension of quotas. But this would be a mistake. "Temporary" textile and clothing protection has persisted for 40 years; continuing protection is likely only to prolong and further distort the adjustment process. The difficult process of adjustment must be started now. Given the role that developed countries have played in creating the scale (if not the fact) of the adjustment challenge, they must now be prepared to contribute to its resolution by covering some of the costs of adjustment. Assistance could help developing countries to move into niche markets or up the value chain and to strengthen their networks of suppliers and clients to meet just-in-time production deadlines. Removal of trade barriers and domestic distortions by developing countries themselves would also help increase competitiveness. Tariff preferences may ease adjustment for some countries in the short term, although restrictive rules of origin will need to be addressed. More helpful and less distortionary temporary breathing space could be provided by all developed countries extending duty-free and quota-free access to all products from the poorest developing countries no later than January 1, 2006.

Services—a major source of gains for developing countries

Liberalization of trade in services, especially of so-called mode 4 (the temporary movement of people to supply services), has been recognized as a major source of gains for developing countries, capable of bringing more benefits to them than perhaps any other part of the Doha Agenda. Services liberalization promises real development gains in efficiency, in the growth potential of the economy, in the export of goods and other services, and in access to basic services to improve the lives of the poor. Done right, services negotiations offer developing countries an opportunity to act in their own economic interest and get paid for it. They also offer the opportunity to manage the world's mounting migration pressures in a much more orderly fashion.

But services gains are not automatic, and producing an outcome that supports development can be a challenge, given the need for regulation to address complex issues of market structure, market failures, and noneconomic objectives. Ensuring that services liberalization results in competition and increases access to services by the poor are key regulatory challenges, requiring increased research and assistance. But with appropriate care to the nature, pace, and sequencing of reform, adjustment—including that related to increased imports of labor-intensive services—can be managed.

The Doha Round must make progress on mode 4 as a high priority. Developing countries should seek to expand access for groups of interest to them (such as contractual service suppliers and intracorporate transferees) and improve the transparency and usability of existing access. Bilateral or plurilateral agreements could also be considered as an interim step. These cover a broader range of workers than mode 4 and provide scope to develop trust and complementary policies (for brain drain, remittance transfers, return, and recognition). Over time, recruitment of workers under these schemes could be opened on a most-favored nation (MFN) basis to any country that can implement the requirements. The WTO would be notified of agreements, and interested WTO members would have the opportunity to indicate their interest in joining or negotiating similar agreements. An MFN waiver would likely be necessary. Bilateral or regional agreements, while a potentially useful interim step, are no substitute in the longer term for bound multilateral commitments under the WTO. WTO commitments remain the best and most effective way to deliver gains to developing countries, and commercially meaningful market access commitments on mode 4 are essential to fulfill the development dimension of the services, and Doha, negotiations.

Contingent protection and standards—avoiding the costs and uncertainties of new barriers

Even if the access issues are addressed, it will be important to ensure that the hard-won gains are not eroded by other policies that recreate trade barriers or create transactions costs and uncertainty. Antidumping is a form of continent protection used disproportionately against the exports of developing countries, with a severe chilling effect on their actual and potential trade—though some developing countries are now also becoming major users of antidumping measures. The Doha Round could help in several ways. The *de minimis* threshold below which developing country exports are immune from antidumping could be raised. Currently, as soon as imports from developing countries emerge from being insignificant, they are restricted by high antidumping barriers. National antidumping laws could also be required to treat all affected domestic interests—import-competing industries, consumers, and users—equally.

Meanwhile, developing countries should not be denied effective market access by their inability to meet ever more and ever higher OECD standards or market entry conditions. Exemptions are unlikely to help, serving only to brand developing country exports as inferior or unsafe and providing no incentive to raise national standards for the benefit of domestic consumers. Where standards are imposed by private buyers, there is even less scope for—or point in—seeking exemptions. Two things are essential if developing countries are not to be left behind: first, assistance to make effective use of the WTO disciplines for Technical Barriers to Trade and Sanitary and Phytosanitary Measures to ensure that standards are not abused for protectionist purposes, and, second, significant assistance to construct the institutional frameworks and infrastructure required to meet legitimate standards. Further, developing countries must be assisted to become more substantively involved in standard-setting processes, and those standard-setting activities themselves need to be oriented toward issues of greater interest to developing countries.

Preferences—to be replaced with equivalent development assistance

Rich countries have often used preferences to divide developing countries and promote their narrower regional, sectoral, and political objectives, often establishing complicated regulations that exclude exports from otherwise eligible countries. The poorest countries have frequently received limited benefits from preference schemes, in part because preferences do nothing to address their multiple supply-side constraints. Benefits are also often gained at the expense of other developing countries, and they are smaller than would be the case with either direct transfers or multilateral liberalization. But the price of preferences is continuing protection in rich countries. MFN liberalization—plus appropriate compensation for countries that may suffer adjustment problems—is likely to be a better path.

Some developing countries may confront possible large losses from preference erosion and will require concrete assistance. Given the history of preference programs, developed countries as a group should pay to replace preferences with equivalent official development assistance, which the recipient governments could use to fund adjustment costs. Making this deal operational should be an explicit part of the Doha Round. Any such assistance should be seen as part of a broader effort needed to help poor countries build and strengthen their ability to use trade beneficially. Specifically in the context of a Doha deal, however, there is a need to accompany global commitments to implement far-reaching trade reforms on an MFN basis with a temporary program to transfer additional resources to developing countries, especially those that will experience preference-erosion losses.

Free trade agreements—imposing high transaction costs

Free trade agreements (FTAs) have a mixed record in achieving real liberalization, especially on the hardest nuts (agricultural subsidies or sensitive products). Benefits may be limited (or achieved at the expense of others) and costs can be high. Unlike at the WTO where developing countries can form effective coalitions, in FTAs they are at a disadvantage in resisting the inclusion of nontrade issues or erosion of their WTO rights (TRIPS+ on patents, especially pharmaceutical patents, and other WTO+ provisions). Multiple FTAs with differing rules of origin impose high transaction costs, particularly on small traders, and divert the limited negotiating resources of poor countries from the pursuit of multilateral liberalization.

Singapore issues—improved trade facilitation promises development gains

Three of the four Singapore issues (competition, investment, transparency in government procurement) have rightly been left off the Doha Round. None meet the three essential tests of whether rules on regulatory issues should be included in the WTO: Are they trade related? Are they in line with broader development priorities? And what is the specific value of a WTO agreement? These issues are not priorities for poor countries and could divert scarce resources from other issues with higher development payoffs. Even where there are development benefits, they may not be best pursued through a WTO agreement.

The remaining Singapore issue, trade facilitation to minimize bureaucratic procedures, promises trade and development gains, but a WTO agreement cannot be business as usual. It should not impose heavy obligations on developing countries and make light promises of assistance. The main value of a WTO agreement on trade facilitation would be as a mechanism for attracting and channeling international assistance. From a development perspective, the best model is one where implementation deadlines could be customized in negotiations with individual countries, with technical and financial assistance packages negotiated and customized as part of a package. A review process involving expert organizations and other developing countries with similar experiences could identify problems early, and negotiated extensions would be possible.

Trade-related intellectual property rights—differing needs by country

The inclusion of intellectual property rights in the WTO has been vigorously debated. Intellectual property laws require a very delicate balance of market forces and public action—a balance unlikely to be the same for all countries. Unfortunately, TRIPS (trade-related aspects of intellectual property rights) obligations have tended to take too little account of levels of development and varying interests and priorities. This was manifestly true of TRIPS coverage of medicines, imposed without due regard for the possible adverse effects on public health. As a result, the TRIPS rules on access to essential medicines have had to be negotiated and renegotiated in a still-acrimonious environment. While the TRIPS Agreement has tried to mitigate the problems of one-size-fits-all by providing for differing implementation periods, countries acceding to the WTO may not even have access to these normal flexibilities.

The TRIPS Agreement does include areas of interest for many developing countries, though the balance of costs and benefits varies by country and according to issue. But the flexibility provided for implementation of TRIPS is still less than sufficient on paper, and even less so in practice. There is a clear case for revisiting more of the rules to examine their impact on developing countries and any additional flexibility required. This remains relevant for access to essential medicines.

Special and differential treatment—making it more effective and operational

All countries will enjoy significant long-term benefits from freer trade. But it is also clear that poorer countries have less capacity to benefit and many will need short-term adjustment support. Developing countries generally have a more limited ability to take advantage of new opportunities and to bear adjustment costs. Special and differential treatment makes sense—and should be made more effective and operational.

There is no compelling case for exemption for rules on traditional trade policies. Additional freedom to use bad policies promises few development gains, and risks harming other developing countries (through subsidy wars). For rules on domestic regulations requiring actual investment of resources, a cost-benefit analysis based on four factors should guide what special and differential treatment to grant and to whom: the extent to which the rules are related to trade (market access), the extent to which they are in line with broader development priorities, the costs of implementation, and the relative costs to others of nonimplementation. Assessments of costs and benefits will vary by issue and level of development of the country.

Where the costs are high and the trade and development benefits minimal, the issue should not be included in the WTO. Where the costs are high, and development benefits only a longer term priority, there is a strong case for extensive—but not eternal—flexibility. Where development benefits are greater or more immediate, a model that calibrates commitments with assistance and gives greater flexibility to countries to determine appropriate implementation periods is appropriate. Where WTO rules promise real and short-term trade and development benefits, concrete technical and financial assistance should be assured—say, through mandatory commitments subject to review and linked to implementation requirements of developing countries.

A trading system limited only to agreements that are in the trade and development interests of all members to implement under the framework of binding multilateral trade rules should be accompanied by special and differential treatment that affords appropriately long and flexible conditions to adjust to trade liberalization and real and substantial aid for trade. Poor countries must be supported in generating the sources of revenue needed to compensate for losses incurred as a result of lowering import duties. They must also be supported in building the human and physical infrastructure they need to benefit from increased market opportunities and in adjusting to erosions of existing trade preferences stemming from multilateral negotiations.

An incremental and temporary "aid for trade fund" commensurate with the size of the task, or significantly ramped-up contributions through such existing channels as the Integrated Framework for Trade-Related Technical Assistance to Least Developed Countries, is needed to support countries in addressing adjustment costs associated with the implementation of a Doha reform agenda. A priority task for the development and trade communities could be the identification of new and existing channels though which this additional funding could most efficiently be made available for relevant, targeted projects in developing countries.

Coherence—adopting sound complementary policies and ramping up aid for trade

If trade liberalization is to contribute to economic growth, expanded trade, and poverty reduction, it must be coordinated with other policies at both the national and international levels. At the national level, policy coherence means adopting sound complementary policies to manage liberalization, as well as ensuring that trade policymaking is appropriately informed by expertise across a range of policy areas. Importantly, export competitiveness must not be pursued in a way that encourages discriminatory or coercive labor practices or adversely affects the natural environment—say, through unsustainable forestry or fisheries practices. At the international level, coherence calls for a significant ramping up of "aid for trade" by the development community (to negotiate, assess, and implement WTO agreements and to design and implement adjustment policies). It also calls for a clear and realistic view of the WTO's role in technical assistance. This assistance for deeper capacity building must be additional to, and not at the expense of, development aid. Trade liberalization requires international negotiations and international assistance, but its benefits and challenges remain fundamentally a question of domestic economic and policy reform. This is particularly important for the Least Developed Countries.

Promoting the export supply side in low-income countries

As indicated at the beginning of this chapter, the Monterrey Consensus called explicitly for market access issues to be complemented by an emphasis on overcoming supply-side constraints (UN 2002a). As the Consensus states:

> We invite multilateral and bilateral financial and development institutions to expand and coordinate their efforts, with increased resources, for gradually removing supply-side constraints; improve trade infrastructure; diversify export capacity and support an increase in the technological content of exports; strengthen institutional development and enhance overall productivity and competitiveness.... Special consideration should be given to Least Developed Countries, landlocked developing countries, small island developing states, African development, transit developing countries and countries with economies in transition (paragraph 36).

Supply-side constraints are most prominent in Least Developed Countries and other low-income countries, which are typically primary commodity exporters with high measures of export concentration. Most nonfuel commodity exporters are located in Sub-Saharan Africa, Latin America, and Central Asia. Many are caught in poverty traps. Low overall productivity and a high concentration of exports in a few primary commodities leave them subject both to frequent commodity price shocks and to a long-term terms-of-trade decline in some commodities.

Primary commodity exporters have tended to experience general economic stagnation over the past few decades. The UN Millennium Project Task Force on Poverty and Economic Development examined the growth performance of developing countries according to export composition (UN Millennium Project 2004a). Focusing attention on non-oil economies, and putting aside the special case of the postcommunist transition economies, the evidence shows that commodity exporters tended to stagnate while exporters of manufactures tended to grow. Among developing countries, only 19 of 41 primary commodity exporters experienced positive economic growth during 1980–98, while 23 of 24 manufactures exporters experienced positive economic growth over the same period. The commodity exporters had average GNP growth of 0.2 percent per person per year, compared with an average growth of 2.7 percent per person per year among the manufacturing exporters. Of course when commodities prices are high, the commodity exporters tend to grow faster, but they are then subject to sharp downturns when highly volatile commodity prices decline.

Export diversification, especially into nontraditional manufactures and services, is not easily achieved in low-income settings. Countries must have adequate infrastructure and human capital to support production and trade in nontraditional sectors. Low-income countries with small populations and with populations living far from sea coasts—and thus far from ports—are especially

Table 14.1

Economic growth rates by population size and location

	Small countries		Large countries	
	Number that grew in GDP per capita (PPP) 1980–98	Average growth in GDP per capita (PPP) 1980–98	Number that grew in GDP per capita (PPP) 1980–98	Average growth in GDP per capita (PPP) 1980–98
Inland populations	24 of 53	–0.2%	10 of 10	2.5%
Coastal populations	15 of 17	1.9%	3 of 4	3.2%

Note: Countries are defined as "small" if they had a population of less than 40 million in 1990 and "coastal" if more than 75 percent of their population lives less than 100 kilometers from the coast.

Source: Maddison 2001; Gallup, Sachs, and Mellinger 2003.

hindered. Since populations of Sub-Saharan countries tend to be both small and located inland, they face special difficulties in overcoming their lack of competitiveness in manufactures exports. Similarly, countries in Central Asia and the Andean region have tended to remain stuck in a few primary commodity exports.

Remote, small economies had a much harder time sustaining economic growth during 1980–98 than those that are either coastal or large (table 14.1). (We consider countries "small" if they had a population of less than 40 million in 1990 and "coastal" if more than 75 percent of their population lives less than 100 kilometers from the coast.) Countries that are both small and noncoastal, as are most countries in Sub-Saharan Africa, experienced negative economic growth in 1980–98. And among non-African small inland countries (with available data) 11 out of 20 grew in GDP per capita (PPP$) during 1980–98.

The implication of these figures is not that geographic barriers are insurmountable for trade and growth. Instead, countries with supply-side constraints (such as long distances from ports) require special supply-side attention. For landlocked and other relatively isolated economies, trade policy priorities need to include regional integration and donor-supported investments in transport infrastructure (trunk roads and ports) and communications infrastructure (mobile telephony and fiberoptic cables for Internet connectivity). We recommend that these investment and policy requirements be addressed as central issues within countries' MDG-based poverty reduction strategies.

Regional and global goods

The UN Millennium Project's core operational recommendation is for the Millennium Development Goals to be implemented at the country level through MDG-based poverty reduction strategies. But for many developing countries the Goals cannot be achieved solely through country-level investments, debt relief, and trade reform. They also require increased investments in regional and global public goods. At the regional level, countries must build cross-country infrastructure, curb pollution, strengthen the management of transboundary ecosystems, and enhance economic and political integration. At the global level the research community must be mobilized to address the specific problems facing developing countries—particularly for health and agriculture in the tropics and subtropics. Other global investments critical for achieving the Millennium Development Goals include efforts to mitigate climate change, protect global fisheries, and maintain biodiversity.

Regional goods

Regional public goods are generally overlooked and underprovided in most parts of the developing world, despite their critical role in promoting development. A country's immediate neighbors tend to be among its most important trading partners. These economic ties can be strengthened through regional cooperation—critical for economic growth and poverty reduction when an economy is landlocked, a small island state, or dependent on neighbors for vital resources, such as food, water, or energy. Regional integration is even more important for countries with very small populations, which must rely on regional rather than national markets to enjoy economies of scale and scope in industry, public administration, and research and development.

Further, the management of transboundary watersheds, desertification, and biodiversity can be addressed only through regional strategies integrated with national plans of action. And since many local conflicts have repercussions on entire regions or are driven by regional tensions, conflict management requires greater regional cooperation to detect conflicts before they erupt and to develop coordinated responses from neighboring countries to end them.

We therefore recommend that the provision of four types of regional goods be supported internationally and integrated into national MDG-based poverty reduction strategies:

- Infrastructure for transport, energy, and water management.
- Coordination mechanisms to manage transboundary environmental issues.
- Institutions to promote economic cooperation, including coordination and harmonization in trade policies and procedures.
- Political cooperation mechanisms for regional dialogue and consensus building, as exemplified by the African Peer Review Mechanism.

Regional infrastructure

Infrastructure for transport, energy, and water resource management underpins economic development. Many countries—particularly those isolated from world markets, such as the landlocked nations in Africa, Central Asia, and Latin America—require investments in transport infrastructure to integrate them more closely with the rest of the world. Rwanda, for instance, can make tremendous investments in its road infrastructure, but the economic returns to those investments will be limited unless Uganda, Kenya, Burundi, and Tanzania make similar investments to facilitate transport to the international shipping ports of Mombasa, Kenya, or Dar-es-Salaam, Tanzania. Economic prospects in Kyrgyzstan, Lao People's Democratic Republic, Paraguay, and Tajikistan, among others, also depend on improved transport infrastructure in neighboring countries for vital access to international trade routes.

As discussed in chapter 10, perhaps nowhere are MDG investment needs for regional transport infrastructure greater than in Africa. Much of the continent's transport infrastructure was designed in the colonial era to transport minerals and other natural resources directly to the nearest port, with virtually no infrastructure to connect African countries. Decades of insufficient maintenance and underinvestment have left transport networks across the continent in extremely poor condition, with an estimated 25 percent of the Trans-African Highway delapidated to the point where it no longer meets design standards (UNECA 2004). The resulting transport costs, several times higher than in other regions, take a tremendous toll on economic development by raising the cost of key economic inputs, such as fertilizers and fuel.

Poor cross-border infrastructure for the transmission of electricity and fossil fuels is another constraint on development (EIA 1999). By connecting national electricity grids and developing subregional electricity pools, countries can share cheap sources of energy—such as the largely untapped potential for hydropower in Africa and parts of Central Asia—reducing costs and increasing reliability. Regional infrastructure investments are also needed to transport fuel. Important examples are the West African gas pipeline, the Baku–Tiblisi pipeline, and an outlet to an international shipping port allowing Bolivia to export its natural gas. The joint development of infrastructure for water management, such as dams, aqueducts, and canals, presents another investment priority.

Regional management of the environment

Many environmental problems require regional solutions. For instance, the Mekong River and the Nile Basin initiatives have improved the multicountry management of transboundary rivers and watersheds. Another promising example is the Amazon Cooperation Treaty, to develop a joint management strategy for the Amazon Basin among all riparian countries. Other major environmental challenges in need of concerted regional responses are combating desertification, managing coastal and freshwater fisheries, and reducing the pollution of air and water. For example, halting the eutrophication of Africa's Lake Victoria will be possible only if Kenya, Tanzania, and Uganda cooperate to reduce nutrient loads in the lake. The Global Environment Facility, which specializes in the financing of regional and global environment strategies, requires more funding to support such regional initiatives in developing countries.

Economic cooperation

Countries need to promote regional economic cooperation to overcome the constraints of small market size and to reap the full benefits of economic specialization. Since developing countries tend to export more to distant developed countries than to developing countries (World Bank 2001), the potential for regional integration among developing countries is tremendous.[1] To promote intraregional trade, countries should continue to reduce tariffs and invest in trade facilitation by simplifying and automating customs procedures, promoting the mutual recognition of standards, and encouraging trade in services (chapter 14). In some cases regional currency unions can further aid intraregional trade by reducing the cost of exchange rate fluctuations and further deepening economic integration.

A second dimension of regional economic cooperation focuses on sharing the high fixed costs of setting up key institutions for development. Universities, research centers, and standards bodies are critical for generating growth, but frequently impossible for small countries to afford. Botswana, for example, despite having one of the highest per capita incomes in Africa, does not have a medical school to train doctors to fight the HIV/AIDS pandemic, since it has a population of only 1.5 million people. Many small island developing states also require regional institutions to help them overcome the constraints of small markets and population.

Third, the example of the European Union, which speaks with one voice in international negotiations over trade, shows that regional economic cooperation can strengthen the international voice of developing countries. By agreeing on common positions and objectives, small countries can reduce the cost of international negotiations and increase the likelihood of successful outcomes on issues like trade and debt relief.

These priorities require strong institutions to coordinate the alignment of customs procedures, the harmonization of standards, and the development of joint infrastructure. Examples are the Association of Southeast Asian Nations (ASEAN), which has been successful at promoting economic cooperation in Southeast Asia, the Caribbean Community (CARICOM) in the Caribbean, the South American Community of Nations in Latin America, and the regional economic commissions in Africa, such as the Economic Community of West African States (ECOWAS) and the Southern African Development Community (SADC).

Political cooperation

Countries can strengthen national policies and promote good governance through regional cooperation. The New Partnership for Africa's Development (NEPAD) has established voluntary procedures for member countries to identify means of strengthening their institutions and policies through the African Peer Review Mechanism. It provides an important new forum for countries to exchange experiences and assist each other in improving their policy environments.

Regional political cooperation can further promote regional peace and security by implementing the recommendations in chapters 7 and 12. Sources of finance for conflict can be cut off through such regional initiatives as the Kimberly Process Certification Scheme. ECOWAS and SADC have helped prevent and end regional conflicts through their peacekeeping activities. Their example should be followed in other regions prone to conflict, but this will require increased training for peacekeepers and support for their missions.

Why are regional goods and integration underfunded?

Birdsall (2004) estimates that only approximately $2 billion of the $65 billion (in 2003 dollars) in official development assistance in 2002 financed regional collaboration and infrastructure. Even for global public goods, where developed countries stand to benefit directly, financing remains dramatically insufficient (Kaul and others 2003).

Birdsall cites two main reasons. First, the cost of coordination among different countries is extremely high, requiring strong regional institutions that do not exist in most parts of the developing world. Many governments in low-income countries are so understaffed and underresourced that they cannot afford to allocate the staff necessary to ensure effective coordination with their neighbors.

Second, the attribution of responsibility is a problem. This relates more to the way donors operate. Bilateral and multilateral agencies tend to allocate funds on the basis of individual country performance and needs—a difficult metric in the case of regional goods. In addition, donors often expect a recipient country to guarantee loan repayments. This approach does not work for regional investments since it is extremely difficult to assign the investment benefit to individual countries. As a result, it becomes nearly impossible to obtain loan guarantees for regional projects from individual countries. To overcome similar problems, regional infrastructure projects in the European Union are justified by their benefits to the entire community and financed from the EU's core budget. Comparable mechanisms could be established among developing countries.

Strengthening the provision of regional goods

How can the provision of regional goods, such as infrastructure, improved environmental management, or economic integration, be improved? Our recommendations focus on two main areas: strengthening regional organizations and making more funding available for regional infrastructure.

Dedicated regional institutions are needed to provide regional public goods and improve coordination among member countries. They must have a clear mandate, be properly staffed and resourced, and be clearly aligned with member countries' national poverty reduction strategies. In most developing regions today this is not the case. To finance necessary investments, organizations like the African Union, CARICOM, and NEPAD urgently require more funding through dedicated or "ring-fenced" sources of financing, such as customs duties, and official development assistance where necessary. Adequate funding must be available to fund the operating costs of organizations in addition to the specific projects that receive most attention today.

Competing responsibilities among regional organizations must be resolved to avoid any unnecessary duplication of effort. While different organizations within one region may provide different regional goods, their mandates must be clearly delineated—not only among them, but also with respect to the national governments of their member states, which will need to gradually transfer some sovereign responsibilities to them.

Direct funding for regional infrastructure must increase substantially. Where regional infrastructure strategies exist, they are often not implemented due to a lack of funding. NEPAD has estimated that more than $8 billion will be required to fund its priority infrastructure investments, eight times the $950 million in total funding for NEPAD-related activities provided by the African Development Bank and the World Bank from 2001 to 2004 (Nduru 2004). Since it can be nearly impossible to receive loan guarantees for regional infrastructure projects from individual countries, the bulk of funding for regional infrastructure critical to achieving the Goals will need to be provided to regional organizations in the form of grants. For this, the concessional financing windows of the World Bank and the regional development banks must increase substantially.[2] Bilateral donors must also increase their financing of regional infrastructure projects. As described in chapter 17, the UN Millennium Project estimates that official development assistance for regional infrastructure will need to rise from $2 billion in 2002 to $11 billion by 2015.

Global goods

Some global responses are required to support countries in achieving the Goals—such as mobilizing global science and technology, curbing climate change, and fighting the degradation of the environment.

Mobilizing global science and technology for the MDGs

Many developing countries need new technologies to address specific needs. There are realistic prospects for developing new vaccines and medicines for malaria, HIV/AIDS, TB, and other killer diseases in poor countries. Improved agricultural varieties and cropping systems can increase the food productivity of rainfed agriculture. Accurate environmental monitoring and forecasting can help focus interventions for the greatest positive impact. Better microbicides and contraceptives can improve sexual and reproductive health for the poor. Many other examples abound for such public goods that, once developed, should be shared broadly to help all countries achieve the Goals.

The international science community—led by national research laboratories, universities, and national academies of science—must play a critical role in developing the global public goods to overcome these constraints. It must bring to bear its tremendous research capabilities to help solve the tough problems facing developing countries—particularly in the tropics.

Global research into areas critical to developing countries, despite several efforts, remains underfunded. The annual operating budget of $400 million for the worldwide network of 15 tropical agricultural research centers known as the Consultative Group on International Agricultural Research (CGIAR) is small in comparison with the combined research and development (R&D) budgets of the world's six largest agrobiotech companies, estimated at roughly $3 billion a year (Evenson 2003). The CGIAR specifically focuses on increasing the agricultural productivity of the poorest rural farmers in the tropics. It has had outstanding success in helping achieve major gains in food security in many parts of the tropical world, particularly in fostering the Green Revolution in Asia. The low budgets of the CGIAR system and national agricultural research centers continue despite considerable evidence of the high social rates of return from R&D on tropical food production.

Likewise, health R&D is limited for diseases affecting the poor, with only 10 percent of global funding used for research into 90 percent of the world's health problems (Global Forum for Health Research 2002). The WHO's

Commission on Macroeconomics and Health recommends that annual funding for R&D on global public goods in health (malaria, AIDS, TB, and nutrition, among other priorities) should be increased to $3 billion by 2007 and $4 billion by 2015, compared with roughly $300 million annually today (WHO 2001). The situation is similar in other areas critical to the needs of poor people. Low-cost sanitation technologies adapted to local cultural preferences, ability to pay, and environmental constraints are notoriously underresearched across the developing world.

Two reasons account for the inattention of global science to the needs of poor countries. First, public investments in research targeted at the needs of the tropics or other developing regions are insufficient due to the resource constraints in developing countries. Second, while private markets in developed countries can produce development-stage science and, to a lesser extent, research-stage science, this is not so in poor countries. No adequate incentives exist for private research to focus on tropical diseases or subsistence and small-scale agriculture, since the poor would be unable to pay for the new medicines, improved plant varieties, or farming techniques. There is simply no commercially attractive market for such products.

These shortcomings have been understood for some time, but the international system has so far not responded adequately. Private research could be mobilized through three tested coordinating mechanisms.

- Ex post prizes have been used frequently to spur innovation. An impressive example, though not related to the MDGs, is the Ansari X Prize, recently awarded for the first commercial flight into space. Similar prizes should be offered for well defined problems, such as developing a new type of vaccine or an improved crop variety (Masters 2002).
- Direct funding of private research has been used successfully by several private foundations, such as the Rockefeller Foundation and the Bill and Melinda Gates Foundation, to promote development-stage research in public health and agriculture. Recently, the privately funded Malaria Vaccine Initiative announced the successful completion of phase 2 of clinical trials for a new malaria vaccine developed jointly with GlaxoSmithKline Biologicals, in partnership with Mozambique's Ministry of Health and the Centro de Investigação em Saude da Manhiça.
- Precommitment purchase agreements, as proposed by Kremer (2002), consist of binding public commitments to buy a product, such as a vaccine against hepatitis, at a minimum price. They enable private companies to plan for a minimum production level, thus removing an important element of risk in the development of product-stage research or in the bulk production of vaccines. The Global Alliance for Vaccines and Immunization (GAVI) and the Vaccine Fund are putting this principle into operation.

In addition to mobilizing private research for the Goals, international donors and foundations need to support more public research on the specific challenges facing developing countries. A preliminary estimate suggests that at least $7 billion a year will be required by 2015 (chapter 17). In addition to the $4 billion for public health research, $1 billion would go toward agriculture, nutrition, and improved natural resource management—by more than doubling the current budget of the CGIAR. Roughly another $1 billion is needed for research toward improved energy technologies. And perhaps $1 billion is needed for greater understanding of interannual, seasonal, and long-term climate change. Improved climate modeling and forecasting tools can help predict changes in precipitation patterns to improve the management of water reservoirs and help farmers adapt their cropping and irrigation techniques accordingly. It can also help improve management of responses to natural disasters.

An international response to climate change

Climate change is a major development issue that needs to be addressed urgently. Unless global warming slows down, the incidence of droughts and floods will likely increase, vector-borne diseases will probably expand their reach, and many ecosystems, such as mangroves and coral reefs, will likely be put under great strain. In short, achievements in the fight against disease, hunger, poverty, and environmental degradation risk being unraveled by climate change (IPCC 2001a–c).

Although, encouragingly, the Kyoto Protocol is now in force, the international response to the gathering threat of climate change has so far been inadequate. Despite the signing of the Protocol, global emissions of greenhouse gases continue to rise. It is therefore necessary that additional measures be implemented to stabilize greenhouse gas concentrations in the atmosphere in the near future. As agreed at Johannesburg, primary responsibility for mitigating climate change and other unsustainable patterns of production and consumption, such as the overharvesting of global fisheries, must lie with the countries that cause the problems—that is, the rich and some of the rapidly growing middle-income countries.[3] The details of how to mitigate climate change go beyond the scope and mandate of the UN Millennium Project. But we stress that urgent action is necessary to agree on binding reductions in the global emissions of greenhouse gases (UN Millennium Project 2005c).

Chapter 16

Getting started in 2005: launching a decade of bold ambition

There is still enough time to meet the Millennium Development Goals—though barely. With a systematic approach and decade-long horizon, many countries now dismissed as too poor or too far off track could still achieve the Goals. The UN Millennium Project argues strongly for introducing a longer term horizon into international development policy, one that focuses on overcoming short-term constraints by scaling up approaches to meet basic needs. But the need for longer term horizons should not be confused with, or detract attention from, the need for urgent action. Without a bold breakthrough in the coming year, a large number of countries that could still achieve the Goals will be consigned to failure.

Major challenges of global policy cooperation need to be addressed in 2005. The Doha trade agenda needs a breakthrough for development. Immediate action is needed to start tackling long-term environmental challenges, such as climate change and fisheries depletion.

The world also needs to move urgently with specific actions of scale-up toward the Goals. Only by acting now can sufficient numbers of doctors or engineers be trained, service delivery capacity strengthened, and infrastructure improved to meet the Goals. To start the decade of bold ambition toward 2015, we recommend a series of worldwide initiatives to kickstart progress, translating the Goals quickly from ambition to action. For all of this, we recommend that they be led by the Secretary-General, with UN system contributions coordinated under the strengthened guidance of the UN Development Group.

Identify fast-track countries

Bold MDG-based investment programs cannot be scaled up in developing countries with extremely poor governance. But the international community has recognized many low-income countries as having strong governance and the potential for much more ambitious investment programs. In 2005 we recommend that these well governed low-income countries be granted "fast-track MDG status" by the international community and receive the massive increase in development assistance needed for them to implement MDG-based poverty reduction strategies.

Several preexisting criteria could be used to help identify the fast-track countries (box 16.1). They include countries that have reached completion point under the HIPC Initiative, those that have qualified for support from the U.S. Millennium Challenge Corporation; those that have acceded to the African Peer Review Mechanism of the New Partnership for Africa's Development; or those with favorable reviews through the World Bank–IMF Joint Staff Assessments of PRSPs. These or other performance-based criteria will yield at least a couple of dozen low-income countries that have reached governance standards sufficient for including them on a fast-track for scaling up MDG-based investments beginning in 2005.

It is with these MDG fast-track countries that the international community, particularly the donor countries, will face the clearest test of their commitment to achieving the Goals. If donors cannot provide the support that these countries require to achieve the Goals, the undertaking to achieve those Goals will be in peril. Moreover, the system for international development assistance needs to consolidate its incentives for countries with weaker governance levels, showing that good performance is indeed rewarded by financial support consistent with the Goals.

Prepare MDG-based poverty reduction strategies

In addition to supporting fast-track countries, we recommend that every interested developing country produce, before the end of 2005, an MDG needs assessment and an MDG-based poverty reduction strategy (chapter 4). We further recommend that the Secretary-General request each resident UN Country Team to assist in this process. Most often the outcome will be a revised version of an existing national strategy, including the Poverty Reduction Strategy Paper (PRSP), where appropriate. It should contain a strategy for enhanced investments at the village, town, and city levels, a financing scenario, and a governance strategy to ensure implementation of the program with minimized corruption, based on fundamental principles of human rights.

The host country should lead and own the effort to design the MDG strategy, drawing in civil society organizations; bilateral donors; the UN specialized agencies, programs, and funds; and the international financial institutions, including the IMF, the World Bank, and the appropriate regional development bank. The contributions of the UN specialized agencies, programs, and funds should be coordinated through the UN Country Team, and the UN Country

Box 16.1
Identifying MDG fast-track countries

The UN Millennium Project recommends that in 2005, the international community designate fast-track status to a significant number of low-income countries that are ready for scale-up. At least four criteria could be used to identify these countries, as listed in the table. One of the first international efforts to reward strong governance with increased foreign assistance was the Heavily Indebted Poor Countries (HIPC) Initiative to reduce debt burdens. As part of the HIPC process, country eligibility to receive debt relief hinges not only on having an extremely high debt burden but also on a positive joint evaluation by the World Bank and IMF of the country's governance quality and economic policies. Countries are granted debt relief when they reach their "completion point," and thus are "recognized by the international community for their satisfactory progress in implementing sound economic and structural policies."

A second mechanism that evaluates and validates strong governance as a precondition to aid disbursement is the U.S. Millennium Challenge Corporation (MCC). The MCC disburses funds only to countries surpassing thresholds for various indicators measuring governance, investment effort in health and education, and economic policies. Seventeen countries have already been deemed eligible for ambitious investment programs. Another 13 "threshold" countries have been assessed as committed to undertaking the reforms necessary to improve policy performance and eventually qualify for MCC assistance.

A third example is the African Peer Review Mechanism (APRM) of the New Partnership for Africa's Development (NEPAD). African Union member states join the APRM to foster the adoption of policies, standards, and practices that lead to political stability, high economic growth, sustainable development, and accelerated subregional and continental economic integration through sharing experiences and reinforcing successful practices, including identifying deficiencies and assessing the needs for capacity building. A key criterion for acceding to the APRM is submitting to periodic peer reviews and facilitating such reviews to ensure that the policies and practices of participating states conform to the agreed political, economic, and corporate values, codes, and standards. As of mid-2004, 23 African countries have signed a Memorandum of Understanding as the first step to accession and submitting their policies and institutions to regular peer review.

The PRSP process offers a fourth mechanism for identifying countries to include on an MDG fast-track. The World Bank and IMF conduct Joint Staff Assessments of the PRSPs, and have given high praise to several low-income countries. For example, "[Mali's] PRSP represents a credible policy framework to reduce poverty, integrating for the first time the country's various poverty-focused programs within the context of a sound macroeconomic framework." Other countries recently praised for having strong PRSPs include Burkina Faso, Ethiopia, Ghana, Mauritania, and Yemen. In addition, the World Bank recently published a paper arguing that developing countries are ready to absorb at least an additional $30 billion in foreign assistance (World Bank 2003c). The study selected a sample of 18 countries that have recently "improved their policies significantly...used aid productively...and continue to have substantial unmet development needs." The paper concludes that a significant increment of aid could be used effectively in all 18 countries.

(continued on next page)

Box 16.1

Identifying MDG fast-track countries

(continued)

* Low-income country.

a. Countries are from www.nepad.org, retrieved on December 20, 2004.

b. Countries are from www.worldbank.org, retrieved on December 20, 2004.

c. Countries are from www.mca.org, retrieved on December 20, 2004.

d. World Bank 2003c.

Source: See notes a–d.

Potential candidates for MDG fast-tracking

	Governance qualification					
Country	African Peer Review Mechanism[a]	HIPC completion point[b]	Millennium Challenge Corporation Qualifier[c]	Millennium Challenge Corporation Threshold[c]	Poverty Reduction Strategy Paper[b]	World Bank Absorptive Capacity Study[d]
Albania				×	×	×
Algeria	×					
Angola*	×					
Armenia			×		×	
Azerbaijan					×	
Bangladesh*						×
Benin*	×	×	×		×	×
Bhutan*					×	
Bolivia		×	×		×	×
Bosnia and Herzegovina					×	
Burkina Faso*	×	×		×	×	×
Cambodia*					×	
Cameroon*	×				×	
Cape Verde			×			
Chad*					×	
Congo, Rep.*	×					
Djibouti					×	
Egypt	×					
Ethiopia*	×	×			×	×
Gabon	×					
Gambia*					×	
Georgia			×		×	
Ghana*	×	×	×		×	
Guinea*					×	
Guyana		×		×	×	
Honduras			×		×	×
India*						×
Indonesia						×
Kenya*	×			×		
Kyrgyzstan*					×	×
Lao PDR*					×	
Lesotho*	×		×			
Madagascar*		×	×		×	×
Malawi*	×			×	×	
Mali*	×	×	×		×	×
Mauritania *		×			×	×
Mauritius	×					
Moldova*					×	
Mongolia*			×		×	
Morocco			×			
Mozambique*	×	×	×		×	×
Nepal*					×	
Nicaragua *		×	×		×	
Niger*		×			×	
Nigeria*	×					
Pakistan*					×	×
Paraguay				×		
Philippines				×		
Rwanda*	×				×	
São Tomé and Principe*				×		
Senegal*	×	×	×		×	
Serbia and Montenegro					×	
Sierra Leone*	×					
Sri Lanka			×		×	
South Africa	×					
Tajikistan*					×	
Tanzania*	×	×		×	×	×
Timor-Leste*				×		
Uganda*	×	×		×	×	×
Vanuatu			×			
Viet Nam*					×	×
Yemen*				×	×	
Zambia*				×	×	

Team should work closely with the international financial institutions. At the headquarters level, the UN Development Group should coordinate the activities among all UN agencies, programs, and funds—with the UNDP Administrator continuing to play a special coordination role.

Launch a global human resource training effort for the Millennium Development Goals

With the design of national strategies, a major worldwide effort in preservice skill training should be launched simultaneously in 2005 to overcome the immediate scale-up constraints in human resources. International agencies and bilateral donors should work with low-income countries to prepare serious strategies and training materials for use at the village and city level. Global champions are needed for this initiative to set targets and confirm financial commitments to train, as first priorities:

- Village specialists in health, soil nutrients, irrigation, land reclamation, drinking water, sanitation, electricity, vehicle repair, road maintenance, and forest management.
- Managers in investment planning, budgeting, computer-based information systems, poverty mapping, and sector needs assessments.
- Teachers, doctors, and other skilled professionals to provide services in education and health.
- Professionals for urban planning and urban infrastructure and services (such as electricity, transport, water, waste management, and industrial zoning) and community development agents to promote local participation, gender equality, and minority rights.

A sharp focus on the short-term training of women and young workers, where appropriate, will provide the added benefits of bringing them into the formal labor market. It will also help them develop a skill base that will contribute to longer term development processes.

Launch the Quick Wins initiatives

We have noted the chance for early breakthroughs in many areas: school attendance, malaria control, school meals, soil nutrients, to name just a few. Each should be championed, and explicit and bold targets should be set on a three-year horizon. For example, it would certainly be possible to make a free antimalaria bed-net available to every African child in an endemic malaria region by the end of 2007. It would also be possible to have every subsistence farmer given the chance to replenish soil nutrients using fertilizers or agroforestry or related techniques by the end of 2007. We have identified a large number of additional Quick Wins (see box 5.1). The relevant UN agencies, together with bilateral agencies and the Bretton Woods institutions, should grasp the opportunity to launch these initiatives in 2005.

Engage the middle-income countries in the challenge of meeting the Goals

Middle-income countries are challenged to complete the process of eradicating extreme poverty within their own countries and to join the ranks of donor countries at the same time. Most large middle-income countries suffer from pockets of poverty that must be targeted for elimination. Grants in support of high debt burdens for heavily indebted middle-income countries can support that process.

But leading middle-income countries—such as Brazil, China, Malaysia, Mexico, and South Africa—also have expertise of direct benefit to the poorer countries. For example, China should help to ensure a steady flow of artemisinin-based antimalaria medicines for Africa in the coming years. It can also assist countries in expanding transport or other infrastructure. Brazil can contribute to development in Lusophone Africa, including the training of Portuguese-speaking professionals. Malaysia can help promote increased competitiveness in labor-intensive manufacturing exports and strengthen science advice mechanisms. South Africa has recently gained unique experience in the rapid scaling-up of infrastructure services for water and electricity in rural areas. It could assist countries in the rest of Africa in designing ambitious investment plans. Many other examples of the tremendous potential for developing country cooperation abound—and should be seized in 2005.

4

The costs and benefits of achieving the Millennium Development Goals

Resources required to finance the Millennium Development Goals

To implement the interventions recommended by the UN Millennium Project Task Forces, as outlined in chapter 5, countries will need to increase public investments in social services, basic infrastructure, and environmental management. Here, we estimate the cost of meeting the Millennium Development Goals (MDGs) at the country level and present a financing strategy based on increased domestic resource mobilization. We also describe the implications for global flows of official development assistance (box 17.1 summarizes key terms and concepts). We conclude by discussing mechanisms for financing the additional aid required to reach the Goals.

To be stressed at the outset is the provisional nature of our estimates of global needs. The estimates are meant to give guidance on the overall volume of aid that will be needed to achieve the Goals, but they should not be confused with the detailed costing that will have to be done country by country—and that will have to be regularly updated with experience and new information. The actual total costs will emerge over time as the sum of individual country costs and overall expenses of global operations. The individual country costs will be known definitively only as country programs are under way and more lessons are learned about scaling up. Here we provide what we believe are the right orders of magnitude.

To summarize our estimates, a typical low-income country in 2006 will need to invest around \$70–\$80 per capita in capital and operating expenditures toward meeting the Goals. Since investments can be scaled up only gradually, the financing will be lower at the beginning of the period and rise to \$120–\$160 per capita toward the end of the period. A rising share of these investments will be financed through domestic resource mobilization, which we project to increase sharply by up to four percentage points of GDP. Still, most low-income countries will experience an MDG financing gap of

Box 17.1
Key terms and concepts

Graduating countries. Countries whose domestic resource mobilization will rise enough to finance all MDG expenditures before 2015. As a result, they will "graduate" from the need for official development assistance (ODA) for direct MDG support.

MDG capacity building. Investments in human resources, including training and management systems for national and local governments as well as NGOs.

MDG investment needs. The capital investments and operating expenditures for basic infrastructure, social services, and improved environmental management required to meet the MDGs, excluding expenditures for capacity building (see below).

MDG financing gap. The portion of a country's MDG investment needs that cannot be financed through domestic resource mobilization by governments and households.

Official development assistance. Grants and concessional loans to developing countries to promote economic development and welfare.

ODA for direct MDG support. The amount of official development assistance that finances MDG investment needs (excluding capacity building) and can be provided to either governments or NGOs.

Other ODA. Flows that do not finance MDG investment needs or MDG capacity building.

10–20 percent of GDP that will need to be financed through official development assistance.

As this chapter describes, only a small share of today's global official development assistance—an estimated $16 billion of $65 billion in 2002 (in 2003 dollars)—supports direct MDG investment needs at the country level. Official development assistance (ODA) for direct MDG support will need to rise to $73 billion in 2006 and $135 billion in 2015 if all countries are to meet the Goals. After adjusting for the fact that several countries will not meet the minimum governance thresholds required to scale up public investments for the Goals, these figures are likely to be lower—$52 billion in 2006 and $110 billion in 2015.[1] In addition to these direct investments on the ground, meeting the Goals also requires capacity building, debt relief, additional early support for the Quick Wins (chapter 5), enhanced support for regional collaboration and infrastructure, global research, and emergency assistance. We estimate that total ODA volumes need to rise to 0.54 percent of rich country GNI in 2015, up from 0.23 percent in 2002 and 0.25 percent in 2003.[2]

By 2006 global official development assistance needs to reach $135 billion, up from $65 billion in 2002 and $69 billion in 2003. Some of the increase will be achieved on the basis of existing commitments made by OECD/DAC member countries. Based on those commitments, ODA in 2006 is to reach approximately $88 billion. Of course, it will be vital that, as ODA increases, it be properly directed at MDG needs.

Several financing mechanisms exist to make this steep rise in development assistance possible, despite short-term donor fiscal constraints. Among them, the International Finance Facility (IFF) stands out as being practical, technically feasible, and fairly straightforward to implement. We encourage all developed countries to support the IFF in 2005, in time to start disbursing funds in 2006. After 2006 the ODA needs continue to rise, and donors should prepare for a continuing scale-up of funding between 2006 and 2015, as country-level investment programs grow in scope. During 2006–15 we project that some countries (such as India) will graduate from the need for ODA. But this will be more than offset by the increased scale of investment programs required in the remaining low-income countries.

Readers are advised to keep in mind that all ODA numbers presented here are in constant 2003 dollars. Inflation and dollar exchange rate depreciation will raise the current dollar amounts of these estimates in the years ahead.

MDG investment needs at the country level

The aggregate ODA figures presented in this chapter are anchored in the preliminary MDG needs assessements that the UN Millennium Project has carried out in Bangladesh, Cambodia, Ghana, Tanzania, and Uganda (box 17.2).[3] The results show that these countries' total MDG investment needs are $70–$80 per capita in 2006, rising to $120–$160 in 2015. Underlying these estimates is the assumption that the scaling up of investment goes hand in hand with optimizing current public expenditures using best practices. We have added $8 per capita in 2006 and $13 in 2015 to account for interventions not originally included in the needs assessments. Additional expenditures will be required for capacity building and for emergency assistance that are not reported in the country results but included in our estimate of global ODA needs.

Per capita MDG investment needs are remarkably similar across the five countries, even though they derive from country-specific coverage data and unit costs. Two reasons account for this low variation.

First, some unit costs are independent of per capita GDP. For example, antiretroviral drugs to treat HIV/AIDS cost several hundred dollars per year regardless of whether per capita income is $100, $300, or $1,000 a year. Likewise, the international market for doctors, a recent product of globalization, leads to more uniform salaries no matter how poor a country is. For this reason we stress the importance of considering MDG investment needs in absolute per capita terms rather than as shares of GDP or national budgets. Since our results suggest that countries will require similar expenditures to meet the Goals, MDG investment needs expressed as a percentage of GDP will be higher in poor countries.

Second, countries will reach similar service coverage or infrastructure stocks to achieve the Goals. In countries with high current coverage, a reduced need for additional capital investments to meet the Goals is partially offset by higher current operating costs. This partial tradeoff between capital investments and operating costs further reduces differences across countries. The remaining variation is driven by differences in unit costs or underlying needs.

Box 17.2

MDG needs assessment methodology

Our approach to estimating resource needs to meet the Millennium Development Goals is guided by four principles.

- First, there is no one-size-fits-all answer to the question, "What will it take to meet the Millennium Development Goals?" The question can be answered only through country-level needs assessments using country-specific coverage data, targets, and unit costs. We recommend this approach for all countries preparing MDG-based poverty reduction strategies.
- Second, instead of using aggregate input-output relationships and unit costs, needs assessments must build on a bottom-up assessment of both capital and operating expenditures. The analyses should also quantify human resource needs and infrastructure requirements for all interventions necessary to meet the Goals.
- Third, operationalizing the Monterrey Consensus requires that domestic resource mobilization by governments and households fund as much of the cost of meeting the Goals as possible. Where MDG investment needs exceed domestic resource mobilization, this MDG financing gap must be covered through official development assistance.
- Fourth, while the Goals cannot be "bought" through more money alone, increased assistance is necessary to meet them. Donors must commit credibly to make sufficient funds available, with actual disbursements contingent on the quality of MDG-based poverty reduction strategies and the credible commitments of countries to undertaking the necessary reforms.

In collaboration with local research institutes in five countries—Bangladesh, Cambodia, Ghana, Tanzania, and Uganda[a]—the UN Millennium Project conducted needs assessments to answer the basic question "What will it take to meet the Goals?" (UN Millennium Project 2004b). To answer this question, we quantified the required investments and operating expenditures as well as the human resources and infrastructure necessary to meet the Goals. A common objection to asking this question is that resources are finite while needs are infinite, as stressed in every introductory economics class. This statement is of course correct, but we define "needs" specifically as the resources required to meet the quantitative, time-bound targets for poverty reduction that the world set for itself in the form of the Millennium Development Goals.

Our needs assessment methodology follows five basic steps summarized here and described in more detail in UN Millennium Project (forthcoming).

- In a first step, we identify all policies and interventions—defined broadly as the provision of services, goods, and infrastructure—necessary to meet the Goals, as described in chapter 5. They include sets of interventions for which no specific Goal exists, but that are nevertheless required to meet the Goals, such as improving access to transport, energy services, and sexual and reproductive health services (appendix 1).
- Second, we identify quantitative targets for each intervention for 2015, such as coverage rates for emergency obstetric care to reduce maternal mortality and the number of teachers, classrooms, and learning materials required to ensure universal primary education and the expansion of secondary education.
- Third, we use transparent investment models to estimate the capital and operating costs of the MDG interventions, including human resources and infrastructure.[b] We project an exponential scaling up of interventions to allow for a gradual expansion of service delivery capacity, as discussed in chapter 6.

a. Bangladesh Institute of Development Studies; Economic Policy Research Center, Uganda; Economic and Social Research Foundation, Tanzania; Institute of Social Statistics and Economic Research, Ghana; and the University of Cambodia.

b. These "open source" investment models can be downloaded at [www.unmillenniumproject.org].

Box 17.2
MDG needs assessment methodology
(continued)

- Fourth, we iteratively revise needs estimates to integrate synergies across intervention areas that would affect overall MDG investment needs. For example, greater access to safe water supply will reduce the incidence of diarrheal diseases and thereby lower health costs.
- In a final step, we develop a financing strategy by matching MDG investment needs with substantially increased domestic resource mobilization to estimate the MDG financing gap.

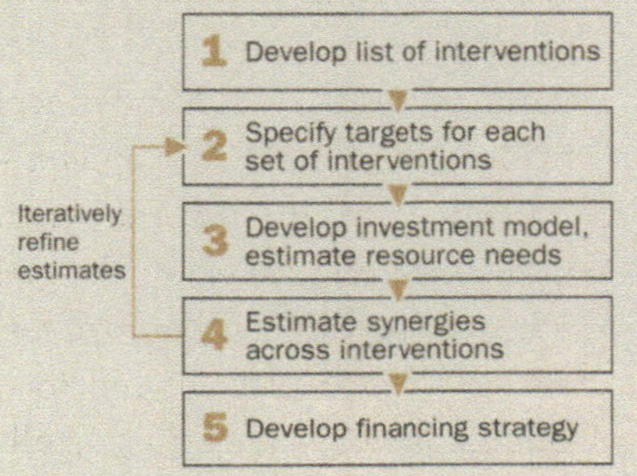

For example, health costs in countries with high HIV prevalence are higher than in low-prevalence countries.

Results of the needs assessment are arranged by MDG outcomes (table 17.1). Under hunger we quantify the MDG investments required to raise the productivity of subsistence farmers as well as nutrition interventions not provided through the primary health system, such as community-based nutrition programs. Other nutrition interventions are included under health. Our education estimates build upon the Education for All estimates by also including secondary school education. Investments in all sectors are targeted to benefit women and young girls. The gender needs assessment addresses additional interventions to combat violence against women, strengthen institutional capacity for promoting gender equality, and raise awareness of sexual and reproductive legal and economic rights.

One of the largest line item costs is for health. It includes the cost of running a health system offering essential medical interventions, such as emergency obstetric care, treatment for the major infectious diseases, and interventions to reduce child mortality. We also include some interventions primarily provided outside the health system, such as preventing major diseases. Consistent with the findings of the WHO (2001), differences in per capita costs are largely driven by HIV prevalence rates. "Improving the lives of slum dwellers" focuses on slum upgrading and providing alternatives to the formation of new slums. Infrastructure services—including domestic water supply, sanitation, electricity, improved cooking fuels, and transport—together account for roughly 35–50 percent of total MDG investment needs. The transport needs assessment includes only

Table 17.1

Per capita MDG investment needs and MDG financing gaps in Bangladesh, Cambodia, Ghana, Tanzania, and Uganda, 2006–15

2003 US$ per capita

Note: Numbers in table may not sum to totals due to rounding. Results describe MDG investment needs excluding expenditures for capacity building. Refer to appendix 3 and UN Millennium Project (forthcoming) for more details.

a. For MDG interventions not yet included in MDG needs assessments due to a lack of data (such as large infrastructure projects, higher education, national research systems, and environmental sustainability). Period average is $10 per capita for each country.

b. Consistent with table 13.2, calculated as net ODA minus technical cooperation, debt relief, aid to NGOs, emergency assistance, and food aid, using data from OECD/DAC (2004d).

Source: Authors' calculations prepared in collaboration with Bangladesh Institute of Development Studies; Economic Policy Research Center, Uganda; Economic and Social Research Foundation, Tanzania; Institute of Social Statistics and Economic Research, Ghana; and the University of Cambodia.

	Bangladesh			Cambodia			Ghana			Tanzania			Uganda		
	2006	**2010**	**2015**	**2006**	**2010**	**2015**	**2006**	**2010**	**2015**	**2006**	**2010**	**2015**	**2006**	**2010**	**2015**
MDG investment needs															
Hunger	2	4	8	4	7	13	3	5	12	4	7	14	3	5	10
Education	11	17	25	15	19	22	17	19	22	11	13	17	14	15	17
Gender equality	2	3	3	2	3	3	2	3	3	2	3	3	2	3	3
Health	13	19	30	14	21	32	18	24	34	24	33	48	25	32	44
Water supply and sanitation	4	5	6	3	5	8	6	7	10	4	5	12	2	3	9
Improving the lives of slum dwellers	2	3	4	3	3	4	2	2	3	3	3	4	2	2	3
Energy	20	19	20	9	13	23	13	15	18	14	15	18	6	10	19
Roads	12	21	31	12	21	31	11	10	10	13	21	31	13	20	27
Other[a]	8	9	13	8	9	13	8	9	13	8	9	13	8	9	13
Total	**74**	**100**	**140**	**71**	**101**	**148**	**80**	**94**	**124**	**82**	**111**	**161**	**75**	**100**	**143**
Sources of financing															
Household contributions	8	10	14	9	13	18	9	11	15	9	11	17	8	9	14
Government expenditures	23	33	49	22	30	43	19	27	39	24	32	46	27	35	48
MDG financing gap	43	56	77	40	58	87	52	57	70	50	67	98	41	56	80
Shortfall of ODA for direct MDG support over 2002 level	**42**	**55**	**75**	**22**	**40**	**69**	**36**	**41**	**54**	**35**	**52**	**83**	**29**	**44**	**68**
For comparison: ODA for direct MDG support, 2002[b]		1			18			16			15			12	

the cost of maintaining and expanding road networks. A more comprehensive assessment must factor in the cost of improving access to transport services as well as expanding ports and other transport infrastructure.

The UN Millennium Project's needs assessment methodology has recently been applied to other countries. The Indian Institute of Management in Ahmedabad collaborated with the Project to carry out detailed MDG needs assessments for three states: Madhya Pradesh, Rajasthan, and Uttar Pradesh, which together contain 28 percent of the Indian population below the poverty line. The researchers estimate that between 2005 and 2015, total annual MDG investment needs for the three states will average $115 per capita (Madhya Pradesh), $110 (Rajasthan), and $113 (Uttar Pradesh) (Dholakia, Kumar, and Datta 2004). These estimates, which exclude the significant costs of HIV/AIDS and transport, are in line with the results in table 17.1.

In Tajikistan, the UN Country Team is leading a detailed needs assessment for the country and has so far produced results for health, education, and water and sanitation. Preliminary estimates there suggest that the 2015 per capita costs of achieving 100 percent primary school enrollment would be $26, achieving universal primary healthcare $39, and reaching 80 percent access to water and sanitation $10. So the total estimated cost of achieving the education, health, and water and sanitation MDGs in Tajikistan would be $75 per capita in 2015. These figures are somewhat higher than the results presented here. (Perhaps unit costs are higher due to the country's rugged terrain and extreme geographical isolation.)

The cross-country consistency of results provides some convenient shorthand for assessing whether poverty reduction strategies and national budgets are consistent with the Millennium Development Goals. For example, the health costs are around $13–$25 per capita in 2006, rising to around $30–$48 in 2015 ($25 in 2006 for countries with high HIV prevalence, rising to up to $48 by 2015). So if an MDG-based poverty reduction strategy includes a health budget of only $5 per capita, as is often the case, it is likely to fall far short of the Goals. Similarly, the scale-up results indicate that any serious MDG investment strategy will have education costs of at least $11–$17 per capita (rising to $17–$25 in 2015), hunger costs of $2–$4 (rising to $8–$14 per year), water and sanitation costs of $2–$6 (rising to $6–$12), energy costs of $6–$20 (rising to $18–$23), and slum upgrading costs of $2–$4. UN Millennium Project (forthcoming) contains a detailed discussion of the sectoral results.

Financing the MDG investments at the country level

To operationalize the Monterrey Consensus, countries need to maximize domestic resource mobilization for the Millennium Development Goals before official development assistance should be called on to fund public investments.

Increasing domestic resource mobilization

We estimate that each of the five countries can increase government spending on the Goals by an ambitious four percentage points of GDP over the next 10 years. In a typical low-income country this corresponds to a rise from 5 percent of GDP to 9 percent, a more than doubling in absolute terms.[4] This expansion is not only necessary—it is achievable through using broad-based revenue sources, such as a value added tax, strengthening tax collection, and redirecting current spending.

Meanwhile, middle-income countries will be able to finance essentially all investments in the Goals without raising government budgets by four percentage points of GDP or resorting to external finance. In some cases, primary surplus requirements for government budgets may need to be adjusted to allow countries to increase public investments for the Goals. Still, modest levels of ODA may be needed to help middle-income countries redress especially difficult "pockets of poverty." Some heavily indebted middle-income countries may also require assistance in refinancing their debt burden.

Households are expected to contribute financially within their means to sectors where the incentive effects of well designed user fees are compatible with the overall policy objectives of ensuring effective and equitable access to basic infrastructure and social services. We project household contributions to investments in agricultural productivity, secondary school education, energy services, domestic water supply, and sanitation (UN Millennium Project forthcoming). In line with the international consensus and the recommendations of the UN Millennium Project, all direct and indirect fees for primary education and basic healthcare are discontinued (UNESCO 2000; WHO 2001).

The need for more official development assistance

Even substantial increases in domestic resource mobilization by governments and households will be insufficient to finance investments of approximately \$120–\$160 per capita by 2015. As a result, each of the five countries will require \$40–\$50 per capita in external finance in 2006, rising to \$70–\$100 in 2015. We stress that no distinction should be made between funding capital and operating costs through official development assistance, since poor countries cannot afford to fund operating expenditures, which account for a large share of total costs in health, education, and other sectors. To maintain macroeconomic stability, external finance to low-income countries will need to be provided in the form of grants (Landau 2004).

In the case of Ghana, direct investments in the Millennium Development Goals need to rise from \$80 per capita in 2006 to \$124 by 2015 (see table 17.1). Even after accounting for a near doubling of domestic resource mobilization between 2006 and 2015, the country's MDG financing gap is projected to rise from \$52 to \$70 per capita.

A step increase in MDG investments

The results for all five countries project a step increase in investment levels during 2006, to be funded largely through official development assistance. A common objection to such a step increase is that countries would not be able to spend the money productively due to constraints in their capacity to scale up public investments. As this report emphasizes, this is a valid concern that will frequently need to be addressed. For several reasons, however, countries like Ghana will be able to absorb the projected step increase in financing. Chief among them are policy changes that should be implemented rapidly with the existing administrative capacity. A significant amount of incremental financial support is often needed simply to abolish fees for primary schools and healthcare for the poor. Public sector salaries across the board will also need to be raised in many countries to improve the performance of public administration and public service delivery. Finally, unemployed teachers and medical staff should be rehired.

As described in chapter 5, several other interventions for achieving the Goals—we call them Quick Wins—can be implemented rapidly by developing countries without the need to invest in prior capacity building. Examples include providing malaria bed-nets, training community health workers, increasing access to water and electricity for schools and healthcare facilities, and maintaining core infrastructure. Since the unmet financing needs in these areas are high, we project that the step increase can be invested during 2006 (figure 17.1). Thereafter, the scaling up of public investments will follow a smoother path to allow for time to remove capacity constraints.

The macroeconomic implications of increasing aid

ODA inflows of up to 20 percent of GDP may raise macroeconomic issues that need to be managed carefully. Since much of the externally financed government spending will be devoted to the nontradable sector, an appreciation of the real exchange rate is likely. But the implications for the "Dutch disease" should not be too quantitatively significant, since much of the official development assistance will be directed to raising supply-side productivity through investments in human capital, agriculture, and infrastructure—not to a consumption boom typically linked to a squeeze in the tradable sector. Further, this strategy is completely different from an oil boom, in which additional revenues are easily dissipated by an irresponsible government. In the current case, the increased resources would be made available in rigorous, monitored investment programs. Appropriate monetary sterilization and other policy tools can further contain real exchange rate appreciation.

Most macroeconomists therefore share the view that the negative macroeconomic implications of increased assistance flows are manageable and are far outweighed by the benefits of scaled-up investments in the Millennium Development Goals—so long as the aid flows are predictable and come in the form of grants (IMF Fiscal Department, personal communication, 2004; Foster and Keith 2003; Prati, Sahay, and Tressel 2003). Likewise, the competitiveness of the private sector is unlikely to be put in jeopardy, as is sometimes feared, since most MDG-related interventions will lower the cost of doing business by improving human capital and infrastructure.

Figure 17.1

Scaling up public investments in the Goals in Ghana

2003 US$ billions

Source: Authors' calculations prepared in collaboration with the Institute of Social Statistics and Economic Research, Ghana.

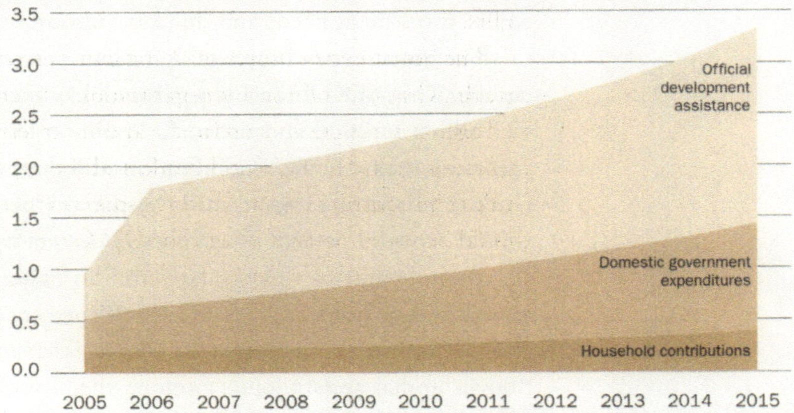

The recent experience of Mozambique underlines this point. Between 1993 and 2002 ODA inflows averaged more than 40 percent of GNI while real growth averaged an impressive 5.5 percent per capita. The ODA inflows were instrumental in achieving dramatic poverty reduction without creating any major appreciation of the real exchange rate or other macroeconomic imbalances. Other economies that have experienced very high inflows of aid over long periods and managed to sustain high growth rates are Botswana after independence, Taiwan (China) in the 1950s, and Uganda in the 1990s (Foster and Keith 2003).

We have nonetheless estimated the likely effect of a real exchange rate appreciation on the external financing required for the Millennium Development Goals. A real exchange rate appreciation of 20 percent over the 10 years from 2006 to 2015, as suggested by Prati, Sahay, and Tressel (2003), would raise the need for external finance by 7–8 percent over the period.[5] Such a modest increase will not significantly alter the results of the needs assessment. Since estimates of real exchange rate appreciation are subject to considerable uncertainty, we have not adjusted our MDG needs assessment for exchange rate effects.

Global ODA needs to meet the Goals

What are the implications of national MDG needs assessments for global ODA volumes? To answer this question we estimate the global MDG financing gap using the country-level results. We then add estimated assistance needs for debt relief, technical cooperation, and other MDG-related expenditures to the MDG financing gap. Finally, we provide an indicative estimate of the total ODA envelope required to finance the Goals through 2015. We include adjustments for countries not qualifying due to inadequate governance and for efficiency gains from improved aid allocation and effectiveness. (Appendix 3 contains a detailed description of the key assumptions.)

The aggregate MDG financing gap

We estimate the global MDG financing gap as the difference between total MDG investment needs and domestic resource mobilization, assuming a rise in government expenditures of up to four percentage points of GDP over the decade. The MDG financing gap for all low-income countries will amount to $73 billion in 2006 and rise to $135 billion by 2015 (table 17.2). Thanks to rising domestic resource mobilization, the share of official development assistance in financing incremental investments (that is, the MDG financing gap) will fall from 59 percent in 2006 to 32 percent by 2015.

Many countries—particularly in Sub-Saharan Africa—will require sustained budget support of more than 10 percent of GDP through 2015 (map 17.1). The map underscores that because of rising incomes several countries, Bolivia, India, and Indonesia among them, will graduate from the need for official development assistance before 2015.

Table 17.2

Cofinancing the MDGs in low-income countries

2003 US$ billions

	2002[a]	2006	2010	2015
MDG financing gap, 2006–15				
MDG investment needs	149	253	348	529
Domestic resource mobilization	137	180	259	394
MDG financing gap	12	73	89	135
Increment over 2002				
MDG investment needs		104	200	380
Domestic resource mobilization		43	122	257
MDG financing gap		61	78	123
Share of increment over 2002 (percent)				
MDG investment needs		100	100	100
Domestic resource mobilization		41	61	68
MDG financing gap		59	39	32

Note: Refer to appendix 3 for more details.

a. For 2002 we report actual expenditures and ODA for direct MDG support.

Source: Authors' calculations.

Map 17.1

MDG financing gap, 2015

Share of GDP (%)

Source: Authors' calculations.

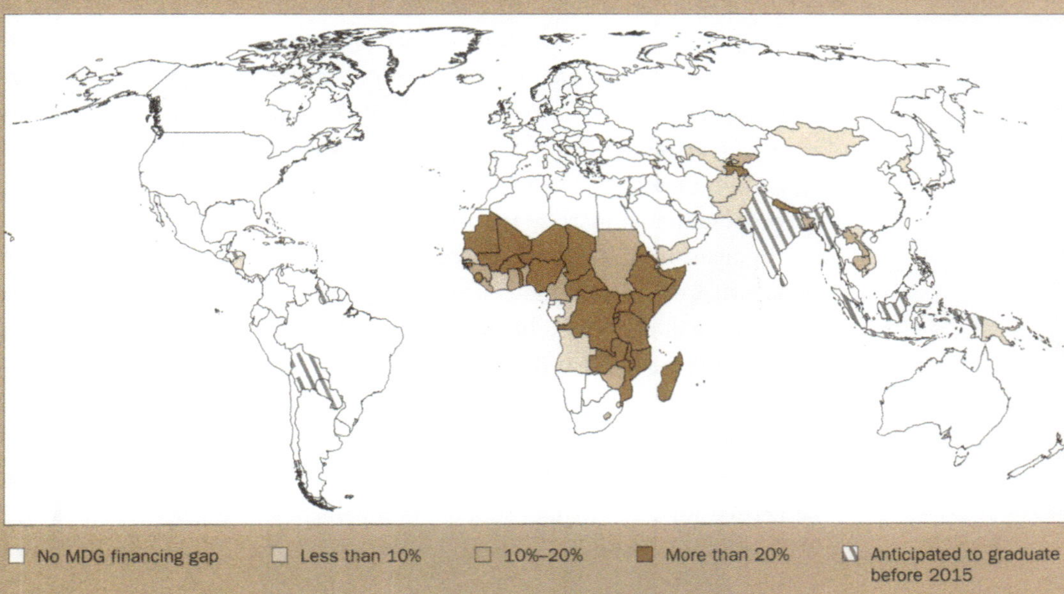

The cost of meeting the Millennium Development Goals in every country

We project that the cost of meeting the Goals in all countries will amount to $121 billion in 2006 and $189 billion in 2015 (table 17.3). This compares with 2002 official development assistance of roughly $28 billion in support of the Goals (out of $65 billion in total ODA).[6] The projections cover the MDG financing gap in all countries as well as the cost of financing capacity building for the Goals, debt relief, and grants in support of heavy debt burdens. We assume that the poorest countries will require outright cancellation of their debt to be able to achieve the Goals (chapter 13). The granting of debt relief should of course be contingent on countries committing themselves to credible strategies for investing the proceeds in the Goals. Countries likely to graduate before 2015 will need grants to finance loan repayments only if the sum of their debt service payments and the MDG investments exceeds domestic resource mobilization.

While middle-income countries will be able to finance most MDG investments through domestic resource mobilization alone, we expect that overcoming entrenched "pockets of poverty" will require international assistance of $10 billion a year in addition to the current level of capacity building supported by bilateral and multilateral agencies. Since our MDG needs assessments have not focused on the MDG investment needs of middle-income countries, this estimate would need to be refined further through detailed country estimates.

We estimate that the $2 billion in current assistance flows for regional infrastructure and cooperation (Birdsall 2004) will need to rise to $11 billion by 2015. As described in chapter 15, additional annual funding of $7 billion is also required by 2015 to expand global research for the Millennium Development Goals. We have included four main areas in our estimates: public health, agriculture and natural resource management, energy technologies, and adaptation to climate change. Investments in many cases should target existing research centers or networks, such as the CGIAR system, which leads agricultural research for the needs of developing countries and whose budget we propose should increase to $1 billion annually.

Implementing the three Rio conventions also requires more funding. We project that the cost of implementing the Convention to Combat Desertification alone will reach $5 billion annually by 2015.[7] Finally, we include the cost of operating the international agencies of the UN system, which must provide enhanced technical support and training programs to assist countries in their pursuit of MDG-based poverty reduction strategies.

In aggregate, the bulk of additional official development assistance will be needed for direct MDG support to low-income countries (see table 17.3). Middle-income countries will also require an increase in net assistance flows—albeit a very modest one.

Implications for total official development assistance

With the cost of meeting the Millennium Development Goals at the country level known, it is possible to ask how the global ODA envelope will need to change to meet them. We estimate that global assistance will need to roughly double from $69 billion in 2003 (and $65 billion in 2002) to $135 billion in 2006, rising thereafter to $195 billion by 2015 (table 17.4). Projected official development assistance is high in absolute terms, but since rich countries'

Table 17.3

Estimated cost of meeting the Millennium Development Goals in all countries

2003 US$ billions

— Not available.

Note: Numbers in table may not sum to totals because of rounding. Refer to appendix 3 for more details.

Source: 2002 data based on OECD/DAC 2004d. Projections for 2006–15 are authors' calculations.

Category	Estimated ODA in 2002	Projected for 2006	Projected for 2010	Projected for 2015
MDG support needs in low-income countries				
MDG financing gap	12	73	89	135
Capacity building to achieve the MDGs	5	7	7	7
Grants in support of heavy debt burden	—	7	6	1
Debt relief	4	6	6	6
Repayments of concessional loans	–5	0	0	0
Subtotal	**15**	**94**	**108**	**149**
MDG support needs in middle-income countries				
Direct support to government	4	10	10	10
Capacity building to achieve the MDGs	5	5	5	5
Repayments of concessional loans	–6	–3	–4	–6
Subtotal	**3**	**12**	**11**	**9**
MDG support needs at the international level				
Regional cooperation and infrastructure	2	3	7	11
Funding for global research	1	5	7	7
Implementing the Rio Conventions	1	2	3	5
Technical cooperation by international organizations	5	5	7	8
Subtotal	**10**	**15**	**23**	**31**
Estimated cost of meeting the MDGs in all countries	**28**	**121**	**143**	**189**

Table 17.4

Plausible ODA needs to meet the Millennium Development Goals

2003 US$ billions

na Not applicable.

Note: Numbers in table may not sum to totals because of rounding. Refer to appendix 3 for more details.

a. Includes assistance that does not contribute directly to the Goals and operating expenditures of donor agencies.

b. Does not include several important ODA needs, such as responding to crises of geopolitical importance (such as in Afghanistan or Iraq), mitigating the impact of climate change, protecting biodiversity and conserving global fisheries, and so on.

Source: 2002 data based on OECD/DAC 2004d. Projections for 2006–15 are authors' calculations.

	Estimated ODA in 2002	Projected for 2006	Projected for 2010	Projected for 2015
Baseline ODA for the Goals in 2002	28	28	28	28
Incremental MDG investment needs	na	94	115	161
Adjustment for countries not qualifying due to inadequate governance	na	–21	–23	–25
Reprogramming of existing ODA	na	–6	–7	–9
Emergency and distress relief	4	4	5	6
Other ODA[a]	34	36	34	35
Total indicative ODA needs for the Goals[b]	**65**	**135**	**152**	**195**
Share of OECD/DAC countries' GNI (percent)	**0.23**	**0.44**	**0.46**	**0.54**
ODA to Least Developed Countries (% of OECD/DAC countries' GNI)	0.06	0.12	0.15	0.22
Absolute increase in ODA required (compared with 2002)	na	70	87	130
Difference between total ODA needs and existing commitments	na	**48**	**50**	**74**

income will grow over the 10 years,[8] the cost of meeting the Goals in all countries with adequate governance corresponds to 0.44 percent of OECD countries' GNI in 2006 and 0.54 percent in 2015 (compared with 0.23 percent in 2002 and 0.25 percent in 2003)—well below the 0.7 percent target that rich countries have committed themselves to (box 17.3).

We emphasize that overall assistance needs are likely to be higher since our estimates cover only investments that contribute directly to achieving the Millennium Development Goals. For example, we exclude official development assistance to stabilize greenhouse gas concentrations, to protect global fisheries, to countries of special geopolitical importance, and so forth. So, total ODA needs will likely approach the 0.7 percent target.

Box 17.3

The 0.7 percent ODA target and the Millennium Development Goals

Although the UN Millennium Project focuses its ODA needs assessments on country-level estimates of the assistance required to achieve the Goals, we do so within the context of developed countries' long-established international target of providing 0.7 percent of their national income as ODA. 2005 marks 35 years since this target was first affirmed by UN member states in a 1970 General Assembly Resolution:

> "In recognition of the special importance of the role that can be fulfilled only by official development assistance, a major part of financial resource transfers to the developing countries should be provided in the form of official development assistance. Each economically advanced country will progressively increase its official development assistance to the developing countries and will exert its best efforts to reach a minimum net amount of 0.7 percent of its gross national product at market prices by the middle of the decade." (UN 1970, paragraph 43)

This first deadline passed. Having fallen from 0.51 percent as a share of donor GNP in 1960 to 0.33 percent in 1970, ODA reached 0.35 percent in 1980. By 1990 ODA was at 0.34 percent and then fell to 0.23 percent in 2002, the same year the 0.7 target was reconfirmed by all countries in the Monterrey Consensus (OECD 2004d).

So far, only five countries have met or surpassed the 0.7 target: Denmark, Luxembourg, the Netherlands, Norway, and Sweden. In the past two years, however, six other countries have committed themselves to specific timetables to achieving the target before 2015: Belgium, Finland, France, Ireland, Spain, and the United Kingdom. Thus nearly half the membership of the OECD's Development Assistance Committee has now set a firm timetable for reaching 0.7 percent. The UN Millennium Project urges all developed countries to follow through on the Monterrey commitment "to make concrete efforts towards the target of 0.7." We urge that "concrete efforts" require a specific timetable, and specifically a timetable before 2015, the target date for the Goals.

The confluence of the 0.7 target and the Goals is an important one. As this report outlines, ours is the first generation in which the world can halve extreme poverty within the 0.7 percent envelope. In 1975, when the donor world economy was around half its current size, the Goals would have required much more than 1 percent of GNP from the donors. Today, after two and a half decades of sustained economic growth in developed countries, the Goals are utterly affordable. No new promises are needed—only a follow-through on commitments already made.

After adding incremental ODA needs (estimated in table 17.3) to baseline ODA, we adjust for countries unlikely to meet the minimum standards of good governance that form the basis of the Monterrey Consensus and are necessary before MDG interventions can be scaled up. To do this, we subtract $21 billion in 2006, or 28 percent of the aggregate MDG financing gap. By 2015 we assume that more countries will have passed this threshold and therefore subtract a smaller financing share for that year ($25 billion, or 19 percent of the MDG financing gap).

Next, we project that greater harmonization and reallocation of existing official development assistance can increase net assistance for the MDGs. By untying aid, aligning official development assistance with government priorities, and shifting away from project implementation toward program funding—as called for in the Rome harmonization agenda—donors can generate major savings. We assume that 20 percent of development assistance that is not currently directed toward the Goals, emergency assistance, or the operation of bilateral agencies can be redirected toward the Goals in 2006. By 2015 the share will rise to 30 percent. This lowers total ODA needs by $6–$9 billion.

We then add official development assistance for emergency assistance to support countries in crisis or experiencing natural disasters. In the face of rising population numbers and the growing impact of climate change, we project that emergency assistance will need to increase by 50 percent over the coming 10 years. A share of this funding should support the UN Immediate Response Account, which is currently underfunded.

The final item, "other ODA," includes a basic extrapolation of various kinds of official development assistance that are not related to the Goals (such as aid to countries of geopolitical importance for needs not covered in our costing). This is not a comprehensive measure of non-MDG needs. It is simply a baseline calculated on the basis of current ODA. We project that actual ODA needs for non-MDG-related goals—such as postwar reconstruction, the consolidation of new democracies, or the mitigation of climate change—will be considerably higher than this line.

The Monterrey Consensus calls for increasing official development assistance to Least Developed Countries to 0.15 to 0.20 percent of rich countries' GNI (UN 2002a). Consistent with this objective, our estimates suggest that aid to Least Developed Countries will increase from currently 0.06 percent to 0.12 percent in 2006 and 0.22 percent by 2015.

Despite some caveats (box 17.4), these estimates highlight three important insights about official development assistance (table 17.5).

- First, the Millennium Development Goals can be met within the 0.7 percent of GNI target that all developed countries endorsed in Monterrey. But due to current shortfalls from that target, ODA volumes need to rise beyond the commitments already made by donors—by $48 billion in 2006.

Box 17.4

Some caveats for the projections of official development assistance needs

The ODA estimation methodology presented here represents a pioneering effort at a bottom-up, integrated, country-level approach to MDG needs assessments. Even so, the results are subject to several sources of uncertainty and should not be interpreted as a definitive point estimate of MDG investment needs. Since only limited data are available on marginal costs and how they change as investments reach greater shares of a population, it is difficult to project the actual costs of service delivery into the future. We also assume that governments and donors alike can provide all investments efficiently. Failing to do so may have a substantial effect on the projected MDG financing gap. And although our analysis does account for many of the most important synergies across intervention areas, only a real-time scaling up of all interventions will show how they interact quantitatively.

The global ODA projections extrapolate results for five countries that have since been validated in a number of other countries. Any such extrapolation is an estimate at best, since a true global needs assessment would require detailed within-country assessments for every developing country. Moreover, we do not account for the possible impact of major events that might affect the cost of achieving the Millennium Development Goals globally or in specific regions—such as major natural disasters, armed conflict, climate change, or major financial crises. Any one of them could substantially alter the results here.

The projections of global official development assistance constitute our best estimate of what donors must be prepared to finance if they are to engage in honest discussions with countries about how to meet the Goals. They lay out the full set of "line items" to be adequately funded to achieve the Goals. Since the focus of the UN Millennium Project has been on quantifying MDG investment needs at the country level, our projections of debt relief required to meet the Goals, the need for enhanced emergency assistance, and regional MDG investment needs cannot substitute for a more detailed analysis.

- Second, the quality of ODA needs to improve substantially. Most incremental aid needs to be provided in the form of budget support or sector-wide approaches to support the scaling up of national programs under the MDG-based poverty reduction strategies. No distinction should be made between aid funding for capital and recurrent costs, because both need to be fully financed to meet the MDGs. This point is critical since current ODA is rarely used to support operating expenses, such as doctors' salaries, preservice training, or the maintenance of core infrastructure.
- Third, new ODA for the Goals must be much better targeted than is currently the case. While middle-income countries require enhanced debt relief and some additional aid, the bulk of official development assistance must focus on low-income countries. Funding for regional infrastructure and collaboration, as well as global scientific research, must also rise sharply.

Financing mechanisms for increasing the ODA envelope

MDG needs assessments call for a step increase in net official development assistance from $69 billion in 2003 to $135 billion in 2006, compared with existing commitments made by OECD/DAC member countries to increase

Table 17.5

Estimated ODA flows and gaps of Development Assistance Committee members based on existing commitments

Constant 2003 US$ billions

Country	Assistance at 0.44 percent of 2006 GNI[a]	Assistance commitment for 2006	Gap in 2006	Assistance at 0.54 percent of 2015 GNI[a]	Assistance commitment for 2015	Gap in 2015
Australia[b]	2.4	1.4	1.0	3.4	1.6	1.8
Austria[c]	1.2	0.9	0.3	1.7	1.1	0.7
Belgium[d]	1.5	2.1	none	2.1	2.7	none
Canada[e]	4.2	2.6	1.7	6.1	3.7	2.4
Denmark[f]	1.0	1.8	none	1.5	2.2	none
Finland[d]	0.8	0.7	0.1	1.1	0.9	0.2
France[g]	8.5	8.8	none	12.3	15.6	none
Germany[c]	11.6	8.4	3.2	16.7	10.0	6.7
Greece[c]	0.9	0.6	0.2	1.3	0.8	0.5
Ireland[h]	0.6	0.8	none	0.9	1.1	none
Italy[c]	7.0	5.1	1.9	10.1	6.1	4.1
Japan[i]	19.9	9.5	10.4	28.7	9.5	19.2
Luxembourg[j]	0.1	0.2	none	0.2	0.3	none
Netherlands[k]	2.4	4.2	none	3.5	5.1	none
New Zealand[l]	0.4	0.2	0.2	0.5	0.2	0.3
Norway[m]	1.1	2.4	none	1.6	2.8	none
Portugal[c]	0.7	0.5	0.2	1.0	0.6	0.4
Spain[g]	4.1	2.9	1.1	5.9	7.5	none
Sweden[n]	1.5	3.2	none	2.1	3.8	none
Switzerland[o]	1.6	1.4	0.3	2.4	1.7	0.6
United Kingdom[p]	9.2	8.5	0.7	13.2	16.8	none
United States[q]	54.5	22.3	32.2	78.7	27.2	51.5
Total	**135.0**	**88.4**		**195.0**	**121.5**	

Note: Numbers in table may not sum to totals because of rounding.

a. Assumes 2 percent real annual GNI growth.
b. 2006 ODA/GNI target held constant through 2015.
c. 2006 ODA/GNI target held constant at 0.33 percent through 2015.
d. 2010 ODA/GNI target of 0.7 percent; commitment level held constant through 2015.
e. Assumes 5.5 percent real annual ODA increase (8 percent nominal increase less 2.5 percent inflation) through 2010; 2010 ODA/GNI target held constant through 2015.
f. 2006 ODA/GNI target of 0.83 percent held constant through 2015.
g. 2012 ODA/GNI target of 0.7 percent through 2015.
h. 2007 ODA/GNI target of 0.7 percent held constant through 2015.
i. 2006 ODA level held constant at $9.5 billion through 2015.
j. 2006 ODA/GNI commitment of 1 percent held constant through 2015.
k. ODA/GNI target of 0.8 percent held constant through 2015.
l. Assumed 2006 ODA/GNI level of 0.26 percent held constant through 2015.
m. 2005 ODA/GNI target of 1 percent held constant through 2015.
n. 2006 ODA/GNI target of 1 percent held constant through 2015.
o. 2010 ODA/GNI target of 0.4 percent held constant through 2015.
p. 2013 ODA/GNI target of 0.7 percent held constant through 2015.
q. Assumes 3 percent real GNI growth to 2006; for 2006 includes over $1.5 billion for the Millennium Challenge Account, nearly $2 billion for the Global AIDS Initiative, increased multilateral aid, and rephased expenditure on reconstruction in Iraq. ODA/GNI share projected to remain constant through 2015.

Source: OECD/DAC forthcoming.

ODA to $88 billion in 2006 (see table 17.5). The shortfall of roughly $48 billion is projected to stay constant through 2010 and may rise to $74 billion by 2015, assuming that assistance volumes increase according to the commitments already made.

Can donors finance the additional increase to raise official development assistance from 0.25 percent of their GNI in 2003 to 0.44 percent in 2006? The most direct way to further increase ODA volumes is of course to allocate increasing shares of national budgets to official development assistance. We recognize, however, that some donor countries face short-term fiscal constraints. While we urge all developed countries to commit to a specific year by which to achieve the 0.7 percent target they have set for themselves, other innovative financing mechanisms may be necessary—as discussed by Atkinson (2004) and Landau (2004). Prominent among recent suggestions are international taxation on financial transactions or carbon emissions, the use of IMF special drawing rights, and the International Finance Facility (IFF) proposed by the United Kingdom. We consider the IFF to be the most advanced and immediately practicable of all of the proposals.

The IFF would be a temporary financing mechanism to at least double development assistance between now and 2015. Importantly, we interpret "doubling" to mean a doubling of the share of ODA in donoı GNP—that is, to reach at least 0.54 percent of donor GNP for the MDGs, rather than simply doubling the current dollar level of aid flows. The IFF will leverage additional money from the international capital markets by issuing bonds, based on legally binding long-term donor commitments. It responds to the need for the rapid scaling up, or "frontloading," of development assistance without placing undue constraints on rich countries' budgets. It also permits donor countries to achieve the overall ODA target of 0.7 percent of GNI by 2015.

In contrast to other proposals, the IFF can be rapidly implemented and does not depend on participation by all high-income countries. It offers the flexibility to align the level of financing with actual assistance needs by adjusting the issuance of bonds. And it enables donors to channel the funds through a range of disbursement mechanisms that can provide high-quality assistance, such as direct budget support, the International Development Association, the development funds of the regional development banks, and the European Development Fund. We encourage all developed countries to support the IFF in 2005, in time to start disbursing funds in 2006.

Immediate ODA needs for 2005

Further to the call to action in chapter 16, as part of the step increase of official development assistance by 2006, we call on donors to ensure that increased assistance is made available in 2005 for the following urgent categories:

- Assistance to countries and international agencies to enable all countries that wish to do so to prepare MDG-based poverty reduction strategies in 2005–06.
- Financing for the Global Fund to Fight AIDS, Tuberculosis, and Malaria, to fund the Quick Wins in HIV/AIDS ("3 by 5"), malaria (bed-nets and effective medicines), and TB control (DOTS).
- Dropping user fees for primary schools and essential health services, rehiring unemployed teachers and medical staff, and raising public sector salaries in developing countries as needed.
- Large-scale training, particularly for community health workers, agricultural extension workers, and community-based experts in infrastructure, to commence in 2005.
- Support for at least one dozen MDG fast-track countries in scaling up MDG-related investments beginning in 2005.

Chapter 18

The benefits: the case for a decade of bold ambition

The Millennium Development Goals lay out a challenging and achievable vision for dramatically reducing extreme poverty by 2015, with tremendous benefits for the entire world. This report presents a roadmap for achieving the Goals and for effecting a dramatic—if partial—global reduction in poverty. The danger of inaction is tremendous, not only in lives lost and opportunities forgone, but also in threatening the security of everyone. The world must act now to begin a decade of bold ambition, starting with a dramatic scaling up of assistance to well governed countries in 2005 and ending with achieving the Goals in 2015.

Dramatic—but only partial—poverty reduction around the world

If the Millennium Development Goals are met, poverty in all its dimensions will be dramatically reduced. A simple analysis helps to tell this story. The difference between extrapolating current trends on a "business as usual" path since the MDG baseline year of 1990 and a simple linear path with every country achieving the Goals (table 18.1)[1] shows that, compared with 2005, if the Goals are met by 2015 approximately 500 million people will be lifted out of extreme poverty and more than 300 million will no longer suffer from hunger. Three hundred and fifty million fewer people will lack access to safe drinking water, and 650 million people otherwise left unserved will enjoy the benefits of basic sanitation to lead healthier and more dignified lives. Over the decade, the lives of 30 million young children will be saved, and more than 2 million women's deaths from pregnancy-related causes will be averted. Not captured in such a table is the environmental degradation reversed, the many millions of children with opportunities for a better future through education, or the hundreds of millions more women and girls who will lead their lives in freedom, with more security and more opportunity—if the Goals are achieved. Behind these achievements lie the lives and hopes of people with new opportunities to end the burden of grinding poverty.

Table 18.1

The benefits of meeting the Millennium Development Goals, by developing region

* Region on track to meet MDG target.

Note: Numbers in table may not sum to totals because of rounding. See appendix 4 for methodology of calculations.

Source: Poverty headcount data from Chen and Ravallion 2004. GDP per capita and child mortality data from World Bank 2004c. Undernourishment data from FAO 2003a. Maternal mortality data from WHO and UNICEF 1996 and WHO, UNICEF, and UNFPA 2003. Water and sanitation data from WHO and UNICEF 2004. HIV/AIDS data from Stover and others 2002. Slum dweller data from United Nations Population Division 2001, 2003a and UN-HABITAT 2003. All population projections from United Nations Population Division 2003a.

Poverty headcount (millions)

Region	2005 estimate	Current trend extrapolated to 2015	MDG scenario for 2015
Eastern Europe and Central Asia	92	88	49
East Asia and the Pacific*	182	0	0
Latin America and the Caribbean	128	123	90
Middle East and North Africa	8	9	4
South Asia*	407	317	317
Sub-Saharan Africa	345	431	198
Total	1,162	968	658

GDP per capita (2003 US$)

Region	2005 estimate	Current trend extrapolated to 2015	MDG scenario for 2015
Europe and Central Asia	2,980	3,827	4,084
East Asia and the Pacific*	1,313	2,139	2,139
Latin America and the Caribbean	3,724	4,104	5,102
Middle East and North Africa	2,447	2,727	3,352
South Asia*	602	980	980
Sub-Saharan Africa	520	509	712

Individuals suffering from undernourishment (millions)

Region	2005 estimate	Current trend extrapolated to 2015	MDG scenario for 2015
Europe and Central Asia	52	61	20
East Asia and the Pacific*	162	65	65
Latin America and the Caribbean*	49	38	38
Middle East and North Africa	32	46	14
South Asia	301	285	228
Sub-Saharan Africa	228	255	155
Total	824	749	520

Child mortality (millions of lives lost)

Region	2005 estimate	Current trend extrapolated to 2015	MDG scenario for 2015
Europe and Central Asia	0.2	0.1	0.1
East Asia and the Pacific	1.1	0.7	0.6
Latin America and the Caribbean*	0.3	0.1	0.1
Middle East and North Africa*	0.4	0.2	0.2
South Asia	3.1	2.0	1.6
Sub-Saharan Africa	4.7	4.7	1.9
Total	9.8	7.9	4.4

Maternal mortality (millions of lives lost)

	2005 estimate	Current trend extrapolated to 2015	MDG scenario for 2015
Total	0.54	0.54	0.15

(continued on next page)

Table 18.1
The benefits of meeting the Millennium Development Goals, by developing region
(continued)

New HIV infections 2002–10 (millions)

Region	Current trend extrapolated to 2010	Expanded response scenario for 2010
Eastern Europe and Central Asia	2.8	1.3
Latin America and the Caribbean	2.3	0.7
Middle East and North Africa	0.9	0.3
South and Southeast Asia	18.5	5.7
Sub-Saharan Africa	21.0	8.8
Total	45.5	16.8

Individuals without access to improved water supply (millions)

Region	2005 estimate	Current trend extrapolated to 2015	MDG scenario for 2015
Europe and Central Asia*	23	10	10
East Asia and the Pacific	388	305	299
Latin America and the Caribbean*	49	16	16
Middle East and North Africa	40	46	26
South Asia*	160	0	0
Sub-Saharan Africa	280	270	230
Total	939	647	581

Individuals without access to improved sanitation (millions)

Region	2005 estimate	Current trend extrapolated to 2015	MDG scenario for 2015
Europe and Central Asia	69	73	32
East Asia and the Pacific*	873	608	608
Latin America and the Caribbean	128	107	103
Middle East and North Africa	80	84	61
South and Southeast Asia	877	770	718
Sub-Saharan Africa	454	531	305
Total	2,481	2,172	1,827

Individuals living in slum conditions (millions)

Region	2001	Current trend extrapolated to 2020	MDG scenario for 2020
Europe and Central Asia	44	47	41
East Asia and the Pacific	237	385	210
Latin America and the Carribean	128	173	116
Middle East and North Africa	61	97	54
South Asia	235	398	207
Sub-Saharan Africa	167	325	144
Total	872	1,425	772

While global aggregates can obscure significant regional variation, table 18.1 highlights the range of benefits the Goals will bring and the acceleration of progress needed. For example, progress toward the child health Goal is well off track, and the benefits of dramatically increasing current progress in child health are extraordinary. Today nearly 10 million children die annually in developing countries. If mortality declines continue at current rates, 8 million children will still die in 2015, most from completely preventable causes. But by reaching the MDG target in 2015, the cumulative number of children's lives saved will be about 30 million, or 20 million compared with the trajectory of progress (figure 18.1).

The benefits of meeting the Millennium Development Goals, compared with business as usual, differ across regions. For example:

- Sub-Saharan Africa will end years of regress across many human development outcomes and make the biggest improvements by achieving the Goals. Its fast-growing population means that although the share of extreme poor will have halved, the numbers will still be large.
- Latin America and the Caribbean will see a substantial reduction in poverty, slums, and the number of new HIV infections.
- East Asia and the Pacific is the only region that has already met the poverty and other targets, but progress is uneven, with many countries falling short. Meeting the Goals will spread progress to laggard countries in the region, generate major improvements for the environment, and reduce the number of slum dwellers.
- South Asia is on track to meet the targets on income poverty and access to improved water supply. By achieving the Goals it will also prevent a large number of new infections of HIV, reduce hunger, cut child mortality, and substantially reduce the number of people living in slums.
- The Middle East and North Africa stands to make great progress in reducing the number of people suffering from undernourishment, living in slums, or without access to safe water and basic sanitation.

Figure 18.1

Under-five mortality

Lives lost each year (millions)

Source: World Bank 2004c; authors' calculations.

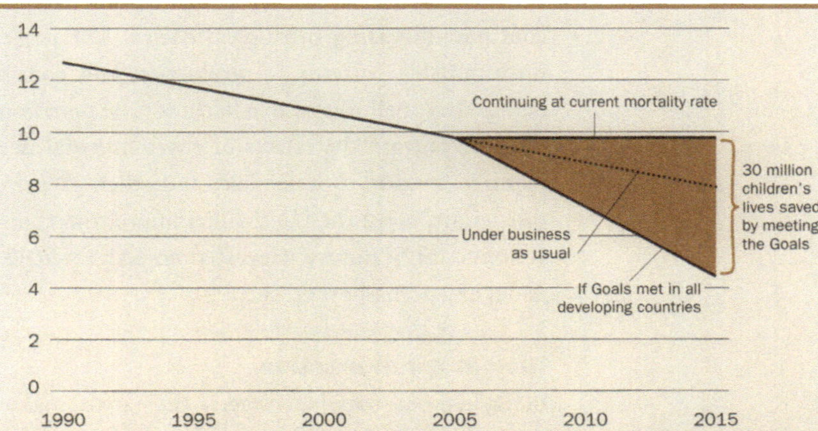

- Substantial reductions in the number of people infected with HIV and without access to sanitation will be achieved in Eastern Europe and Central Asia, where the Goals also offer a framework for reducing hunger, recently on the rise across the region.

We emphasize that the paucity of available data gives only a partial picture. In particular, progress in gender equality, universal primary education, and environmental sustainability are not reported in table 18.1. Achieving all of these Goals will further transform the lives of hundreds of millions of people and families.

The need for ongoing assistance beyond 2015

Table 18.1 underscores a deep truth about the Goals. They aim to cut poverty by half, rather than to eliminate it. They are a midstation to ending absolute poverty, but they are definitely not the final destination. Even if the Goals are achieved in every country, extreme poverty will remain a major issue requiring ongoing attention. As just one example, even if the sanitation target is met, as many as 1.8 billion people will still need to resort to open defecation or use inadequate facilities. Well over half a billion people will still live on less than a dollar a day.

For these people, those for whom even the Goals are not enough, sustained global partnership will be needed. While a scale-up of high-quality development assistance will allow many countries to graduate from the need for large-scale external budget support by 2015, many countries will still require ongoing support equal to 10–20 percent of their GDP to graduate from external assistance sometime later—likely by 2025 (see map 17.1). Until then, and to eventually make development assistance obsolete, sustained ODA will be crucial. To that end high-income countries will need to maintain support at close to 0.7 percent of their GNI for some period beyond 2015. By 2015 extreme poverty can be cut by half. By 2025 extreme poverty can be substantially eliminated.

Responding to other long-term global challenges will also be crucial for development. For instance, climate change poses a major threat to small island states threatened by rising sea levels and to countries that will experience increases in climatic variability. The frequency of drought is likely to continue increasing in tropical Africa. The depletion of global fisheries and unsustainable patterns of production and consumption must be addressed before they spell irreparable harm for the poor countries that can least protect themselves from the effects of environmental degradation. Foresight will be required to address these issues and others. We do not focus on all of them in this report, given the UN Millennium Project's mandate to produce a plan of action to 2015. But we stress that no plan for 2015 should overlook the serious issues extending beyond then.

The danger of inaction

In laying out tangible targets, the Goals make explicit the most obvious costs of inaction—in terms of lives and opportunities lost. They also form a centerpiece for the world's security agenda. As the Secretary-General's High-Level Panel on Threats, Challenges, and Change and many others have noted, human development and environmental management are intricately linked to peace and security. Only by reducing poverty and improving environmental management over the coming decades can a rise in conflicts and state failures be averted. If the Goals are not met, millions will die who would otherwise live. Countries that would be stable will descend into conflict. And the environment will continue to be degraded. In short, many crises we face today will only be more pronounced and expensive to resolve in 10 years unless the world starts investing in the MDGs straight away.

At stake is the credibility and functioning of the international development system. Without a breakthrough in 2005, if poor countries that have met their Monterrey commitments are not effectively supported in pursuing an MDG-oriented strategy, the already dwindling faith in international commitments to reduce poverty will likely vanish. In that case, the Goals will not be met and the trust in rich countries' sincerity to support well governed countries in need of external assistance will be deeply, if not fatally, damaged.

If we do not act now, the world will live without development goals—seriously threatening the stability of poor countries, but also endangering the interests of a rich world that cannot insulate itself from the consequences of extreme poverty, instability, disease, and environmental degradation.

A decade of bold ambition

Meeting the Goals in all countries poses a tremendous challenge to the international community, but with the best efforts on all sides, they can be achieved. To ensure success, in 2005 the world must start training to build capacity, improve policies, and deliver the investments needed to meet the Millennium Development Goals. This effort will need to be sustained at the global, national, and local levels over the next 10 years. Only by acting now can sufficient numbers of doctors or engineers be trained, service delivery capacity strengthened, and infrastructure improved to meet the Goals.

Fortunately the costs of achieving the Goals are entirely affordable and well within the promises of 0.7 percent of GDP made at Monterrey and Johannesburg. The required doubling of annual official development assistance to $135 billion in 2006, rising to $195 billion by 2015, pales beside the wealth of high-income countries—and the world's military budget of $900 billion a year. Indeed, the increased development assistance for the Goals will only amount to one half of one percent of rich countries' combined income.

Of course money alone will not be sufficient to meet the Goals, but increased investments need to accompany institutional reform, trade liberalization, good policies, and increased efforts to improve the accountability of governments. Unless more financial assistance is available effectively, low-income countries and their development partners will not be able to have serious and honest discussions about the reforms and investments required to meet the Goals. For countries where such reforms and good policies are in place, high-income countries should, in the spirit of the Monterrey Consensus, make good on their commitment to increasing ODA for them to work toward achieving the Millennium Development Goals.

The risks and benefits of achieving the Goals will be shared across the entire world, so genuine international leadership and responsibility will need joint action on both development and security policy. In line with the High-Level Panel's recommendation, developed countries aspiring to permanent seats on the UN Security Council, for example, should be prepared to fulfill the commitment to 0.7 percent of GNP in ODA by 2015 as part of their leadership responsibilities.

Urgent action is needed if we are to usher in a decade of bold ambition to achieve the Millennium Development Goals. Developing countries need to make every effort to mobilize around the Goals. Rich countries need to ask themselves if they should be more concerned, as many of them are today, with pointing fingers at the responsibilities of poor countries than with meeting their own commitments. In 2005 the world needs desperately to follow through on its commitments, taking quick practical steps at scale before the Goals become impossible to achieve. If we fail to invest now, it will be a very long way to the next Millennium Summit in the year 3000.

Appendix 1

Millennium Development Goals interventions by area

Intervention area 1: investments in rural development

Investments in rural development include interventions in poverty and hunger reduction; domestic water supply, sanitation, and water management infrastructure; rural transport; and rural energy services.

Poverty and hunger reduction

Increasing agricultural productivity

Investments in soil health. Combinations of mineral fertilizers, agroforestry (use of trees to replenish soil nutrients), green manures, cover crops, return of crop residues, and soil erosion control, as appropriate, depending on soil characteristics, partly financed by market-oriented smart vouchers to food-insecure farmers.

Small-scale water management. Development of water management techniques and structures, pumps, drip irrigation, wells, and the like, as appropriate, partly financed by market-oriented smart vouchers to food-insecure farmers.

Improved inputs. Provision of seeds of improved varieties of crops, pastures, and trees, as well as improved breeds of livestock and fish,* with delivery systems accessible to food-insecure farmers, such as community tree nurseries.

*Farm diversification.** Incentives to farmers to diversify to high-value livestock, vegetables, and tree products, once they are food-secure.

Extension services. Strengthening of extension services with village-level paraprofessionals that have a strong participatory approach and up-to-date knowledge of soil health, small-scale water management, improved germplasm, high-value products, and other ecologically sound agricultural techniques.

* *Interventions not included in the MDG needs assessment (chapter 17)*

Agricultural research. Increased investments in national research systems for agriculture and natural resource management to 2 percent of agricultural GDP.

Special interventions to reach women farmers. Recruitment and training of women extension workers; provision of inputs (seeds, fertilizers, implements) targeted to reach women; promotion of women's property rights to land, water, trees, and fisheries, and access to information on agriculture, nutrition, marketing, finance, and environmental protection.

Linking farmers to markets

*Storage, marketing, and agroprocessing facilities.** Construction of warehouses to reduce postharvest losses, construction of market spaces, provision of training and equipment to encourage small-scale agroprocessing industries in rural areas, supporting shifts to high-value farming and skill building, supporting rural input traders, and providing access to market information.

*Agrodealer networks.** Fostering local agrodealers to sell fertilizers, seeds for agroforestry, green manure, water management equipment, and improved seeds, redeeming smart vouchers and receiving training from extension workers.

*Support to farmer associations.** Investments to support farmer and rural laborer associations to organize to improve negotiating price outcomes and access to markets, with emphasis on cell pones and internet access modeled after the "biovillages" in South India and the Hunger Project's "epicenters" in Africa.

*Access to credit.** Extension of the formal banking system and provision of microcredit services.

Nutrition

Nutrition for infants, pregnant women, and nursing mothers. Promotion of mother- and baby-friendly community initiatives, including exclusive breastfeeding for first 6 months and complementary feeding with continuing breastfeeding for infants ages 7–24 months. HIV-positive mothers should use replacement feeding when it is acceptable, feasible, affordable, sustainable, and safe. Provision of sufficient calories, protein, and micronutrients to pregnant women and nursing mothers, supported by nutrition extension workers and using locally produced food to the extent possible.

Nutrition for undernourished children under five years. Complementary feeding, including fortified and blended foods with take-home rations supported by nutrition extension workers.

Nutrition for school-going children. Provision of balanced school meals with locally produced foods at the primary and secondary levels.

Addressing hidden hunger. Reduction of vitamin A and iron, zinc, and iodine deficiencies by increasing the production and consumption of micronutrient-rich foods, particularly local fruits, vegetables, livestock products, iodized salt, and fortified foods from local products (such as India Mix); special attention to nutrition needs of the above groups and people living with HIV/AIDS; support to research on biofortification of food.

*Emergency food assistance**

Early warning systems. Strengthening of early warning systems to cope with natural disasters.

Emergency response. Direct food aid to areas where droughts, floods, earthquakes, and civil wars threaten the acutely hungry with starvation.

Social safety nets. Investments in social safety net solutions such as food for work, cash for work, community grain banks, and environmental rehabilitation to mitigate shocks and reduce longer term food security risks, using locally produced food to the extent possible.

Domestic water supply, sanitation, and water management infrastructure

Domestic water supply

Water supply infrastructure. Provision and operation of infrastructure for water supply (such as standpipes, boreholes, dug wells, or rainwater harvesting), including water treatment as necessary.

Water management

*Water storage and other infrastructure for water management.** Construction and operation of water storage infrastructure for drinking water supply, agricultural water use, and hydropower; extension of large-scale water harvesting.

*Integrated water resources management.** Protection and allocation of water resources to agricultural, domestic, and industrial uses, as well as environmental needs based on comprehensive assessment of renewable and nonrenewable water resources.

*Hydrological monitoring.** Operation and extension of hydrological monitoring systems.

Sanitation

Sanitation infrastructure. Construction and operation of sanitation facilities (simple pit latrines, ventilated improved pit latrines, septic tanks, flush toilets, and the like), including emptying of pits and safe disposal of sullage.

Building awareness. Targeted awareness-building measures accompanying the provision of new sanitation infrastructure to ensure the informed choice of technology options and proper use by all household members.

Hygiene education	Awareness campaigns (in primary schools, through community-based organizations, media, and so on) to promote hygienic behavior, with particular focus on hand washing and personal hygiene, as well as appropriate use of sanitation facilities and safe water storage.
	Rural transport
Transport infrastructure	*District roads.* Upgrading and construction of paved secondary or district roads.
	Feeder and community roads. Upgrading and construction of small paved roads connecting villages and farmers to the nearest district road.
	*Footpaths.** Extension and improvement of footpaths connecting individual rural farmers to feeder roads.
	Road maintenance. Institutional structure and funding arrangements for adequate road maintenance (such as dedicated road funds).
Transport services	*Vehicle supply.** Investments in supply and distribution systems for bicycles and motorized vehicles.
	*Other interventions.** Deregulation of transport market to increase competition. Support to small-scale transport entrepreneurs to reduce barriers to market entry.
	Rural energy services
Thermal energy	*Improved cooking stoves.* Distribution and maintenance or replacement of appropriate cooking stoves (ceramic stoves, liquid petroleum gas stoves (LPG), ethanol stoves, charcoal stoves, and the like).
	Modern cooking fuels. Strengthening of distribution and production systems for modern fuels (such as liquid petroleum gas, ethanol, dimethylsulfoxide, and kerosene), including safe containers.
Electricity	*Off-grid electric power systems and batteries.* Provision of diesel generators, hybrid systems, or solar home systems together with necessary wiring to schools, hospital, clinics and health centers, and other community facilities. Provision of batteries and charging stations to remote rural communities.
	Electric power generation capacity. Extension, upgrading, and maintenance of generation capacity (thermal energy plants, hydropower, or geothermal, as appropriate) to supply electricity grids.

Electric power grid. Extension of grid through high-voltage lines, medium- to low-voltage lines (including end-user connections), and other related infrastructure (such as transformer stations).

Provision of basic machinery for food processing and other motive power needs.

Intervention area 2: investments in urban development and slum upgrading

Investments in urban development and slum upgrading include interventions in urban hunger, slum upgrading, urban transport, energy services, domestic water supply and sanitation, environmental management, and industrial development.

Slum upgrading and urban planning

Slum upgrading

Housing. Incremental improvements to and construction of housing.

Infrastructure for slum upgrading. Upgrading and extension of roads and sidewalks, street lighting, storm drainage, and communications infrastructure within slums. (See below for domestic water supply, sanitation, and energy services.)

Tenure

*Security of tenure.** Improving the security of tenure through legislation against forced eviction and through legitimized occupancy or formal title.

*Enforcement of improved land tenure legislation.** Legal protection and enforcement of slum dwellers' rights.

City-wide urban planning and management

Urban infrastructure. Planning of urban infrastructure (roads, footpaths, sidewalks, street lighting, stormwater drainage, bus lanes, and other transport infrastructure). Providing health and educational systems without social or economic barriers for the urban poor.

Basic services. Provision of basic services (such as refuse collection and solid waste disposal, policing and security, and fire protection).

Urban transport

Transport infrastructure

Infrastructure for mass transport. For example, bus lanes.

Urban roads. Upgrading, construction, and maintenance of urban roads.

*Footpaths.** Extension and improvement of footpaths within cities.

Transport services

Mass transport system. Operation of bus, rail, and other mass transport systems.

Small-scale transport providers. Reduce barriers to market entry for small-scale transport providers and ensure uniform safety and regulatory standards.

Urban energy services

Thermal energy systems

Improved cooking stoves. Distribution and maintenance or replacement of appropriate cooking stoves (ceramic stoves, liquid petroleum gas stoves (LPG), ethanol stoves, charcoal stoves, and the like).

Modern cooking fuels. Strengthening of distribution and production systems for modern fuels (such as liquid petroleum gas, ethanol, dimethylsulfoxide, and kerosene), including safe containers.

Electricity

Electric power generation capacity. Extension, upgrading, and maintenance of electric power generation capacity (thermal energy plants, hydropower, or geothermal, as appropriate) to supply electric power grids.

Electric power grid. Extension of electricity grid through high-voltage lines, medium- to low-voltage lines (including end-user connections), and other related infrastructure (such as transformer stations).

Domestic water supply and sanitation

Water supply

Water supply infrastructure. Provision and operation of infrastructure for water supply (such as household connections, standpipes, or boreholes), including water treatment as necessary.

Trunk water infrastructure. Maintenance and extension of trunk infrastructure for urban water supply, including treatment facilities and reservoirs.

Other water management infrastructure

Storm drainage and flood control measures. Extension and rehabilitation of storm drainage infrastructure, including conversion of sanitation infrastructure to serve as storm drainage.

Sanitation

Sanitation infrastructure. Construction and operation of sanitation facilities (simple pit latrines, ventilated improved pit latrines, septic tanks, flush toilets, and the like) and sewers, including emptying of pits and safe disposal of sullage.

Sewage treatment. Construction and operation of simple sewage and other wastewater treatment facilities (such as waste stabilization ponds or other forms of primary treatment) where needed in dense urban settlements or because of specific environmental concerns (such as eutrophication of freshwater lakes).

Awareness building. Targeted awareness-building measures accompanying the provision of new sanitation infrastructure to ensure the choice of the adequate technology option and proper use by all household members.

Hygiene

Hygiene education. Awareness campaigns (in primary schools, through community-based organizations, media, and so on) to promote hygienic behavior, with particular focus on hand washing and personal hygiene, as well as appropriate use of sanitation facilities and safe water storage.

Urban environmental management

Pollution control

*Air pollution control.** Adoption and enforcement of regulatory standards and investments in pollution abatement technologies.

*Water pollution control.** For example, industrial wastewater treatment to complement sewage treatment, as necessary.

*Solid waste and soil pollution control.** Construction and maintenance of technically sound landfills.

Urban industrial development

Private sector development

*Industrial promotion.** Supportive policies, including tax concessions and grants, as well as provision of additional infrastructure for development of manufacturing and service industries.

*Export processing zones.** Provision of export processing zones, industrial parks, and other designated areas for private sector development.

Urban hunger

Food production

Urban agriculture. Promote urban and periurban food production, particularly of root and tuber crops, bananas, fruit trees, vegetables, and small-scale livestock.

Nutrition

Nutrition for infants, pregnant women, and nursing mothers. Promotion of mother- and baby-friendly community initiatives, including exclusive breastfeeding for first 6 months and complementary feeding with continuing breastfeeding for infants ages 7 to 24 months. Provision of sufficient calories, protein, and micronutrients to pregnant women and nursing mothers, supported by nutrition extension workers. Universal access to reproductive and sexual health services is also needed to ensure that women are able to delay first pregnancy and properly space births to avoid cumulative nutritional deficits and reduce the risk of complications for themselves and their children.

Nutrition for undernourished children under five years. Complementary feeding, including fortified and blended foods, with take-home rations supported by nutrition extension workers.

Nutrition for school-going children. Provision of balanced school meals with locally produced foods at the primary and secondary level.

Addressing hidden hunger. Reduction of vitamin A and iron, zinc, and iodine deficiencies by increasing the production and consumption of micronutrient-rich foods, particularly local fruits, vegetables, livestock products, and iodized salt and fortified foods from local products (such as India Mix); special attention to nutrition needs of the above groups and people living with HIV/AIDS; support to research on biofortification of food.

Emergency food assistance

Early warning systems. Strengthening of early warning systems to cope with natural disasters.

Emergency response. Direct food aid for areas where droughts, floods, earthquakes, and civil wars threaten the acutely hungry with starvation.

Social safety nets. Investments in social safety nets such as food for work, cash for work, community grain banks, and environmental rehabilitation to mitigate shocks and reduce longer term food security risks.

Intervention area 3: investments in the health system

Investments in the health system include interventions in child and maternal health; prevention, care, and treatment of HIV/AIDS, TB, and malaria; access to essential medicines; measures to strengthen health systems management and health services delivery; and sexual and reproductive health.

Health

Child health

Neonatal integrated package. Clean delivery, newborn resuscitation, prevention of hypothermia, kangaroo care (skin-to-skin contact), antibiotics for infection, tetanus toxoid, breastfeeding education (including education on replacement feeding for HIV-positive mothers), and hygiene education.

Integrated management of childhood illness plus immunization. Integrated approach to reduce child mortality, illness, and disability, which includes both preventive and curative elements to address leading causes of child mortality such as oral rehydration therapy and antibiotics for diarrheal disease, antibiotics for acute respiratory infection, care for measles, antimalarials for malaria, and nutritional supplements for malnutrition plus immunization.

Maternal health

Emergency obstetric care. Rapidly accessible treatment for delivery complications such as eclampsia, hemorrhage, obstructed labor, and sepsis. Emergency obstetric care requires functioning referral systems and well equipped and staffed district hospitals.

Skilled attendance, clean delivery, and postpartum care. Presence of trained and registered midwives, nurses, nurse-midwives, or doctors at birth with ability to diagnose and refer emergent complications as well as postpartum care (including counseling on nutrition, family planning, and parenthood skills*).

Antenatal care. Routine care during pregnancy, including preventive and curative interventions such as blood pressure and weight monitoring, treatment of infections, nutrition and smoking counseling, intermittent preventive treatment for malaria, and antiretrovirals for HIV-positive women to prevent mother-to-child transmission of HIV.

Safe abortion services. Access to postabortion care, access to abortion counseling and, where permitted by law, safe abortion services.

HIV/AIDS prevention

Improved linkages. Effective joint programming between reproductive health and HIV/AIDS programs.*

Behavior change programs. Programs to encourage safer sexual behavior, including condom social marketing, peer-based education, mass media campaigns, work-based programs, and school-based HIV education.

Control of sexually transmitted diseases. Routine screening and effective treatment of sexually transmitted diseases (such as syphilis, gonorrhea, and chlamydia).

Voluntary counseling and testing. Pre- and post-test counseling and HIV testing.

Harm reduction for injecting drug users. Actions to prevent transmission of HIV and other infections that occur through sharing of nonsterile injection equipment and drug preparations; specific programs include provision of sterile syringes and needles and drug substitution treatment.

Prevention of mother-to-child transmission. Prevention of transmission of HIV from infected women to their infants during pregnancy, labor, and delivery, as well as during breastfeeding (that is, replacement feeding when it is acceptable, feasible, affordable, sustainable, and safe); includes short-term antiretroviral prophylactic treatment; infant feeding, counseling, and support; and the use of safer infant feeding methods.

Blood safety interventions. Measures to reduce the risk of receiving infected blood through a transfusion, including HIV antibody screening, protocols to avoid unnecessary blood transfusions, and policies to exclude high-risk donors.

HIV/AIDS care and treatment

Antiretroviral therapy. Combination drug therapy to treat AIDS.

Treatment of opportunistic infections. Treatment of any infection caused by a microorganism that would not normally cause disease in a healthy individual.

Orphan support. Provision of support to orphans to minimize the impact of AIDS on their lives; includes school fee support, community support, and support to extended families.

TB

DOTS. Internationally recommended TB control strategy combining five elements: political commitment, microscopy services, drug supplies, surveillance and monitoring systems, and use of highly efficacious regimes with direct observation of treatment.

Adaptation of treatment to high-prevalence TB/HIV and multidrug-resistant (MDR) TB settings. Integration with HIV diagnosis and treatment for high HIV-prevalence settings; use of effective diagnostics and treatment protocols for areas with MDR TB.

Malaria

Insecticide-treated bed-nets. Provision of antimosquito bed-nets that are treated with insecticide, providing a physical and chemical barrier to mosquitoes, shortening the mosquito's life span, and thus reducing incidence of malaria.

*Indoor residual spraying.** Periodic spraying of indoor surfaces with insecticide to reduce malaria transmission.

Artemisinin combination treatment. Combination of drugs used to treat first-line-drug-resistant falciparum malaria, which is now widespread in Africa.

*Larviciding, drainage, and house improvement.** Measures designed to reduce mosquito breeding. Focal use of these measures is especially important in urban areas.

Access to essential medicines

Interventions to ensure availability, affordability, and appropriate use. Incentives to direct research and development processes toward appropriate medicines for developing countries; establishment of national essential medicines lists (including preventive, curative, and reproductive health commodities, equipment, and supplies); ensuring reliable procurement and distribution systems; prequalifying quality suppliers and procurement and distribution facilities; monitoring systems to assure drug quality; elimination of user fees for essential medicines; programs to improve the way drugs are prescribed, dispensed, and used, including public media campaigns and education of providers.

Health systems

Multiple interventions to strengthen health system. Human resource training and salary enhancement, improving management capacity, enhancing monitoring and evaluation, strengthening quality control, strengthening medical information systems, increasing capacity for research and development, enhancing community demand, and improving infrastructure.

Sexual and reproductive health (elements of integrated programming not covered under maternal health, child health, HIV/AIDS programs)

Counseling on contraception and birth spacing. Information and education on benefits and methods of family planning and birth spacing; appropriate follow-up on method satisfaction, consistent and correct use of method, and options for appropriate method switching.

Universal access to contraception. Program to ensure universal access to family planning choices, including effective modern contraceptive methods, and to guarantee reliably available and affordable supplies and choice among methods.

*Age-appropriate sexuality education and services (especially for adolescents).** School and community-based education programs, mass media education programs, youth-friendly information and service delivery, beneficiary-driven programming to meet the information and service needs of diverse adolescent groups (including married adolescents), and programs to educate parents to improve adolescent reproductive health.

Prevention and treatment of sexually transmitted infections. Programs to detect and treat sexually transmitted infections (such as syphilis, gonorrhea, and chlamydia) and other reproductive tract infections that can increase the risk of HIV/AIDS and infertility and affect the choice of appropriate contraceptive methods.

*Outreach to men to increase participation and support in reproductive health.** Counseling and information services for men to address their reproductive health needs, support the decisions of their partners, and change gender and relationship norms to ensure greater gender equality; prevent gender violence and harmful traditional practices and promote collaborative decisionmaking; information and services for reproductive health in the army and police forces, including efforts to combat gender violence.

Intervention area 4: investments in education at all levels

Investments in education at all levels include interventions in primary and postprimary education and adult literacy.

Primary and postprimary education[1]

Demand-side incentives. Elimination or reduction of school fees, conditional cash transfers to parents, school feeding (and take-home food rations where needed), school health programs such as deworming and iron supplementation, targeted subsidies to girls, and vulnerable populations such as ethnic groups or HIV/AIDS orphans, provision of school material such as textbooks and uniforms, and so on.

*Local control and management.** Systems to involve parents in school management: parent committees, school-based management, financing, auditing, and expenditure management systems that are consistent with more local control.

*Information/assessment.** Provision of transparent information regarding resources, greater access to information through school report cards, better data systems, and better learning outcomes.

*Improving and evaluating learning outcomes.** Learning evaluation systems that assess acquisition of skills and knowledge, and learning outcomes.

Special packages to make schools safe for girls. Training teachers and administrators in gender sensitivity, hiring female teachers, and investing in gender-sensitive infrastructure such as latrine facilities.

*Special packages for children with disabilities.** Investments in infrastructure, special training for teachers, specific outreach and retention efforts, and separate performance assessments.

*Special packages for education in conflict and postconflict situations.** Community participation to increase coverage of children affected by conflict and efforts involving private institutions and NGOs to create a participatory and culturally and environmentally sensitive learning environment through training of teachers and relevant learning material.

Adult literacy [1]

Adult literacy for women. Providing informal educational opportunities to uneducated and/or illiterate mothers of young children, particularly in settings where there are pockets of undereducated women, such as ethnic minority/indigenous communities, and in areas where parental literacy is a constraint on children's enrollment and completion.

Other core interventions for primary and postprimary schooling

Infrastructure. Provision of schools, including classrooms, furniture, transportation, and other facilities such as libraries, laboratories, and sports facilities, where needed for primary and postprimary schooling.

Teachers. Recruitment of teachers, with provision of incentives (such as adequate salaries and housing in rural areas where applicable) and ensuring adequate preservice and in-service training.

Curriculum reform. Implementation of curriculum reform, where necessary, to improve education content, quality, and relevance, with a focus on vocational and informal training as necessary to prepare students for transition to work and to adulthood.

Higher education

Extension and maintenance of higher education system, with a particular focus on science and engineering education.

Intervention area 5: investments in gender equality

Investments in gender equality include interventions for sexual and reproductive health, access to property rights and work, security, participation and institutional reform, and data collection and monitoring.

Gender equality

Sexual and reproductive health

Universal access to sexual and reproductive health information and services and protection of reproductive rights. (Service packages described under health interventions above.) Legislation* and awareness campaigns to protect the rights of individuals and couples to plan their families; to ensure access to sexual and reproductive health information and services; to discourage early marriage (at ages posing health risks), female genital mutilation, and other traditional harmful practices; and to expand access to safe abortions (where permitted by law) and review the legal status of abortion in order to improve public health while respecting national sovereignty, cultural values, and diversity.*

Access to property rights and work

Equal access to and treatment at work. Provision and enforcement of equal opportunity legislation* and legislation promoting gender-sensitive policies, such as provision of maternity and dependent care leave and training,* and support programs for women entrepreneurs and young girls training to transition to work (including care centers for young children to ensure early childhood development).

*Equal access to property rights.** Legislation and administrative support to provide and protect women's equal rights to property and other inherited and acquired assets.

Security

Security for girls and women from violence. Legislation and administrative actions to protect women against violence,* promotion of awareness of women's right to seek redress, protection from perpetrators of violence (through access to shelters, services, and so on), and mechanisms to dispense justice to perpetrators.

Participation and institutional reform

Political representation. Mechanisms* (such as quotas and reservations) to allow for adequate representation at all levels of government, along with adequate training.

*Involvement of women's groups at the community level.** Recognition of and support to women's groups organized at the community level to encourage women to be partners in the design and delivery of public services.

National women's machineries. Legislative and financial support to national women's machineries (defined by the United Nations as "a single body or complex organized system of bodies, often under different authorities, but recognized by the government as the institution dealing with the promotion of the status of women").

*Data collection and monitoring**

Gender-disaggregated data. Collection of gender-disaggregated statistics on health, education outcomes, access to assets and infrastructure, conditions of work and employment, political representation, and gender-specific violence.

Intervention area 6: investments in environmental sustainability

Investments in environmental sustainability include interventions in improved environmental management, integration of environmental sustainability into sector strategies, strengthening regulatory and institutional mechanisms, and monitoring.

Improved environmental management

Natural resource management

*Soil management and prevention of desertification.** Implementation of soil erosion control (by wind and water) by planting windbreaks and cover crops; improvements in soil fertility with agroforestry systems, cover crops, and conservation of ground and surface water.

*Forest management.** Implementation of sustainable forest management techniques, forest plantations in appropriate areas to satisfy demand for forestry products, and tree seedlings and other measures to support afforestation.

*Watershed management.** Promotion of reforestation and afforestation to protect selected water catchment areas.

*Management of coastal ecosystems and fisheries.** Elimination of destructive technologies (for example, dynamite and cyanide, bottom trawling); design and implementation of fisheries rebuilding plans to restore depleted fish populations to target levels (biomass at maximum sustainable yield); implementation of a representative network of fully protected marine and coastal areas to restore fisheries.

*Management of freshwater resources and ecosystems.** Institution of Integrated Water Resources Management plans; promotion of reforestation to protect selected catchment areas; increasing efficiency of cropping systems; and monitoring of wells and groundwater-dependent systems.

Integration of environmental sustainability into sector strategies

Technical support

Advisory mechanisms. Strengthening institutions for environmental management (such as ministries and environmental protection agencies) to provide technical support to the development of sector strategies.

Impact assessments

*Environmental impact assessments.** Carry out strategic environmental impact assessments for large-scale infrastructure projects and other development strategies that are likely to have a major impact on the environment.

Strengthening regulatory and institutional mechanisms

Property rights

*Access to tenure and rights.** Local ownership of natural resources, including common property and provision of access rights.

Regulation of pollution

*Pollution control.** Development and implementation of pollution control standards.

Market-based strategies

*Reformation of tax laws.** Taxation of environmental "bads" (such as pollution and degradation), and appropriate carbon tax systems.

*Transformation of market incentives.** Revision of subsidies in forestry and fisheries that cause overexploitation of these resources, design of agricultural subsidy programs to prevent overuse, development of an internationally credible system of certification of raw natural resource materials.

Monitoring and enforcement

*Environmental monitoring systems.** Better dissemination and use of existing environmental monitoring and assessments at national and local levels; provision of funds, technical support, and tools for countries to undertake monitoring, data collection, and harmonization based on established standards (based on core set of indicators).

*Enforcement of environmental regulation.** Strengthening systems for monitoring environmental pollution to help enforce regulation for pollution control.

Intervention area 7: investments in science, technology, and innovation

Investments in science, technology, and innovation include interventions in science and technology institutions and information and communication technologies.

Science, technology, and innovation

Science and technology institutions

*Science and technology advice.** Creation of independent body charged with providing scientific advice and technology forecasting to policymakers.

*Science and technology research.** Extension and maintenance of centers of excellence for scientific research, including the financing of research at universities.

*Science parks and business incubators.** Establishment of science parks and incubators for technology-based companies.

Information and communication technologies

*Telecommunications infrastructure.** Provision of telecommunications infrastructure, including international and trunk fiber infrastructure; provision of connectivity to hospitals and schools.

Millennium Development Goals interventions by target

Income poverty (Goal 1, target 1)

Agriculture. Increasing agricultural productivity directly raises the incomes of the rural poor and generates rural jobs.

Nutrition. Better nutrition contributes to human capital accumulation and improved labor productivity.

Education. Education increases human capital, which contributes to economic growth. Education is linked to lower fertility rates, which are in turn linked to increases in economic growth per capita.

Gender equality. Awareness of and access to reproductive health rights and services enable and empower women to plan their families, leading to lower fertility rates and reduced poverty. Empowerment through access to work, property rights, political representation, and safety from violence leads to increased participation of women in economic activity.

Health. Improved health has pervasive direct and indirect effects on raising both the level and the growth rate of income.

Environment. Many poor people depend on natural resources for their livelihoods. Improving natural resource management can sustain or even raise their incomes.

Water and sanitation. Improved water supply for productive activities can raise economic growth through agriculture and the urban manufacturing and service sectors.

Slum upgrading and urban planning. Providing security of tenure can improve labor market participation and access to credit markets. Urban infrastructure, including transport systems, is necessary for establishing manufacturing and service industries.

Science and technology. Science and technology institutions improve technological learning in society and improve the adoption of technology by the private sector. Higher education can open new employment opportunities.

Energy. Access to electricity, motive power, and improved thermal energy systems is necessary for manufacturing, service, or cottage industries.

Transport. Roads, railroads, and ports lower transport costs and thereby increase the real incomes of the poor. In urban areas improved transport infrastructure supports manufacturing and service industries, contributing to employment.

Hunger (Goal 1, target 2)

Agriculture. Increasing agricultural productivity through investments in soil health, water management, extension services, and research increases food availability.

Rural incomes and access to markets. Improved access to credit, storage facilities, processing, and value-added technologies can help raise incomes, together with access to markets, farmer cooperatives/associations, and physical market spaces.

Nutrition. Nutrition interventions are needed for vulnerable populations and to ensure micronutrient intake. Direct food assistance in food-scarce areas alleviates short-term hunger.

Education. Education leads to more productive farming and better management of nutritional needs.

Gender equality. Land rights allow women to increase agricultural production. Increased access to work and higher incomes enable women to purchase adequate food for themselves and their families. Equal access to productive inputs increases plot yields.

Health. Reducing parasitic and infectious disease burden improves nutrition levels. Birth spacing protects maternal and child nutrition and health.

Environment. Improved water resource management and protection of water catchment areas can raise crop yields. Biodiversity protection sustains pollination and seed dispersal mechanisms necessary for agricultural production.

Water and sanitation. Safe drinking water reduces the incidence of diarrheal diseases, which contribute to malnourishment. Drinking water supplied through wells and boreholes can help irrigate fields during droughts. Access to sanitation reduces the incidence of diarrheal disease and thereby increases nutrient uptake. Integrated water resources management sustains adequate water supply for agriculture. Water storage and water management infrastructure improve water management for agriculture.

Slum upgrading and urban planning. Slum upgrading and accompanying interventions help raise incomes and reduce urban hunger. Improved transport infrastructure lowers the cost of food products, further reducing hunger.

Science and technology. Increased agricultural research is critical for improving seed varieties, cropping systems, pest control, and water management to increase agricultural productivity, thus reducing hunger. Increased access to higher education can help increase the number of agricultural extension workers. Information and communications technology improves farmers' market information, raising agricultural production.

Energy. Improved access to electricity and liquid fuels can power diesel pumps for irrigation, facilitate agricultural mechanization, and power agroprocessing machinery, thus increasing agricultural output and reducing hunger. Improved energy services lower transportation and marketing costs, which reduces food prices. Access to improved cooking fuels is necessary to ensure safe cooking of food.

Transport. Footpaths, roads, and improved transport services lower the cost of agricultural inputs, increase farmgate prices, and facilitate marketing, which can increase agricultural production. Improved transport infrastructure reduces postharvest losses through accelerated transport of products to markets.

Primary education (Goal 2, target 3)

Education interventions. Provide demand-side incentives to retain children in school, management systems to increase parental involvement and school-based management, increased transparency and information, evaluation of learning outcomes, special packages to attract girls and children with disabilities and those in conflict and postconflict situations, adult literacy for women in particular, together with other core interventions such as building schools, providing trained teachers, and developing appropriate curricula, these will increase enrollment and retention of children at primary and postprimary levels.

Agriculture. Agricultural interventions to improve soils, seeds, and water management raise rural incomes and reduce the time young children spend in the field, freeing them to attend school.

Nutrition. Nutrition interventions for infants improve cognitive development and improve learning outcomes in the future.

Gender equality. Maternal education contributes strongly to higher primary enrollment.

Health. Improved health enhances educational outcomes by improving cognitive abilities and attendance rates. AIDS prevention and treatment reduce the disease's impact on teacher attendance and attrition. AIDS prevention and treatment reduce the number of orphans, who are less likely to complete primary education. Reproductive health services reduce the withdrawal of girls from school related to sibling care burdens caused by unplanned pregnancies or due to adolescent pregnancy.

Environment. Improved natural resource management can free up children's time and increase school attendance.

Water and sanitation. Improved access to water frees up children's time, thus allowing them to attend school. Improved health through sanitation and hygiene reduces school absenteeism. Installing girls' toilets in schools can increase girls' enrollment and completion rates.

Slum upgrading and urban planning. Security of land tenure and a fixed address are often necessary for children to be allowed to attend school.

Science and technology. Higher education is essential to training secondary school teachers and provides additional incentives to complete primary and secondary school. Information and communications infrastructure can improve the quality of education.

Energy. Access to electric power and improved cooking fuels lowers time spent by children (especially girls) collecting fuelwood, thus facilitating school attendance. Improved access to liquid fuels is necessary to render mechanized school transport more affordable. Electrification permits children to read and study for longer hours, thus improving school outcomes.

Transport. Improved transport infrastructure and services increase incentives for teachers to work in rural areas, reducing the time it takes for children to reach school and allowing them to travel farther, raising enrollment rates, reducing the time required for households to fetch fuel and water and to carry out other tasks, and lowering the opportunity cost of children attending school.

Gender equality (Goal 3, target 4)

Gender interventions (not specified elsewhere). Reduce violence against women, improve their property and other rights, ensure full access to reproductive health services, including contraception, and improve women's participation in decisionmaking processes.

Agriculture. Improved soils, seeds, and water provision can reduce the time girls spend in the field, freeing them to attend school. Providing fuelwood on farms through agroforestry trees decreases women's labor in search of firewood. Providing small-scale water management such as water harvesting decreases women and girls' transport burden to fetch water. Increased agricultural production increases the incomes of women farmers.

Nutrition. Nutrition interventions for girls (infants and children) lead to better health and education outcomes.

Education. Education contributes to increased employment opportunities, improved decisionmaking, and empowerment of women more broadly.

Health. Family planning services facilitate employment and social participation opportunities for women, strengthen partner relationships, and provide a greater sense of well-being and agency. Access to emergency obstetric care in the event of pregnancy and delivery complications saves women's lives.

Environment. Women benefit disproportionately from improved management of natural resources, including through time saving and reduced transport burden.

Water and sanitation. Improved access to water generates time savings for women and girls. Improved access to water reduces the need to carry heavy loads over long distances, thus improving women's health. Increasing access to toilets reduces women's exposure to harassment and improves personal hygiene and well-being.

Slum upgrading and urban planning. Women benefit disproportionately from slum upgrading, since it reduces their transport burden and time-poverty, improves their health, and provides them with additional income-generating opportunities.

Energy. Improved access to electricity and fuels reduces the time-poverty of women and lowers their daily transport burden. Improved access to energy creates additional employment opportunities for women.

Transport. Improved transport infrastructure and services reduce women's time-poverty and transport burden. Roads improve communication and lower transactions costs, thus increasing employment opportunities for women.

Child mortality (Goal 4, target 5)

Health interventions. Provide the neonatal integrated package of interventions, immunization, the integrated management of childhood illness, and the range of preventive approaches (such as mass distribution of insecticide-treated bednets). Family planning can delay first births and reduce very short and very long birth intervals, thereby improving child health outcomes. Strengthening health systems will also be critical to achieving this Goal.

Agriculture. Increased rural incomes and food availability lead to improved health outcomes.

Nutrition. Nutrition interventions for pregnant women lead to higher birthweight, an important determinant of child survival. Exclusive breastfeeding, complementary feeding after six months (HIV-positive mothers should use replacement feeding when it is acceptable, feasible, affordable, sustainable, and safe), and micronutrient supplementation reduce child mortality.

Education. Postprimary education increases the age of marriage, lowers fertility rates, and increases care seeking for child illnesses. Adult literacy programs increase awareness of the causes and prevention of child mortality.

Gender equality. Women's empowerment leads in multiple ways to greater awareness of child health issues.

Environment. Reducing pollution of water and air can lower child morbidity and mortality.

Water and sanitation. Access to clean water, sanitation, and improved hygiene reduce the incidence of waterborne disease.

Slum upgrading and urban planning. Slum upgrading, improved urban infrastructure, and access to basic services (including solid waste disposal) can reduce exposure to pollutants and thereby reduce child mortality rates. Road curbing and street lighting can reduce traffic deaths.

Science and technology. Information and communications technology improves diffusion of hygiene education and thereby lowers child mortality. Access to higher education increases the supply of health workers.

Energy. Reducing indoor air pollution through improved cooking fuels and stoves decreases respiratory infections. Improved access to energy allows households to boil water, thus reducing incidence of waterborne diseases.

Transport. Improved transport infrastructure increases access to healthcare clinics and services and reduces costs for healthcare workers to serve rural areas.

Maternal mortality (Goal 5, target 6)

Health interventions. Ensure access to emergency obstetric care, skilled birth attendance and clean delivery, antenatal care and postpartum counseling, as well as safe abortion (where permitted by law). Access to family planning can reduce the number of unwanted and ill-timed pregnancies, reducing the lifetime exposure to the risk of maternal mortality and preventing recourse to abortion. Strengthening health systems will be critical to achieving this Goal.

Agriculture. Increased rural incomes and food intake lead to improved health outcomes.

Nutrition. Nutrition interventions, such as adequate caloric intake and iron supplementation for women of reproductive age, reduce risk during pregnancy and childbirth.

Education. Postprimary education increases the age of marriage, contraceptive use, and access to prenatal care and safe delivery, all of which reduce maternal mortality. Adult literacy programs increase awareness of the causes and prevention of maternal mortality.

Gender equality. Women's empowerment leads to greater effective demand for family planning services, prenatal care, and safe delivery.

Water and sanitation. Running water and sanitation facilities are essential for provision of prenatal care and emergency obstetric care. Access to sanitation and hygienic behavior improve women's health.

Slum upgrading and urban planning. Slum upgrading and security of land tenure improve women's access to health systems and emergency obstetric care.

Science and technology. Information and communications technology is critical for providing adequate access to emergency obstetric care. Access to higher education increases the supply of health workers.

Energy. Improved access to energy services improves communication and transport, which are critical for emergency obstetric care. Modern energy services reduce costs for healthcare workers serving in rural areas.

Transport. Feeder roads and emergency transport are critical for providing timely access to emergency obstetric care. Improved transport infrastructure reduces the cost for healthcare workers serving in rural areas.

HIV/AIDS (Goal 6, target 7)

Health interventions. Provide comprehensive HIV/AIDS prevention programs, orphan support, voluntary counseling and testing, harm reduction for drug users, prevention of mother-to-child transmission, antiretroviral treatment, and treatment of opportunistic infections. Linking reproductive health and HIV/AIDS program efforts can increase effectiveness, coverage, and efficiency of service delivery. Strengthening health systems will be critical to achieving this Goal.

Agriculture. Increased agricultural incomes improve access to prevention and treatment.

Nutrition. Adequate nutrition can improve survival and quality of life for people with HIV. Nutritional supplementation programs for people with HIV improve antiretroviral adherence.

Education. With education, people are less likely to contract HIV and more likely to use health services effectively.

Gender equality. Women's empowerment leads to greater effective demand for HIV/AIDS prevention and treatment, including the ability to negotiate safe sexual practices.

Water and sanitation. Improving access to clean water and sanitation improves the nutritional status of people with HIV.

Slum upgrading and urban planning. Slum upgrading and security of land tenure improve access to HIV/AIDS treatment and prevention.

Science and technology. Scientific research can improve diagnosis and treatment of HIV. Information and communications technology is critical for media-based HIV prevention. Access to higher education increases the supply of health workers.

Energy. Electricity and modern energy services support functioning health clinics and hospitals. Modern energy services increase incentives for healthcare workers to work in rural areas.

Transport. Improved transport infrastructure and services facilitate treatment and prevention of HIV/AIDS. Improved transport infrastructure reduces costs for health workers serving in rural areas.

Malaria and other major diseases (Goal 6, target 8)

Health interventions. For malaria, comprehensive use of insecticide-treated bed-nets, indoor residual spraying where appropriate, effective malaria treatment (using artemisinin combination therapies, where indicated), epidemic control measures, and promotion of new diagnostics, drugs, and vaccines. Strengthening health systems will be critical to achieving this Goal.

For TB, expansion of DOTS, DOTS-plus for multidrug-resistant TB, adaptation of TB treatment in high HIV prevalence settings, and promotion of new diagnostics, drugs, and vaccines. Strengthening health systems will be critical to achieving this Goal.

Agriculture. An increase in agricultural incomes improves access to and information on ways of preventing and treating malaria and TB.

Education. Education and literacy programs increase awareness of ways to prevent and treat malaria and TB.

Gender equality. Women's empowerment leads to greater effective demand for insecticide-treated bed-nets and effective malaria and TB treatment.

Environment. In some instances environmental control can contribute to containing malaria and TB.

Water and sanitation. Improved water management in urban areas can contribute to containing mosquito breeding sites and transmission.

Slum upgrading and urban planning. Improving housing and urban water management infrastructure can reduce the incidence of malaria and especially TB (which has a higher rate of transmission in overcrowded slum conditions). Slum upgrading improves access to appropriate malaria and TB treatment.

Science and technology. Research is necessary to develop new drugs and diagnostics for malaria and TB. Access to higher education increases the supply of health workers.

Energy. Electricity and modern energy services improve healthcare. Modern energy services reduce cost for healthcare workers serving in rural areas.

Transport. Improved transport infrastructure and services reduce the cost of distributing bed-nets and essential health services including malaria and TB treatment. Improved transport infrastructure increases incentives for health-care workers to work in rural areas.

Access to essential medicines (Goal 8, target 17)

Health interventions. Improve supply and distribution systems for essential medicines together with strengthening quality control, quality assurance, and programs to promote rational use. Strengthening health systems will be critical to achieving this Goal.

Agriculture. An increase in agricultural incomes makes all medicines more affordable.

Education. Education and literacy programs increase access to and appropriate use of essential medicines.

Gender equality. Women's empowerment leads to greater effective demand for essential medicines of good quality, including reproductive health commodities and supplies.

Science and technology. Research can generate new essential medicines and increase the effectiveness of existing ones.

Transport. Improved transport infrastructure and services lower the cost of essential medicines and improves access.

Reverse loss of environmental resources (Goal 7, target 9)

Environmental interventions. Improve management of natural resources through market mechanisms, strengthened regulation and enforcement, and investments in the management of critical ecosystems.

Agriculture. Investments in soil health replenish soils and prevent further land degradation. Labor-intensive agricultural production is an alternative to slash-and-burn and to the deforestation that results. Agroforestry and other organic incomes increase agro-biodiversity and sequester carbon. Small-scale water management can restore water tables and reduce runoff. Food-for-work programs can help restore degraded ecosystems.

Gender equality. Equal access to property rights allows women, as primary users, to manage natural resources in a sustainable manner.

Health. Access to family planning services reduces total fertility rates to levels people desire, thus mitigating population pressures on the environment.

Water and sanitation. Improved sanitation and sewage treatment can reduce environmental pollution. Integrated water resources management can maintain ecosystem functioning. Hydrological monitoring systems can help protect aquifers and freshwater ecosystems from excessive withdrawals.

Slum upgrading and urban planning. Slum upgrading and improved urban water and waste management infrastructure reduce environmental pollution.

Science and technology. Research can improve natural resource management (including management of freshwater ecosystems and wetlands, and biodiversity conservation).

Energy. Access to modern cooking fuels reduces demand for biomass, thus reducing pressure on marginal lands and forests. Improved energy services reduce indoor labor and outdoor air pollution as well as carbon emissions.

Water and sanitation (Goal 7, target 10)

Water and sanitation interventions. Provide, operate, and maintain water and sanitation infrastructure and services in conjunction with behavior change programs to improve household hygiene.

Agriculture. Small-scale water management increases water availability for rural farmers.

Education. Education and literacy programs improve hygiene and help ensure proper operation of water and sanitation facilities.

Gender equality. Political representation allows women to ensure that access to water is a priority in local decisionmaking.

Environment. Improved management of wetlands, water catchment areas, and freshwater ecosystems is critical for ensuring access to drinking water. Control of industrial pollution improves drinking water quality.

Slum upgrading and urban planning. Slum upgrading reduces water pollution and improves drinking water quality. Improved urban infrastructure ensures the separation of sewage from drinking water supplies.

Science and technology. Research can help improve sanitation and water management techniques. Access to higher education increases the supply of trained workers to design and manage water supply and sewer infrastructure.

Energy. Electricity and improved access to modern fuels are necessary to power water supply infrastructure and water treatment systems.

Transport. Improved transport infrastructure and services facilitate the provision, operation, and maintenance of water supply and sanitation systems. Improved transport systems reduce the costs of providing hygiene education through community workers.

Improve the lives of slum dwellers (Goal 7, target 11)

Urban investments and slum-upgrading. Scale up slum-upgrading with the support of improved urban planning and investments in core urban infrastructure as well as basic services.

Agriculture. Investing in urban agriculture increases agricultural productivity and the incomes of slum dwellers.

Nutrition. Nutrition interventions improve the health outcomes of urban populations.

Education. Education and literacy programs improve the employment prospects of slum dwellers.

Gender equality. Equal access to property rights, political representation, and security for girls and women allows women living in slums to improve their lives and the lives of their families.

Health. Access to preventive and curative health services, including sexual and reproductive health information and services, reduces the burden of ill health for slum dwellers.

Environment. Improving solid waste disposal and water treatment can improve health outcomes.

Water and sanitation. Improved access to water supply and sanitation services can reduce household expenditure on water. Sewage treatment can further improve health outcomes in urban areas. Storm water drainage systems are improved through sanitation infrastructure, thus minimizing the risk of flooding.

Science and technology. Higher education provides new employment opportunities for the urban poor. Information and communications technology reduces the cost of income-generating activities.

Energy. Improving access to electricity and modern fuels lowers indoor air pollution. Access to electricity and modern fuels can lower household expenditure on energy services, thus raising incomes. Improved energy services lower the cost of urban transport.

Transport. Improved rural transport infrastructure and services reduce the cost of food in urban areas, thus increasing disposable incomes of slum dwellers. Improved urban transport infrastructure is critical to enhancing income-generating opportunities as well as access to social services. Proper sidewalks and curbing are critical to reduce traffic deaths.

Information and communications technology (Goal 8, target 18)

Direct interventions. Strengthen science advisory mechanisms, invest in higher education and research, promote private sector development, and improve access to communications technologies.

Agriculture. Increased agricultural incomes improve access to information and communications technology.

Education. Postprimary education prepares students for ability to provide, use, and manage information and communications technology.

Energy. Electricity is necessary to power information and communications technology applications and to operate research institutions.

Appendix 3

Assumptions underlying the resource estimates in chapter 17

Throughout, we have used the OECD/DAC deflator to rebase estimates to 2003 U.S. dollars.

Notes to table 17.1—Per capita MDG investment needs and MDG financing gaps

UN Millennium Project (forthcoming) contains a detailed description of the national needs assessments by the UN Millennium Project and explains the technical assumptions underlying the estimates for MDG investment needs and sources of financing. It also presents more detailed country-level results including a breakdown by operating and capital expenditures.

ODA for direct MDG support in 2002. This is calculated by subtracting the following items from net official development assistance: assistance for emergency and distress relief, food aid, and technical cooperation, and ODA channeled through NGOs (OECD/DAC 2004e). We assume that 75 percent of this residual assistance directly supports investments for achieving the Goals and add 40 percent of the ODA channeled through NGOs (see table 13.2).

Notes to table 17.2—Cofinancing the MDGs in low-income countries

MDG investment needs. We estimate aggregate MDG investment needs across low-income countries using a two-step approach. First, the unadjusted MDG investment need in each country is calculated. Second, we adjust for the relative price level in each country.

MDG investment needs cover the following areas: hunger and agriculture, primary education, secondary education, adult literacy, gender equality, health, water supply and sanitation, improving the lives of slum dwellers, energy services, and roads. To take into account the variation in per capita investment needs across countries, we identified the key drivers of variation in the sample of five countries and used them to adjust for differences in needs across countries. For example, since health interventions will be more expensive in countries with high rates of HIV prevalence, higher per capita costs are assigned to countries that have high rates of HIV prevalence. Likewise, MDG investment needs in the road sector are driven largely by the current stock of paved roads, which is therefore used to scale costs. In other areas, such as primary education or water supply, cross-country variation is relatively low, so we assume uniform per capita MDG investment needs.

In the next step, MDG investment needs are adjusted to the relative price level in each country to account for differences in salaries, materials costs, and other prices. In the absence of sufficiently disaggregated price indicators that would allow a disaggregation of tradables and nontradables, the analysis focuses on changes in the overall level of prices. It is assumed that the adjusted cost of MDG investments scales linearly with a country's price level, defined as the ratio of GDP per capita expressed in international U.S. dollars to GDP per capita in purchasing power parity–adjusted dollars. The price level depends on GDP expressed in international dollars, according to

$$price(GDP_{USD}) = a * GDP_{USD} \wedge b$$

The constants *a* and *b* have been estimated using 2002 GDP data for all countries (World Bank 2004c).

Domestic resource mobilization. We first project GDP per capita in 2006 by extrapolating the five-year average growth rate from 1997 to 2002 forward to 2006. It is assumed that as a result of the MDG investments, real per capita growth rates accelerate in regions that have experienced slow growth (table A3.1). Price levels are adjusted for each year as described above.

Government expenditures on the Goals in 2006 are estimated by income group and expressed as a percentage of GDP. We project that the share of national income devoted to public expenditures on social services and infrastructure rises with national income (table A3.2). In some countries where domestic resource mobilization in 2002 deviates substantially from the income group's average, these assumptions were modified.[1] As discussed in chapter 17,

Table A3.1

Annual per capita GDP growth assumptions for 2006–15, by region

Percent

Region	Projected growth in per capita GDP, 2006–15
East Asia and Pacific	5.0
Europe and Central Asia	3.2
Latin America and the Caribbean	3.2
Middle East and North Africa	3.2
South Asia	5.0
Sub-Saharan Africa	3.2

Source: Authors' calculations.

Table A3.2

Estimated government resource mobilization

Government expenditures for the MDGs as a share of GDP (percent)

Source: Authors' calculations.

	2006	2015
Least Developed Countries (per capita GDP less than $450)	5	9
Low-income countries (per capita GDP $450–$734)	7	11
Lower-middle-income countries (per capita GDP $735–$2,935)	9	13
Upper-middle- and high-income countries (per capita GDP greater than $2,935)	10	14

government resource mobilization is projected to rise by up to four percentage points of GDP between 2006 and 2015.

As in the country-level needs assessments, partial cost recovery from households is restricted to secondary education, water and sanitation, and energy services. We assume that people below the poverty line will not pay any user fees.

MDG financing gap. This is calculated by subtracting projected domestic resource mobilization from projected MDG investment needs. For 2002 we estimate that up to $12 billion was provided as ODA for direct MDG support. This amount does not include technical cooperation for capacity building or other investments that have not been estimated in the MDG needs assessments.

Notes to table 17.3—Estimated cost of meeting the MDGs in all countries

MDG support needs in low-income countries

MDG financing gap. Referenced from table 17.2.

Capacity building to achieve the Goals. As recommended by the OECD/DAC it is assumed that 60 percent of technical cooperation provided to low-income countries in 2002, or $5 billion, directly supports achieving the Goals—largely through capacity building (OECD/DAC 2004e). Also included in this line item is 60 percent of all ODA currently provided through NGOs. We estimate that to support the scaling up of MDG interventions, donor support for capacity building in low-income countries will need to rise by 50 percent to $7 billion. As described in the text, major human resource training efforts need to be prioritized in the scaling up of MDG interventions.

Grants in support of heavy debt burden. Countries that require official development assistance to meet the Goals, but that will graduate from assistance before 2015 ("graduating countries"), may nevertheless require grants to support the repayment of heavy debt burden. We assume that such grants will be provided if a country's domestic resource mobilization is insufficient to finance debt service payments for nonconcessional loans in addition to all MDG-related expenditures.[2] We use a three-year average of debt repayments for 1999 to 2002 (World Bank 2004b) as a proxy for projected annual debt service payments for 2006 to 2015. Aggregate debt service payments will need to be reduced by an estimated $7 billion in 2006, falling to $1 billion by 2015, to ensure that all countries can finance MDG investments.

Debt relief. Based on the criterion for debt sustainability proposed in chapter 13, countries that require substantial ODA transfers to finance the Goals through 2015 are eligible for debt cancellation. Using Global Development Finance data, outstanding debt stocks for all "nongraduating countries" are estimated for 2002 to amount to $174 billion in concessional loans and $63 billion in nonconcessional loans (World Bank 2004b). Both stocks are assumed to have stayed constant through 2005 and to be canceled over the 10 years from 2006 to 2015. Under the rules of the OECD/DAC, only the cancellation of nonconcessional loans can be booked as debt relief that counts toward official development assistance. We make the simple assumption that the stock of nonconcessional debt is written down in equal installments of $6.3 billion a year.

Repayments of concessional loans. The 2002 flow of loan repayments is estimated on the basis of OECD/DAC data (2004a). We assume that loan repayments from all nongraduating countries that receive full debt cancellation will be set to zero since their domestic resource mobilization is too low to finance the MDG investments.

MDG support needs in middle-income countries

Direct support to government. Some middle-income countries with particularly severe pockets of poverty will likely have investment needs that cannot be financed through domestic resource mobilization alone. We suggest that $10 billion will be required in official development assistance for direct MDG support in these countries.

Capacity building to achieve the Goals. Technical cooperation for the Goals, estimated at 60 percent of total technical cooperation in 2002, will concentrate on assisting countries in regions that are currently underserved, such as urban slums or remote rural areas (OECD/DAC 2004e). Also included in this line item is assistance currently provided directly to NGOs. The budget for technical cooperation is expected to stay constant through 2015.

Repayments of concessional loans. Loan repayments by middle-income countries in 2002 are estimated on the basis of data from OECD/DAC (2004a). World Bank (2004b) data on outstanding loans suggests that middle-income countries are unlikely to require outright debt relief to finance the Goals. Some countries will require a reduction in their loan service payments to make the investments necessary for meeting the Goals. Using the same approach as for low-income countries we estimate that repayments of concessional loans may need to be reduced by some $3 billion in 2006, compared with 2002, and only $0.2 billion by 2015.

MDG support needs at the international level

Regional cooperation and infrastructure. These needs are projected to require an extra $9 billion a year by 2015, in addition to investments financed through national budgets. The estimate of current assistance for regional cooperation and infrastructure is based on Birdsall (2004).

Funding for global research. Official development assistance for research in 2002 has been estimated by OECD/DAC (table A3.3) (OECD/DAC 2004e). They are expected to rise to $7 billion by 2015, focusing on public health, agriculture and natural resource management, low-cost and sustainable energy technologies, and adaptation to long-term climate change in developing countries. Projected investments in public health are based on recommendations by the WHO Commission on Macroeconomics and Health (2001). The remaining projections are made by the UN Millennium Project.

Implementing the Rio conventions. Current assistance for implementing the environment conventions agreed to in Rio in 1992 is estimated on the basis of OECD/DAC (2002). Unfortunately, few cost estimates exist for implementing these and other environmental agreements. We have included the cost of implementing the Desertification Convention based on UNEP (1991). The estimates have been revised downward by focusing on preventive and corrective measures and excluding the cost of rehabilitation. To limit the possibility of a double-counting of interventions included in the agricultural component of our national MDG needs assessments, we reduce the cost of preventive measures by 50 percent.

Table A3.3

Breakdown of funding for global research

2003 US$ billions

	2002	2006	2010	2015
Public health	0.3	2	4	4
Agriculture and natural resource management	0.4	1	1	1
Low-cost and sustainable energy technologies	0.1	1	1	1
Adaptation to long-term climate change in developing countries	0.1	1	1	1
Total	**1.0**	**5**	**7**	**7**

Note: Numbers in table may not sum to totals due to rounding.

Source: 2002 data based on OECD/DAC Secretariat, personal communication, 2004. Projections for public health from WHO 2001. All other projections are authors' calculations.

Other available cost data for the environment point to major discrepancies in the estimates. For example, the cost of maintaining biodiversity ranges from $23 billion to $45 billion a year depending on the source (Balmford and others 2002; Zhou 2003). Since neither cost estimate is underpinned by an operational strategy for maintaining biodiversity, we have not included the cost of biodiversity protection in the table. Likewise, no robust estimates exist for the cost of adaptation to climate change in developing countries.

Technical cooperation by international organizations. The 2002 ODA flows to UN agencies, funds, and programs are estimated at $5 billion, based on OECD/DAC (2004a). We assume that funding for these organizations will need to increase by 50 percent over the coming 10 years. The bulk of this funding will go toward improving the technical advisory capacities of the UN system and for providing extensive training programs.

Notes to table 17.4—Plausible ODA needs to meet the MDGs

Baseline ODA for the Goals in 2002. Equal to 2002 ODA for direct MDG support only, as in table 17.3.

Incremental MDG investment needs. Calculated as the difference between the estimated cost of meeting the Goals (as in table 17.3) and baseline assistance.

Adjustment for nonqualifying countries due to inadequate governance. To adjust for nonqualifying countries we construct an aggregate governance indicator by calculating the mean of five variables measuring control of corruption, government effectiveness, quality of institutions, regulatory quality, and the rule of law (Kaufmann, Kraay, and Mastruzzi 2003). We then make the simplified assumption that countries need to score within one standard deviation below the mean of this indicator to qualify for ODA for direct MDG support. Of course, actual decisions on whether MDG support can be provided at the scale necessary to achieve the Goals must be made case by case.

Based on this assumption, total ODA needs in 2006 will be reduced by approximately $21 billion to account for countries with inadequate governance. If no currently nonqualifying countries were to qualify for assistance by 2015, this adjustment factor would rise to $42 billion. It is assumed that a number of countries will improve their governance to qualify for ODA for direct MDG support, so that the adjustment factor rises to only $25 billion in 2015.

Reprogramming of existing ODA. It is assumed that 20 percent of the following ODA items can be reprogrammed toward the Goals in 2006: technical assistance not directed toward the Goals, development food aid, and other official development assistance that does not directly support the Goals. These items amounted to $30 billion in 2002. We assume that by 2015, 30 percent can be reprogrammed toward supporting the Millennium Development Goals.

Emergency and distress relief. This is an integral part of financing the Goals. In 2002, emergency assistance amounted to $4 billion (OECD/DAC 2004a). Even if the Goals are met, the need for emergency assistance will rise since an important share of needs are not currently met.[3] In addition, rising population numbers and the effects of long-term climate change will increase the incidence and severity of natural disasters, as well as their impact. It is projected that total emergency assistance, excluding the cost of peacekeeping and security operations, will rise by 50 percent by 2015.

Other ODA. This line item contains all assistance not included elsewhere in the table. In particular, we account for the cost to bilateral agencies of effectively managing higher ODA flows. Data on operational and administrative costs to donors in 2002 are taken from OECD/DAC (2004d). We assume that the cost of managing incremental ODA amounts to 2 percent of the bilateral aid that is not channeled through multilateral organizations, estimated at 70 percent of total ODA volumes (OECD/DAC 2004e).

Percentage of OECD countries' GNI. We assume that the 2002 GNI of all OECD countries ($28 trillion in 2003 U.S. dollars) grows in real terms at 2 percent per capita per year to reach $36 trillion (in 2003 US dollars) in 2015.

ODA to Least Developed Countries. In reference to the Monterrey target for ODA to Least Developed Countries, we estimate total ODA needed for direct MDG support and MDG capacity building required for these countries. The estimates include the adjustment for nonqualifying countries due to inadequate governance.

Methodology and data sources for projecting progress toward the Goals in table 18.1

The following discussion describes the methodology used to determine the number of lives affected by meeting the Millennium Development Goals (table 18.1). In the first scenario, labeled "current trend extrapolated to 2015," outcomes for 2005 and 2015 are estimated for each indicator by extrapolating population-weighted regional averages from 2002 figures using population projections (United Nations Population Division 2003a) and trend data for 1990–2002 (unless otherwise noted).

The "MDG scenario for 2015" is created by assuming that, subsequent to 2005, countries accelerate progress to meet the Goals in 2015 or continue on their current trajectory if the historical rate of progress exceeds what is needed to achieve the Goals. In the text, the number of lives described as improved or saved by achieving the Goals is calculated as the difference between outcomes under the MDG scenario in 2015 and the 2005 estimates.

Indicator-specific notes

Poverty headcount

Poverty headcounts are calculated as the percentage of people with incomes below $1.08 a day in 1993 PPP (purchasing power parity) dollars. Regional trends are extrapolated from data for 1990 and 2001 using estimates and regional definitions from Chen and Ravallion (2004).

GDP per capita

GDP per capita statistics are based on data from *World Development Indicators* (World Bank 2004c). All GDP data are adjusted to 2003 dollars using the OECD/DAC deflator[1] and divided by population numbers to obtain per capita GDP. Growth rates for both scenarios differ by region. In East Asia and the Pacific and in South Asia, growth rates over the past decade have been on track to achieve the poverty headcount Goal assuming standard elasticities between growth and poverty reduction. In these regions, we assume that per capita incomes will continue to grow at 5 percent under both scenarios. For the other regions, we extrapolate current trends by using the average growth rate that the region experienced from 1990 to 2002, except in Europe and Central Asia, where we extrapolate growth rates since 1997 (as a more representative post-transition trend over five years). We project the MDG scenario for these slower growing regions by assuming an acceleration of real per capita growth to 3.2 percent.

Undernourishment

Undernourishment trends by region are calculated using population-weighted country data from 1990 and 2000 from FAO (2003a) for all regions except Europe and Central Asia, where data from 1994 to 2000 are used. Trends are extrapolated to 2005 and 2015 to estimate undernourishment under the assumption that current trends continue. The MDG scenario is calculated assuming that the share of undernourished people halves in each country between 1990 and 2015.

Child mortality

Regional under-five mortality rates are calculated using population-weighted national mortality rates (per 1,000 live births) for 1990 and 2002 and extrapolating them through 2005 and 2015. We then multiply these estimates by projected birth rates (United Nations Population Division 2003a) to calculate children's lives lost on trends. The MDG scenario is calculated by projecting a fall in mortality by two-thirds from the 1990 rate and interpolating linearly between 2005 and 2015. The number of lives lost under the MDG scenario was calculated by applying this trajectory to projected birth rates, assuming that countries that have exceeded progress required to meet the target will continue on their historic trend.

Maternal mortality

Insufficient national data are available to estimate regional averages for maternal mortality ratios. We therefore use global totals. According to statistics compiled from UNFPA, UNICEF, and WHO, the global maternal mortality ratio did not change between 1995 and 2000 (400 deaths per 100,000 live births) (WHO and UNICEF 1996; WHO, UNICEF, and UNFPA 2003). We assume further that it remains unchanged through to 2005. Since the worldwide rate was 430 per 100,000 live births in 1990, achieving the Goal would require reducing the worldwide rate to 108 (a 75 percent reduction). So under the MDG scenario the maternal mortality ratio will fall linearly from 400 in 2005 to 108 in 2015. If current trends continue, the maternal mortality ratio will stay constant at 400 through to 2015. We calculate the number of deaths under the two scenarios, and then take the difference between the outcome under the MDG scenario in 2015 (of meeting the maternal mortality ratio target of 108 deaths per 100,000 live births) and the 2005 estimates to determine the number of lives that would be saved.

New HIV infections prevented

Data and calculations are from Stover and others (2002), who project the cumulative number of new HIV infections between 2002 and 2010 in a baseline trajectory and compare it with an expanded response scenario, as outlined in the July 2001 United Nations General Assembly Special Session (UNGASS). The difference between these two scenarios yields the infections averted by an expanded response by 2010. Although the expanded response scenario is not extended through 2015 and uses assumptions that differ slightly from those recommended by the UN Millennium Project Working Group on HIV/AIDS, it illustrates what an MDG scenario for HIV/AIDS might look like.

Water and sanitation

We use data on access to improved water supply and sanitation from the WHO/UNICEF Joint Monitoring Program (WHO and UNICEF 2004). As for other indicators, we extrapolate current trends, estimate an MDG scenario, and calculate the difference between the outcomes under the MDG scenario in 2015 and the 2005 estimates to determine the number of lives affected between 2005 and 2015.

Individuals living in slum conditions

We use country-level statistics for urban populations in 2001 (United Nations Population Division 2003a) and the proportion of urban populations living in slums (UN-HABITAT 2003) to determine the number of slum dwellers today. The number of slum dwellers in 2020 given current trends is estimated by holding constant the share of urban populations living in slums and multiplying it with projected urban populations in 2020 (United Nations Population Division 2001, 2003a). The MDG scenario is estimated by holding constant the total number of slum dwellers in 2001, assuming that the formation of new slums will be halted, and subtracting a further 100 million people in accord with target 11. The 100 million were distributed across the regions on a pro rata basis.

Notes

Overview

1. Extreme poverty figures are for $1.08 per day in 1990 and 2001.

2. Scheduled for publication in early 2005.

3. Language often fosters confusion in distinguishing between inputs and outcomes. "Health," for instance, describes both a sector (or ministry) and an outcome resulting from a complex set of inputs across sectors. "Education" is similar. "Hunger," in contrast, is a complex outcome but not a sector or ministry.

Chapter 2

1. The poverty headcount ratio is the proportion of the national population whose incomes are below the official threshold(s) set by the national government.

2. Tropical Sub-Saharan Africa as defined in Sachs and others (2004).

3. Several other indicators of absolute poverty were attempted for inclusion in this map but not used due to data sparseness. Unless otherwise stated, all data are from UNDESA (2004).

4. For this section, we follow the regional groupings and use regional data as presented in UNDESA (2004). [http://millenniumindicators.un.org/unsd/mi/mi_worldmillennium_new.asp]. However, we include Armenia, Azerbaijan, and Georgia with the CIS countries of Europe and include Iran in Western Asia.

Chapter 3

1. Using many different growth model specifications, Sala-i-Martin, Doppelhofer, and Miller identify tropical land area, coastal population density, and malaria prevalence as among the more robust determinants of economic growth. See Sala-i-Martin, Doppelhofer, and Miller (2004).

2. A detailed study on Africa's unique challenges and the investments and financing it needs to break the poverty trap is Sachs and others (2004).

3. This is the critique perhaps best represented by Easterly (2001).

Chapter 4

1. It should also include gender expertise, which is too frequently dropped in real-time activities.

Chapter 5

1. Essential health services related to the Goals include preventive and curative health, reproductive health, environmental health, and nutrition.

2. These interventions cannot work in isolation: they need to be accompanied by appropriate policy reforms. These include increased allocations of the national budget to agriculture, linking nutrition with agriculture food security and the right to adequate food and funding it appropriately, empowering women with property rights to local resources, strengthening agricultural research, and removing internal and regional barriers to trade in agricultural inputs and products (UN Millennium Project 2005d).

3. See for example, World Energy Council, UN DESA, and UNDP (2000). Energy services for the Goals are discussed in more detail in a forthcoming UN Millennium Project background paper by Vijay Modi (2004).

4. From a public health perspective, it is important to reduce recourse to abortion by promoting family planning services and levels of unsafe abortion by reviewing the laws regulating abortion to ensure that they address risks to health as well as cultural values and diversity and institutional capacity.

5. Social protection consists of health insurance, disability through work, unemployment insurance, child maintenance, social security, and old-age pensions.

Chapter 6

1. Tools for the job are also an important part of capacity and include consumables or commodities required for service delivery. These are part and parcel of the recommendations in chapter 5 and in the task force reports. Here we focus on the first three components of capacity.

2. See, for example, the case studies and international study of the PARIS21 Task Team on Improved Statistical Support for Monitoring Development Goals, available at www.paris21.org.

3. Deininger and Mpuga (2004) conclude: "We find that the policy change [abolition of user fees] improved access and reduced the probability of sickness in a way that was particularly beneficial to the poor. Although the challenge of maintaining service quality remains, aggregate benefits are estimated to be significantly larger than the estimated shortfalls from the abolition of user fees."

Chapter 7

1. Many researchers have advanced this point, with important early evidence presented by Knack and Keefer (1995), Mauro (1995), and Sachs and Warner (1995).

2. A number of recent studies have analyzed the relationship between governance and income in detail. For example, Radelet (2004) constructs an adjusted governance indicator controlling for levels of income. It shows that many low-income countries perform very well on governance relative to their level of income.

3. Glaeser and others (2004) also make this point.

4. See, for example, "The Right of Everyone to the Highest Attainable Standard of Physical and Mental Health" (UN 2004d).

5. Article 25 of the United Nations International Covenant on Civil and Political Rights states only that "Every citizen shall have the right and the opportunity, [without discrimination] and without unreasonable restrictions: (a) to take part in the conduct of public affairs, directly or through freely chosen representatives ...." This statement of the right to participate must be seen in the context of other rights recognized in the Universal Declaration, which give substance to it. They include the right to freedom of speech, the right to nondiscrimination, the rights to assembly and association, the right to a free press, and so on.

6. See UN (1998) on "The role of national human rights institutions in the protection of economic, social and cultural rights." (E/C.12/1998/25); and see UN (2002e) on "The role of independent national human rights institutions in the promotion and protection of the rights of the child."

7. World Development Report 2005 draws this analysis from World Bank (2004a).

Chapter 8

1. Civil society organizations refer to a broad group of organizations and actors including but not limited to community-based organizations, nongovernmental organizations, business associations, think tanks, social movements, religious organizations, women's rights movements, grassroots and indigenous people's movements, and voluntary organizations.

Chapter 9

1. The government holds primary responsibility to create incentives for informal entrepreneurs to enter the formal economy—by enabling access to credit, better technologies, and other inputs, by simplifying registration systems, and by providing access to training, as described in chapter 7.

Chapter 10

1. This chapter draws extensively from "Ending Africa's Poverty Trap," Sachs and others (2004).

2. A regression (not reported here) of several different governance indicators on log GDP per capita PPP and a dummy for tropical Sub-Saharan Africa results in a statistically insignificant coefficient for the dummy variable, indicating that Africa is not governed worse, after controlling for income.

3. Other evidence suggests, for example, that between 1986 and 1988 long-distance freight tariffs in francophone West Africa were more than five times higher than comparable tariffs in India, Pakistan, and Viet Nam (Rizet and Hine 1993). A more recent study suggests that long-distance freight rates in Tanzania are three times higher than in Indonesia (Hine and Ellis 2001).

Chapter 11

1. UN Department of Economic and Social Affairs, Statistics Division 2004; based on data provided by the United Nations Children's Fund, World Health Organization, and Food and Agriculture Organization of the United Nations. The numbers represent unweighted averages for Least Developed Countries in the regions.

2. The 14 priority areas included: climate change and sea-level rise; natural and environmental disasters; management of wastes; coastal and marine resources; freshwater resources; land, energy, tourism, and biodiversity resources; national institutions and

administrative capacity; regional institutions and technical cooperation; transport and communication; science and technology; and human resource development (UN 1994).

3. The index for each hazard is created based on population distribution, severity of each hazard, and hazard mortality rate data for a 20-year period from 1981 to 2000. See Dilley and others (2005) for further explanation. The index is based on authors' calculations, by multiplying the country's percentage of population exposed to different severities of each hazard to create a 0–10 index where a higher number indicates more severe hazard exposure or a larger percentage of the population exposed.

Chapter 12

1. Based on Fearon and Laitin (2003) dataset. Note that 3 percent was the average annual risk of a new conflict experience by Sub-Saharan African countries in the 1990s.

2. When the basic issues in a particular conflict are threats to identity, cultural or religious, not the economic disadvantage of some regions or groups, more political measures may have to be undertaken.

3. By structural indicators we refer to economic, social, environmental and geographic descriptors of groups and communities; by behavioral indicators we refer to individual and group decisionmaking indicators, including protests, strikes, and riots.

4. For further evidence on the lingering public health effects of civil wars, see Ghobarah, Huth, and Russett (2004).

5. Related proposals have already been voiced in other contexts. See, for example, International Peace Academy (2004).

Chapter 13

1. For example, Harrison, Klugman, and Swanson (2003) show the lack of PRSP alignment with the Millennium Development Goals. Their table 5, for example, indicates that 7 of 12 recent PRSPs do not even include targets that can be compared with the hunger Goal. Similarly, 7 of 12 do not have targets relevant to the education Goal.

Chapter 15

1. For example, in 1999 only 1 percent of exports from Least Developed Countries went to other LDCs, with other developing countries accounting for another 28 percent (UNCTAD 2001).

2. For example, the Inter-American Development Bank allocates a mere $10 million of a $6 billion portfolio to grants-based funding for regional infrastructure (Birdsall 2004).

3. The Johannesburg Plan of Implementation states: "All countries should promote sustainable consumption and production patterns, with the developed countries taking the lead" (paragraph 14).

Chapter 17

1. Subtracting $21 billion from the MDG financing gap of $73 billion in 2005 yields $52 billion. Similarly, subtracting $25 billion from $135 billion yields $110 billion.

2. Analysis for this chapter was conducted using 2002 data. Aggregate ODA figures for 2003 were released just prior to publication of this report and are thus included in the discussion here.

3. The *Handbook for Best Practice to Meet the MDGs,* scheduled for publication in early 2005, describes the needs assessments in more detail.

4. The difficulty of estimating the potential for increasing resource mobilization by the government is reflected in conflicting comments received on earlier drafts of the UN Millennium Project's needs assessment results. The World Bank judged an increase of four percentage points in Uganda as too low, while the IMF thought the same estimate was too high (World Bank 2003e; IMF 2003).

5. This assumes that 25 percent of all MDG-related investments will go toward tradables, as suggested by the UN Millennium Project needs assessments.

6. All data on 2002 ODA flows are based on OECD/DAC (2004a). Additional information has been provided by the OECD/DAC secretariat. We are particularly grateful to Brian Hammond for his invaluable support. Additional details on the analysis are provided in appendix 3.

7. Unfortunately, no robust cost estimates are available for implementing the other two Rio conventions. Table 17.3 thus does not include the cost of protecting biodiversity and implementing the UN Framework Convention on Climate Change.

8. The OECD/DAC uses a projection of 2 percent a year real growth in members' GNI—from $30 trillion in 2005 to $36 trillion in 2015.

Chapter 18

1. The estimation methodology and data sources used are summarized in appendix 4.

Appendix 1

1. These primary and postprimary education categories focus on priorities highlighted by the UN Millennium Project Task Force on Education, although they are not an exhaustive list of all task force interventions.

Appendix 3

1. The following assumptions were made based on public expenditure data: Pakistan and Viet Nam 8–12 percent, Indonesia 9–13 percent, China and India 14–16 percent.

2. Under the terms of the OECD/DAC, concessional loans do not qualify for debt relief that counts toward ODA.

3. For comparison, in 2003 only 53 percent of appeals for emergency aid made through the UN's Consolidated Appeals Process in 2003 were funded (excluding Iraq), even if one assumes that some funding requests may be too high.

Appendix 4

1. Available online at http://www.oecd.org/dataoecd/43/43/1894330.xls.

References

Abdullah II. 2004. "Address to World Economic Forum." Presented at Annual Meeting, January 23, Davos, Switzerland. [Retrieved on November 30, 2004, from www.kingabdullah.jo/press_room/speechpage.php?ki_serial=231&menu_id=607&lang_hmka1=1].

African Agricultural Market Information Network. 2004. "Regional Inputs." [Retrieved on March 8, 2004, from www.afamin.net/regionalenglish/reg_mis_en.asp].

Alston, P. 2004. "A Human Rights Perspective on the Millennium Development Goals." Background paper for the UN Millennium Project Task Force on Poverty and Economic Development. UN Millennium Project, New York.

Andreassi, T. 2003. "Innovation in Small and Medium Enterprises." *International Journal of Entrepreneurship and Innovation Management* 3 (1/2): 99–106.

Annan, K. 2004. "Africa's Green Revolution—A Call to Action." Address to the High-level Seminar on Innovative Approaches to Meet the Hunger Millennium Development Goal in Africa, July 5, Addis Ababa. [Retrieved on December 7, 2004, from www.un.org/News/Press/docs/2004/sgsm9405.doc.htm].

Asian Development Bank. 2003. "Millennium Development Goals in the Pacific: Relevance and Progress." Manila.

Atkinson, A. B. 2004. "New Sources of Development Finance: Funding the Millennium Development Goals." WIDER Policy Brief 10. United Nations University, World Institute for Development Economics Research, Helsinki. [www.wider.unu.edu/publications/policy-brief/PB10.pdf].

Bajpai, N., J. D. Sachs, and N. Volavka. 2004. "Reaching the Millennium Development Goals in South Asia." Background paper for the UN Millennium Project Task Force on Poverty and Economic Development. UN Millennium Project, New York.

Balmford, A., A. Bruner, P. Cooper, R. Costanza, S. Farber, R. E. Green, M. Jenkins, P. Jefferiss, V. Jessamy, J. Madden, K. Munro, N. Myers, S. Naeem, J. Paavola, M. Rayment, S. Rosendo, J. Roughgarden, K. Trumper, and R. K. Turner. 2002. "Ecology—Economic Reasons for Conserving Wild Nature." *Science* 297:950–53.

Barnes, D. F., K. Openshaw, K. Smith, and R. van der Plas. 1994. *What Makes People Cook with Improved Biomass Stoves?* World Bank Technical Paper 242. Energy Series. Washington, D.C.

Barrios Salvador, L. Bertinelli, and E. Strobl. 2003. "Dry Times in Africa: Rainfall and Africa's Growth Performance." Center for Operations Research and Econometrics (CORE) Discussion Paper 2003/61. Université catholique de Louvain, Belgium.

Barro, R. J. 1999. "Determinants of Democracy." *Journal of Political Economy* 107 (0): S158–83.

Barro, R. J., and J. W. Lee. 2000. "International Data on Educational Attainment: Updates and Implications." CID Working Paper 42. Center for International Development. [www.cid.harvard.edu/ciddata/ciddata.html].

Bernstein, S. 2004. "A Proposal for Including a Measure of Unmet Need for Contraception and Adolescent Fertility or Early Marriage Levels as Indicators of the Reproductive Health Component of Gender Equality." UN Millennium Project, New York. [http://unstats.un.org/unsd/mi/techgroup/subgroups/IAEG submission on unmet need v3.pdf].

Bhalla, S. 2002. *Imagine There's No Country: Poverty, Inequality, and Growth in the Era of Globalization.* Washington, D.C.: Institute for International Economics.

Bill and Melinda Gates Foundation. 2003. *Annual Report 2003.* Seattle, Washington.

Birdsall, N. 2004. "Underfunded Regionalism in the Developing World." CGD Working Paper 49. Center for Global Development, Washington, D.C.

Blair, T. 2004. "Address to African Union." October 7, Addis Ababa. [Retrieved on November 30, 2004, from www.pm.gov.uk/output/Page6452.asp].

Bogg, L., D. Hengjin, W. Keli, C. Wenwei, and V. Diwan. 1996. The Cost of Coverage: Rural Health Insurance in China." *Health Policy and Planning* 11: 238–52.

Bruton, G., D. Ahlstrom, and K. Yeh. 2003. "Understanding Venture Capital in East Asia: The Impact of Institutions on the Industry Today and Tomorrow." *Journal of World Business* 39 (1): 72–88.

Burnside, C., and D. Dollar. 2000. "Aid, Policies, and Growth." *American Economic Review* 90 (4): 847–68.

Bush, G. W. 2002. "Address to the Inter-American Development Bank." March 14, Washington, D.C. [Retrieved on November 30, 2004, from www.whitehouse.gov/news/releases/2002/03/20020314-7.html].

Caixa Economica Federal, Instituto Polis, UN-HABITAT Office for Latin America and the Caribbean. 2002. "The Statute of the City: New Tools for Assuring the Right to the City in Brasil." [www.polis.org.br/publicacoes/download/arquivos/statute_of_the_city.pdf]

Carter, S., and B. Currie-Alder. n.d. "Scaling Up Natural Resource Management: Insights from Research in Latin America." International Development Research Center, Ottawa.

Casterline, J., and S. W. Sinding. 2000. "Unmet Need for Family Planning in Developing Countries and Implications for Population Policy." *Population and Development Review* 26 (4): 691–724.

Charlton, A. 2004. "Why Is There So Little Foreign Investment in Most Developing Countries: Vertical FDI in a Multi-country World." Background Paper for the UN Millennium Project. UN Millennium Project, New York.

Chen, S., and M. Ravallion. 2004. "How Have the World's Poorest Fared since the Early 1980s?" Policy Research Paper 3341. World Bank, Washington, D.C.

Chirac, J. 2004. "Scaling Up Poverty Reduction—A Global Learning Process." Address for the opening of the World Bank Conference, May 26, Shanghai. [Retrieved on November 30, 2004, from www.elysee.fr/cgi-bin/auracom/aurweb/search/file?aur_file=discours/2004/UK040526.html].

Chocce, G. R. 2003. "Necessary Conditions for Venture Capital Development in Latin America: The Chilean Case." *International Journal of Entrepreneurship and Innovation Management* 3(1/2): 139–50.

Christiansen, L., C. Scott, and Q. Wodon. 2002. "Development Targets and Costs." In J. Klugman, ed., *A Sourcebook for Poverty Reduction Strategies*. Vol. 2. World Bank, Washington, D.C.

CIESIN (Center for International Earth Science Information Network), Columbia University. 2002. "National Aggregates of Geospatial Data: Population, Landscape and Climate Estimates (PLACE)." Palisades, N.Y. [http://sedac.ciesin.columbia.edu/plue/nagd/place.html].

———. 2004. "Gridded Population of the World (GPW)." Version 3. Palisades, N.Y. [http://sedac.ciesin.columbia.edu/gpw].

———. 2005a (forthcoming). *Global Subnational Infant Mortality Rates [Dataset]*. Palisades, N.Y.

———. 2005b (forthcoming). *Global Subnational Rates of Child Underweight Status [Dataset]*. Palisades, N.Y.

Clapham, A. 2001. "The Question of Jurisdiction under International Criminal Law over Legal Persons: Lessons from the Rome Conference on an International Criminal Court." In M. Kamminga and S. Zia-Ziarifi, eds., *Liability of Multinational Corporations under International Law*. Boston: Martinus Nijhoff.

Clemens, M., S. Radelet, and R. Bhavnani. 2004. "Counting Chickens When They Hatch: The Short-Term Effect of Aid on Growth." Working Paper 44. Center for Global Development, Washington, D.C.

Collier, P., and A. Hoeffler. 2002. "Greed and Grievance in Civil Wars." Working Paper 2002-01. Centre for the Study of African Economies, Oxford, UK.

———. 2004. "The Challenge of Reducing the Global Incidence of Civil War." Copenhagen Consensus Challenge Paper. Copenhagen Consensus, Copenhagen.

Commission on Human Security. 2003. *Human Security Now*. United Nations. New York.

Cooke, R., B. Dickens, and M. Fathalla. 2003. *Reproductive Health and Human Rights: Integrating Medicine, Ethics, and Law (Issues in Biomedical Ethics)*. Toronto: Oxford University Press.

Correa, S. 1999. "ICPD: Moving Forward in the Eye of the Storm." Development Alternatives with Women for a New Era, Suva.

CSR (Corporate Social Responsibility) Platform. 2003. "CSR Frame of Reference." Amsterdam.

da Silva, L. I. L. 2004. "Statement at the 59th Session of the UN General Assembly." September 21, 2004, New York. [Retrieved on November 30, 2004, from www.brazil.org.uk/page.php?cid=1887].

de Rato y Figaredo, R. 2004. "Africa Needs International Help to Turn Promises into Progress." *Business Day*, November 12.

Deaton, A. 2003. "Data for Monitoring the Poverty MDG." Princeton University, Research Program in Development Studies, Princeton, N.J.

Deininger, K., and P. Mpuga. 2004. "Economic and Welfare Effects of the Abolition of Health User Fees: Evidence from Uganda." Policy Research Working Paper 3276. World Bank, Washington, D.C.

DFID (Department for International Development). 2004. "Improving Health in Malawi: a Sector-Wide Approach Including Essential Health Package and Emergency Human Resources Programme." Programme Memorandum. London.

Dholakia, R., A. Kumar, and S. Datta. 2004. "Millennium Development Goals Needs Assessment at State Level in India: Madhya Pradesh, Rajasthan, Uttar Pradesh." Background paper for the UN Millennium Project Task Force on Poverty and Economic Development. UN Millennium Project, New York.

DHS (Demographic and Health Surveys). 2004. "Demographic and Health Surveys (DHS) Database." [www.measuredhs.com].

Dilley M., R. Chen, U. Deichmann, A. Lerner-Lam, and M. Arnold with J. Agwe, P. Buys, O. Kjekstad, B. Lyon, and G. Yetman. 2005. "Natural Disaster Hotspots: A Global Risk Analysis." World Bank, Washington, D.C.

Doppelhofer, G., R. I. Miller, and X. Sala-i-Martin. 2000. "Determinants of Long-Term Growth: A Bayesian Averaging of Classical Estimates (BACE) Approach." NBER Working Paper 7750. National Bureau of Economic Research, Cambridge, Mass.

Durand-Lasserve, A., and L. Royston, eds. 2002. *Holding Their Ground: Secure Land Tenure for the Urban Poor in Developing Countries.* Earthscan: London.

Easterly, W. 1999. "The Ghost of Financing Gap: Testing the Growth Model of the International Financial Institutions." *Journal of Development Economics* 60 (2): 423–38.

———. 2001. *The Elusive Quest for Growth: Economists Adventures and Misadventures in the Tropics.* Cambridge, Mass.: MIT Press.

EI (Earthwatch Institute) (Europe), IUCN (International Union for Conservation of Nature and Natural Resources), and WBCSD (World Business Council for Sustainable Development). 2002. *Business and Biodiversity: The Handbook for Corporate Action.* Geneva.

EIA (Energy Information Administration). 1999. *Energy in Africa.* Washington, D.C.

Ellis, S., and J. L. Hine. 1989. "The Provision of Rural Transport Services." Sub-Saharan Africa Transport Policy Program Working Paper 37. World Bank, Washington, D.C.

Evenson, R. 2004. Department of Economics, Yale University, personal communication with the authors.

Evenson, R. E., and D. G. 2003. *Crop Variety Improvement and Its Effect on Productivity: The Impact of International Research.* Wallingford, UK: CABI Publishing.

FAO (Food and Agriculture Organization). 2003a. *The State of Food Insecurity in the World.* Rome.

———. 2003b. "FAO Statistical Databases." Rome.

———. 2004. *The State of Food Insecurity in the World.* Rome.

Faye, M., J. McArthur, T. Snow, and J. Sachs. 2004. "The Challenge Facing Landlocked Developing Countries." *Journal of Human Development* 5 (1): 31–68.

Fearon, J. D., and D. D. Laitin. 2003. "Ethnicity, Insurgency, and Civil War." *American Political Science Review* 91 (1): 75–90.

Finger, J. M., and P. Schuler. 2000. "Implementation of Uruguay Round Commitments: The Development Challenge." *The World Economy* 23:511–26.

Foster, M., and A. Keith. 2003. "The Case for More Aid. Final Report to the Department for International Development." Mick Foster Economics Ltd., Essex, UK. [www.odi.org.uk/PPPG/cape/seminars/may04papers/Foster_Case_for_Increased_Aid_Summary.pdf].

Freedom House. 2003. *Freedom in the World 2003.* New York. [www.freedomhouse.org/research/index.htm].

Gallup, J., J. Sachs, and A. Mellinger. 2003. "Geography and Economic Development." *International Regional Science Review* 22 (2): 179–232.

GEMS (Global Environmental Monitoring and Research Center). 1995. *Environmental Data Report.* London: Blackwell Publishers.

Ghebreyesus, T. A., T. Alemayehu, A. Bosman, K. H. Witten, and A. Teklehaimanot. 1996. "Community Participation in Malaria Control in Tigray Region Ethiopia." *Acta Tropica* 61 (2): 145–56.

Ghebreyesus, T.A., K. H. Witten, A. Getachew, K. O'Neill, A. Bosman, and A. Teklehaimanot. 1999. "Community Based Malaria Control in Tigray, Northern Ethiopia." *Parasitologia* 41:367–71.

Ghobarah, H. A., P. Huth, and B. Russett. 2004. "The Post-War Public Health Effects of Civil Conflict." *Social Science and Medicine* 59:869–84.

Glaeser, L., R. LaPorta, F. López-de-Silanes, and A. Shleifer. 2004. *Do Institutions Cause Growth?* NBER Working Paper 10568. Cambridge, Mass.: National Bureau of Economic Research.

Global Forum for Health Research. 2002. "10/19 Report on Health Research 2001–2002." Geneva. [www.globalforumhealth.org/Files Upld/36.pdf].

Grameen Trust. 2004. "Scaling Up Microfinance in Millennium Project Pilot Projects." Background note for the UN Millennium Project. Dhaka.

Gray, C., and D. Kaufmann. 1998. "Corruption and Development." *Finance and Development* 35 (1): 7–10.

Gray, C., J. Hellman, and R. Ryterman. 2004. *Anti-Corruption in Transition 2: Corruption in Enterprise-State Interactions in Europe and Central Asia 1999–2002.* Washington, D.C.: World Bank.

Greene, M., M. Mehta, J. Pulerwitz, D. Wulf, A. Bankole, and S. Singh. 2005. Forthcoming. "Involving Men in Reproductive Health: Contributions to Development." Background paper for the UN Millennium Project. UN Millennium Project, New York.

Government of Kazakhstan and United Nations Country Team. 2002. *Millennium Development Goals in Kazakhstan.* Almaty.

Gwatkin, D., S. Rutstein, K. Johnson, R. Pande, and A. Wagstaff. 2003. *Initial Country-Level Information about Socio-Economic Differences in Health, Nutrition and Population.* 2nd ed. Washington, D.C.: World Bank.

Harrison, M., J. Klugman, and E. Swanson. 2003. *Are Poverty Reduction Strategies Undercutting the Millennium Development Goals? An Empirical Review.* Washington, D.C.: World Bank.

Heckman, J. 1999. *Policies to Foster Human Capital.* NBER Working Paper 7288. Cambridge, Mass: National Bureau of Economic Research. [http://nber.org/papers/w7288.]

Henao, J., and C. Baanante. 1999. "Nutrient Depletion in the Agricultural Soils of Africa." 2020 Brief 62. International Food Policy Research Institute, Washington, D.C. [www.cgiar.org/ifpri/2020/briefs].

Hertel, T., and W. Martin. 2000. "Liberalizing Agriculture and Manufactures in a Millennium Round: Implications for Developing Countries." *The World Economy* 23 (April): 455–69.

Hine, J. L., and S. D. Ellis. 2001. "Agricultural Marketing and Access to Transport Services." Transport Research Laboratory. Wokingham, UK.

Homer-Dixon, T. 1994. "Environmental Scarcities and Violent Conflict: Evidence from Cases." *International Security* 16 (1): 4–40.

Humphreys, M., and A. Varshney. 2004. "Violent Conflict and the Millennium Development Goals: Diagnosis and Recommendations." Background paper for the UN Millennium Project Task Force on Poverty and Economic Development. UN Millennium Project Indigenous Peoples Forum 2004, New York.

IDA (International Development Association) and IMF (International Monetary Fund). 2003. "Republic of Mali: Poverty Reduction Strategy Paper Joint Staff Assessment." [http://poverty.worldbank.org/files/Mali_PRSP_JSA.pdf].

ILO (International Labour Organization). 2003. "Working Out of Poverty." Report of the Director-General at the International Labor Conference, 91st Session, June 3–19, Geneva.

IMF (International Monetary Fund). 2003. Personal communication. Washington, D.C.

———. 2004. Personal communication. Washington, D.C.

International Alert. 2004. "Building Institutional Capacity for Conflict Sensitive Practice." [Retrieved on November 18, 2004, from www.international-alert.org/pdf/pubdev/institutional_capacity_ngos.pdf].

International Peace Academy. 2004. "Building Effective Partnerships: Improving the Relationship between Internal and External Actors in Post-Conflict Countries." The WSP Peacebuilding Forum Conference October 7, New York.

International Rescue Committee and Ministry of Health and Sanitation, Sierra Leone. 2001. "Mortality in Kenema District in Sierra Leone: A Survey Covering January 2000–2001." Sierra Leone.

IWHC (International Women's Health Coalition). 2004. [Retrieved on November 20, 2004, from www.iwhc.org].

IPCC (Intergovernmental Panel on Climate Change). 2001a. *Climate Change 2001: The Scientific Basis. Contribution of Working Group I to the Third Assessment Report of the Intergovernmental Panel on Climate Change.* Eds.: J. T. Houghton, Y. Ding, D. J. Griggs, M. Noguer, P. J. van der Linden, X. Dai, K. Maskall, and C. A. Johnson. Cambridge: Cambridge University Press.

———. 2001b. *Climate Change 2001: Impacts, Adaptation, and Vulnerability, Contribution of Working Group II to the Third Assessment Report of the Intergovernmental Panel on Climate Change.* Eds.: J. J. McCarthy, O. F. Canziani, N. A. Leary, D. J. Dokken, and K. S. White. Cambridge: Cambridge University Press.

———. 2001c. *Climate Change 2001: Mitigation, Contribution of Working Group III to the Third Assessment Report of the Intergovernmental Panel on Climate Change.* Eds.: B. Metz, O. Davidson, R. Swart, and J. Pan. Cambridge: Cambridge University Press.

IPU (Inter-parliamentary Union). 2004. "Women in National Parliaments Data Tables." Geneva. [www.ipu.org/wmn-e/classif.htm].

Jadresic, A. 2000. "A Case Study on Subsidizing Rural Electrification in Chile." In *Energy Services for the World's Poor.* Washington, D.C.: World Bank.

Joint Learning Initiative. 2004. *Human Resources for Health—Overcoming the Crisis.* Cambridge, Mass.: Harvard University Press.

Kammen, D. M. 1995. "From Energy Efficiency to Social Utility: Improved Cookstoves and the Small Is Beautiful Model of Development." In J. Goldemberg and T. B. Johansson, eds., *Energy as an Instrument for Socio-economic Development.* New York: United Nations Development Programme.

Kaufmann, D. 2004. Background Note for the UN Millennium Project Task Force on Poverty and Economic Development. UN Millennium Project, New York.

Kaufmann, D., A. Kraay, and M. Mastruzzi. 2003. "Governance Matters III: Governance Indicators for 1996–2002." Policy Research Working Paper 3106. World Bank, Washington, D.C. [www.worldbank.org/wbi/governance/pubs/govmatters3.html].

Kaufmann, D., A. Kraay, and P. Zoido-Lobaton. 2002. "Governance Matters II—Updated Indicators for 2000/01." Policy Research Working Paper 2772. World Bank, Washington, D.C.

Kaul, I., P. Conceicao, K. L. Goulven, and R. U. Mendoza, eds. 2003. *Providing Global Public Goods. Managing Globalization.* Oxford University Press.

Kazakhstan, Government of, and United Nations Country Team–Kazakhstan. 2002. *United Nations Millennium Development Goals in Kazakhstan.* [www.undp.kz/library_of_publications/center_view.html?id=64&back=1].

Kidane, G., and R. H. Morrow. 2000. "Teaching Mothers to Provide Home Treatment of Malaria in Tigray, Ethiopia: A Randomized Trial." *The Lancet* 356:550–5.

Kiszewski, A., A. Mellinger, A. Spielman, P. Malaney, S. E. Sachs, and J. Sachs. 2004. "A Global Index Representing the Stability of Malaria Transmission." *American Journal of Tropical Medicine and Hygiene* 70 (5): 486–98.

Kjaerum, M. 2003. "National Human Rights Institutions Implementing Human Rights." In M. Bergsmo, ed., *Human Rights and Criminal Justice for the Downtrodden: Essays in Honour of Asbjørn Eide.* Leiden, The Netherlands: Martinus Nijhoff.

Knack, S., and P. Keefer. 1995. "Institutions and Economic Performance: Cross-Country Tests Using Alternative Measures." *Economics and Politics* 7 (Nov.): 207–27.

Koenig, M. A., M. B. Hossain, and M. Whittaker. 1997. "The Influence of Quality of Care upon Contraceptive Use in Rural Bangladesh." *Studies in Family Planning* 28 (4): 278–89.

Koizumi, J. 2004. "A New United Nations for the New Era." Address to the 59th Session of the General Assembly of the United Nations, September 21, New York. [Retrieved on November 30, 2004, from www.kantei.go.jp/foreign/koizumispeech/2004/09/21address_e.html].

Kreimer, A., M. Arnold, and A. Carlin, eds. 2003. *Building Safer Cities: The Future of Disaster Risk.* Washington, D.C.: World Bank.

Kremer, M. 2002. "A Purchase Commitment for Vaccines." In I. Kaul, K. L. Goulven, and M. Schnupf, eds. *Global Public Goods Financing: New Tools for New Challenges, A Policy Dialogue.* New York: UN Development Programme, Office of Development Studies.

Landau, J. P. 2004. "Les Nouvelles Contributions Internationales. Rapport au Président de la République." La Documentation Française, Paris.

Levine, R., and M. Kinder. 2004. *Millions Saved: Proven Successes in Global Health.* Center for Global Development, What Works Working Group. Washington, D.C.

Limao, N., and A. J. Venables. 1999. "Infrastructure, Geographical Disadvantage and Transport Costs." London School of Economics, Department of Economics, London.

Liu, Y., K. Rao, and W. Hsiao. 2003. "Medical Expenditures and Rural Impoverishment in China." *Journal of Health, Population and Nutrition* 21 (3): 216–22.

Lowell, L., and A. M. Findlay. 2001. "Migration of Highly Skilled Persons from Developing Countries: Impact and Policy Responses." Department for International Development, International Labour Office, London.

Malhotra, M. 2004. "Lessons: Scaling Up Successful Efforts to Reduce Poverty." World Bank Institute, Washington, D.C.

Maddison, A. 2001. *The World Economy: A Millennial Perspective.* Paris: Organisation for Economic Co-operation and Development. [Retrieved on March 2003, from www.theworldeconomy.org/about.htm].

Masters, W. 2002. "Research Prizes: A Mechanism to Reward Agricultural Innovation in Low-Income Regions." *AgBioForum* 5 (4): 1–5.

Mauro, P. 1995. "Corruption and Growth." *The Quarterly Journal of Economics* 110 (3): 681–712.

Maxx, D., R. Chen, U. Deichmann, A. L. Lerner-Lam, and M. Arnold, and others. 2005. *Natural Disaster Hotspots: A Global Risk Analysis.* Washington, D.C.: World Bank.

MDG Technical Support Centre. 2004. "Africa's Green Revolution: A Call to Action." Proceedings of a high-level seminar convened by the government of Ethiopia and the UN Millennium Project, July 5. Addis Ababa.

Miguel, E., S. Satyanath, and E. Sergenti. 2004. "Economic Shocks and Civil Conflict: An Instrumental Variables Approach." *Journal of Political Economy* 112 (August): 725–53.

Mitchell, T. D., M. Hulme, and M. New. 2002. "Climate Data for Political Areas." *Area* 34 (1): 109–112.

Mkapa, B. 2003. As cited in BBC Worldwide Monitoring, January 10, Dar-es-Salaam.

———. 2004. "A Better Way to Help the Least Developed Countries." *International Herald Tribune*, July 13.

ML Infomap Pvt. Ltd. 2003. "Tahsil Digital Map of India." New Delhi.

Modi, V. 2004. "Energy Services for the Millennium Development Goals." Background paper for UN Millennium Project Task Force on Poverty and Economic Development. UN Millennium Project, New York.

Muirhead, S. 2004. *The 2004 Corporate Contributions Report. An Analysis of the Giving Patterns of 232 Major Corporations in 2003*. Research Report R-1355-04-RR. The Conference Board. New York.

Mutti, J. H. 2003. "Foreign Direct Investment and Tax Competition." Institute for International Economics, Washington, D.C.

Nabarro, D., S. Colombo, and A. Griekspoor. 2004. "The Ultimate Challenge: Sustaining Life (and Realizing the Health MDGs) in Fragile States." Paper presented at the High Level Forum on Health-Related MDGs, December 2–3, Abuja.

National Bureau of Statistics of China. 2003. "China Statistical Yearbook 2003." China Statistical Press, Beijing.

Ndong, I., R. M. Becker, J. M. Haws, and M. N. Wegner. 1999. "Men's Reproductive Health: Defining, Designing and Delivering Services." *International Family Planning Perspectives* 25 (January Suppl.): S53–S55. [Retrieved on December 10, 2004, from www.agi-usa.org/pubs/journals/25s5399.html].

Nduru, M. 2004. "Development–Southern Africa: NEPAD Highlights Infrastructure Needs." *IPS (Inter Press Service) News Agency*. July 10. [www.ipsnews.net/interna.asp?idnews=24578].

Obasanjo, O. 2004. "Statement at the 59th Session of the United Nations General Assembly." September 23, 2004, New York. [Retrieved on November 30, 2004, from www.un.org/webcast/ga/59/statements/nigeng040923.pdf].

OECD/DAC (Organisation for Economic Co-operation and Development, Development Assistance Committee). 2002. *Aid Targeting the Objectives of the Rio Conventions 1998–2000*. Paris.

———. 2004a. *The DAC Journal Development Co-operation Report 2003*. Vol. 5 (1). Paris. [www.oecd.org/findDocument/0,2350,en_2649_33721_1_119687_1_1_1,00.html].

———. 2004b. "International Development Statistics CD-ROM, 2004 Edition." Paris.

———. 2004c. "Security System Reform and Governance: Policy and Good Practice." Policy Brief. Paris.

———. 2004d. "CRS Database." [Retrieved on December 7, 2004, from www.oecd.org].

———. 2004e. Personal communication. Washington, D.C.

———. Forthcoming. *Development Co-operation Report 2005*. Paris.

OHCHR (Office of the United Nations High Commissioner for Human Rights). 2004. "The Right of Everyone to the Enjoyment of the Highest Attainable Standard of Physical and Mental Health." Resolution 2004/27. April 16, New York.

Okejiri, E. 2000. "Foreign Technology and Development of Indigenous Technological Capabilities in the Nigerian Manufacturing Industry." *Technology in Society* 22 (2): 189–99.

ORC-Macro. 2004. Demographic and Health Surveys (DHS) Database. Data for various years. [www.measuredhs.com].

Oxfam. 2002. "Rigged Rules and Double Standards: Trade, Globalization, and the Fight against Poverty." Oxford, UK.

———. 2004. "Stitched Up: How Rich-Country Protectionism in Textiles and Clothing Trade Prevents Poverty Alleviation." Briefing Paper. Oxford, UK.

Pariani, S., D. M. Heer, and M. D. van Arsdol. 1991. "Does Choice Make a Difference to Contraceptive Use? Evidence from East Java." *Studies in Family Planning* 22 (6): 384–90.

PARIS21 Task Team. "Case Studies and International Study on Improved Statistical Support for Monitoring Development Goals." [www.paris21.org].

Prati, A., R. Sahay, and T. Tressel. 2003. "Is There a Case for Sterilizing Foreign Aid Inflows?" Working paper preliminary draft presented at the 18th annual European Economic Association Congress and the 58th Econometric Society European Meeting, August 20–24, Stockholm. [www.eea-esem.com/papers/eea-esem/2003/2499/Sterilizing-Aid.pdf].

PRB (People's Republic of Bangladesh). 2002. "Briefing Paper." Ministry of Health and Family Welfare, Department of Public Health and Engineering (DPHE) and Director General Health Services (DGHS). Dhaka. [http://phys4.harvard.edu/~wilson/arsenic_project_ground_water.html].

PRIO/Uppsala University (International Peace Research Institute, Centre for the Study of Civil War, and Uppsala University, Department of Peace and Conflict Studies). 2004. Armed Conflict Dataset. [www.prio.no/cwp/ArmedConflict/].

PROFAMILIA Colombia. [Retrieved on December 7, 2004, from www.profamilia.org.co/profamilia/english/INDEX.HTM].

PRS Group. 2003. *International Country Risk Guide.* East Syracuse, N.Y. [Retrieved in January, 2003, from www.prsgroup.com/icrg/icrg.html].

Publish What You Pay Campaign. [Retrieved on December 7, 2004, from www.publishwhatyoupay.org/english/].

Radelet, S. 2004. "Aid Effectiveness and the Millennium Development Goals." Working Paper 39. Center for Global Development, Washington, D.C.

Reinikka, R., and J. Svensson. 2004a. "Local Capture: Evidence from a Central Government Transfer Program in Uganda." *Quarterly Journal of Economics* 119 (2): 1–28.

———. 2004b. "The Power of Information: Evidence from a Newspaper Campaign to Reduce Capture." Working Paper 3239. World Bank, Washington, D.C.

Reno, W. 1995. *Corruption and State Politics in Sierra Leone.* Cambridge: Cambridge University Press.

Rizet, C., and J. Hine. 1993. "A Comparison of the Costs and Productivity of Road Freight Transport in Africa and Pakistan." *Transport Reviews* 13 (2): 151–65.

Sachs, J. D. Forthcoming. "Globalization and Patterns of Economic Growth." In Michael M. Weinstein, ed., *Globalization: What's New?* New York: Columbia University Press and Council on Foreign Relations.

Sachs, J. D., and Andrew Warner. 1995. "Economic Reform and the Process of Global Integration." *Brookings Papers on Economic Activity* 1995 (1): 1–118.

Sachs, J. D., J. McArthur, G. Schmidt-Traub, M. Kruk, C. Bahadur, M. Faye, and G. McCord. 2004. "Ending Africa's Poverty Trap." *Brookings Papers on Economic Activity* 2004 (1): 117–216.

Sala-i-Martin, X., G. Doppelhofer, and R. Miller. 2004. "Determinants of Long-Term Growth: A Bayesian Averaging of Classical Estimates (BACE) Approach." *American Economic Review* 94 (4): 813–35.

Sambanis, N. 2004. "Poverty and the Organization of Political Violence: A Review and Some Conjectures." Paper prepared for Brookings Trade Forum 2004, May 13–14, Washington, D.C.

Satterthwaite, D. 2004. "The Under-Estimation of Urban Poverty in Low and Middle-Income Countries." Working Paper on Poverty Reduction in Urban Areas 14. International Institute for Environment and Development, London. [www.iied.org/docs/urban/urbpov_wp14.pdf].

Schröder, G. 2001. "Poverty Reduction—A Global Responsibility. Program of Action 2015." Foreword. [www.gm-unccd.org/FIELD/Bilaterals/Ger/Halving.pdf]

Shi, Y. 2001. "Technological Capabilities and International Production Strategy of Firms: The Case of Foreign Direct Investment in China." *Journal of World Business* 36 (2): 184–204.

Singh, S., J. E. Darroch, M. Vlassoff, and J. Nadeau. 2004. *Adding It Up: The Benefits of Investing in Sexual and Reproductive Health Care.* New York: The Alan Guttmacher Institute and UN Population Fund.

Starkey, P., S. Ellis, J. Hine, and A. Ternell. 2002. "Improving Rural Mobility—Options for Developing Motorized and Nonmotorized Transport in Rural Areas." Technical Paper 525. World Bank, Washington, D.C.

Stoorvogel, J. J., E. M. A. Smaling, and B. H. Janssen. 1993. "Calculating Soil Nutrient Balances in Africa at Different Scales." *Fertilizer Research* 35:227–335.

Stover, J., N. Walker, G. P. Garnett, J. A. Salomon, K. A. Stanecki, P. D. Ghys, N. C. Grassly, R. M. Anderson, and B. Schwartländer. 2002. "Can We Reverse the HIV/AIDS Pandemic with an Expanded Response?" *The Lancet* 360:73–77.

Transparency International. 2004. *Global Corruption Report 2004.* London: Pluto.

Uganda AIDS Commission. 2002. "Overview of HIV/AIDS Coordination." [www.aidsuganda.org/pdf/overview_of_coordination.pdf].

Uganda Ministry of Health. 2003. "Status of Emergency Obstetric Care (EmOC) in Uganda. A National Needs Assessment of EmOC Process Indicators." Kampala.

UK Department of Health. 2004. "Code of Practice for the International Recruitment of Healthcare Professionals." [www.dh.gov.uk/assetRoot/04/08/88/50/04088850.pdf].

UN (United Nations). 1970. "Resolution 2626 (XXV)." United Nations General Assembly. October 24, New York.

———. 1994. "Barbados Programme of Action: Programme of Action for the Sustainable Development of Small Island Developing States." [www.un.org/documents/ga/conf167/aconf167-9.htm].

———. 1995. *Report of the International Conference on Population and Development: The Programme of Action.* [www.unfpa.org/icpd/docs/icpd/icpd_eng.pdf].

———. 1998. "The Role of National Human Rights Institutions in the Protection of Economic, Social and Cultural Rights." General Comment 10. E/C.12/1998/25. Committee on Economic, Social and Cultural Rights, Geneva.

———. 2000. "United Nations Millennium Declaration." A/RES/55/2, Section II. New York.

———. 2001. "Millennium Development Goals: Armenia. Status of Implementation." Office of the United Nations Resident Coordinator. Yerevan.

———. 2002a. "Report of the International Conference on Financing for Development." A/CONF.198/11. New York.

———. 2002b. "Meeting the Millennium Poverty Reduction Targets in Latin America and the Caribbean." United Nations Economic Commission for Latin America and the Caribbean, Instituto de Pesquisas Económicas Aplicadas, and UN Development Programme, Santiago.

———. 2002c. "Review of Progress in the Implementation of the Programme of Action for the Sustainable Development of Small Island Developing States." Report of the Secretary-General. E/CN.17/2004/1. Economic and Social Council, New York.

———. 2002d. "Women, Peace and Security." Study submitted by the UN Secretary-General pursuant to Security Council Resolution 1325 (2000). New York.

———. 2002e. Committee on the Rights of the Child. General Comment 2. "The Role of Independent National Human Rights Institutions in the Promotion and Protection of the Rights of the Child." CRC/GC/2002/2.

———. 2003. "Almaty Programme of Action: Addresssing the Special Needs of Landlocked Developing Countries within a New Global Framework for Transit Transport Cooperation for Landlocked and Transit Developing Countries and the Almaty Ministerial Declaration." [www.un.org/special-rep/ohrlls/imc/Almaty%20Programme%20of%20Action.htm.]

———. 2004a. "A More Secure World: Our Shared Responsibility." Report of the Secretary General's High-Level Panel on Threats, Challenges and Change. New York.

———. 2004b. "Implementation of the United Nations Millennium Declaration: Report of the Secretary-General." A/59/282. New York.

———. 2004c. "Indicators for Assessing Progress towards the 2010 Target: Status and Trends of Linguistic Diversity and Numbers of Speakers of Indigenous Languages." UNEP/CBD/AHTEG-2010-Ind/I/INF/7. Convention on Biodiversity, New York.

———. 2004d. "The Right of Everyone to the Enjoyment of the Highest Attainable Standard of Physical and Mental Health." Note by the Sectetary-General. United Nations General Assembly. A/59/422. October. New York.

———. 2004e. "10 Principles of the UN Global Compact." n.d. [Retrieved on September 5, 2004, from www.unglobalcompact.org/Portal/Default.asp].

UNAIDS (Joint United Nations Programme on HIV/AIDS). 2004. *2004 Report on the Global AIDS Epidemic.* Geneva. [www.unaids.org/bangkok2004/report_pdf.html].

UNAIDS (Joint United Nations Programme on HIV/AIDS) and WHO (World Health Organization). 2003. *AIDS Epidemic Update.* Geneva.

———. 2004. "Global HIV/AIDS Online database." Geneva. [www.who.int/GlobalAtlas/home.asp].

UNCTAD (United Nations Conference on Trade and Development). 2001. *Duty and Quota Free Market Access for LDCs: An Analysis of Quad Initiatives.* Geneva and London. [www.unctad.org/en/docs/poditctabm7.en.pdf].

———. 2003. *Handbook of Statistics.* Geneva.

———. 2004. "Export Performance and Its Determinants: Supply and Demand Constraints." Geneva.

UNDESA (UN Department of Economic and Social Affairs), Statistics Division. 2004. "World and Regional Trends." *Millennium Indicators Database.* [http://millenniumindicators.un.org/unsd/mi/mi_goals.asp]

UNDP (United Nations Development Programme). 2001. *United Nations Development Goals Cambodia 2001.* Phnom Penh. [www.undp.org/mdg/Cambodia.pdf].

———. 2002. "Justice and Security Sector Reform: BCPR's Thematic Approach." Working Paper. New York.

———. 2003a. *The Millennium Development Goals in Arab Countries.* New York. [www.undp.org/mdg/Arab_RegionalReport_english.pdf].

———. 2003b. "Millennium Development Goals Progress Report." Bishkek.

———. 2003c. *Progress toward the Millennium Development Goals in Tajikistan 2003.* Dushanabe. [www.undp.org/mdg/Tajikistan_report.pdf].

———. 2003d. *Human Development Report 2003—MDGs: A Compact among Nations to End Human Poverty.* New York.

———. 2004a. "The Achievement of the Millennium Development Goals in the Caribbean Community." [www.undp.org/rblac/targets/Regional%20Report%20on%20the%20Achievement%20of%20the%20MDGs%20in%20the%20Carib1.pdf]

———. 2004b. *Human Development Report 2004: Cultural Liberty in Today's Diverse World.* New York.

———. 2004c. *Unleashing Entrepreneurship: Making Business Work for the Poor.* Report of the Commission on the Private Sector and Development to the UN Secretary-General. New York.

———. 2004d. *Thailand's Response to HIV/AIDS: Progress and Challenges.* Bangkok.

UNDP (United Nations Development Programme) Albania. 2004. *Albania National MDG Report.* Tirana. [www.undp.org.al/?elib,659].

UNDP (United Nations Development Programme) Ethiopia. 2004. "Note to UN Millennium Project." Addis Ababa.

UNECA (United Nations Economic Commission for Africa). 2004. *Assessing Regional Integration in Africa.* ECA Policy Research Report 135. Addis Ababa. [www.uneca.org/aria/ARIA English_full.pdf].

UNEP (United Nations Environment Program). 1991. "Status of Desertification and Implementation of the United Nations Plan of Action to Combat Desertification." Nairobi.

UN ESCAP (United Nations Economic and Social Commission for Asia and the Pacific and the United Nations Development Programme). 2003. "Promoting the Millennium Development Goals in Asia and the Pacific." ST/ESCAP/2253. [www.unescap.org/].

UNESCO (United Nations Educational, Scientific and Cultural Organization). 2000. "Dakar Framework for Action 2000." [www.unesco.org/education/efa/ed_for_all/dakfram_eng.shtml;

———. 2004. *Education for All Global Monitoring Report 2004.* Paris.

UNFPA (United Nations Population Fund). 2004. *State of the World Population Report 2004: The Cairo Consensus at Ten: Population, Reproductive Health and the Global Effort to End Poverty.* New York.

UN-HABITAT. 2003. *The Challenge of Slums: Global Report on Human Settlements 2003.* Earthscan: London. [www.unchs.org/global_report.asp].

UNICEF (United Nations Children's Fund). 2002. *State of the World's Children 2002.* New York.

UNIFEM (United Nations Development Fund for Women). 2000. *Progress of the World's Women.* New York.

———. 2002. *Progress of the World's Women. Gender Equality and the Millennium Development Goals.* New York.

United Nations Country Team and government of Armenia. 2001. "Millennium Development Goals: Status of Implementation." Yerevan. [www.undp.org/mdg/Millennium goalsARMENIA2001.doc].

United Nations Country Team and government of Bulgaria. 2003. *Millennium Development Goals Report for Bulgaria.* Sofia: UNDP. [www.undp.bg/en/publications.php?content=yes&ID=2&PHPSESSID=d7032e68416fc971a39a5a1f00761e3a2003].

United Nations Country Team and government of Romania. 2003. *Millennium Development Goals Report.* Bucharest. [www.un.ro/pdf/MDGR_ENG.zip].

United Nations Country Team in Georgia. 2004. "Millennium Development Goals in Georgia." Tiblisi. [www.undp.org.ge/news/Georgiamdg.pdf].

United Nations Country Team and government of Thailand. 2004. *Thailand Millennium Development Goals Report 2004.* Bangkok.

United Nations Population Division. 2001. *World Population Prospects: The 2001 Revision.* Percentage of the Population at Mid-Year Residing in Urban Areas by Major Area, Region and Country, 1950–2030. Department of Economic and Social Affairs, New York.

———. 2003a. "World Population Prospects: The 2002 Revision." Population Database. Department of Economic and Social Affairs. [http://esa.un.org/unpp/].

———. 2003b. Levels and Trends in Contraceptive Use Series. Department of Economic and Social Affairs, New York.

———. 2004. *Review and Appraisal of the Progress Made in Achieving the Goals and Objectives of the Programme of the International Conference on Population and Development: The 2004 Report.* Department of Economic and Social Affairs. New York.

UN Millennium Project. 2004a. "An Enhanced Strategy for Reducing Extreme Poverty by the Year 2015." Interim Report. Task Force on Poverty and Economic Development. New York.

———. 2004b. "Millennium Development Goals Needs Assessment: Background Paper to 'Ending Africa's Poverty Trap.'" Working paper. New York. [www.unmillennium-project.org/html/backgroundpaper.shtm].

———. 2005a. *Combating AIDS in the Developing World.* Report of the Task Force on HIV/AIDS, Malaria, TB, and Access to Medicines, Working Group on HIV/AIDS. New York.

———. 2005b. *Coming to Grips with Malaria in the New Millennium.* Report of the Task Force on HIV/AIDS, Malaria, TB, and Access to Medicines, Working Group on Malaria. New York.

———. 2005c. *Environment and Human Well-Being: A Practical Strategy.* Report of the Task Force on Environmental Sustainability. New York.

———. 2005d. *Halving Hunger: It Can Be Done.* Report of the Task Force on Hunger. New York.

———. 2005e. *Health, Dignity, and Development: What Will It Take?* Report of the Task Force on Water and Sanitation. New York.

———. 2005f. *A Home in the City.* Report of the Task Force on Slum Dwellers. New York.

———. 2005g. *Innovation: Applying Knowledge in Development.* Report of the Task Force on Science, Technology, and Innovation.

———. 2005h. *Investing in Strategies to Reverse the Global Incidence of TB.* Report of the Task Force on HIV/AIDS, Malaria, TB, and Access to Medicines, Working Group on TB. New York.

———. 2005i. *Prescription for Healthy Development: Increasing Access to Medicines.* Report of the Task Force on HIV/AIDS, Malaria, TB, and Access to Medicines, Working Group on Access to Medicines. New York.

———. 2005j. *Taking Action: Achieving Gender Equality and Empowering Women.* Report of the Task Force on Education and Gender Equality. New York.

———. 2005k. *Toward Universal Primary Education: Investments, Incentives, and Institutions.* Report of the Task Force on Education and Gender Equality. New York.

———. 2005l. *Trade for Development.* Report of the Task Force on Trade. New York.

———. 2005m. *Who's Got the Power? Transforming Health Systems for Women and Children.* Report of the Task Force on Child and Maternal Health. New York.

———. Forthcoming. *The Handbook for MDG Best Practice.* New York.

UN OHRLLS (United Nations Office of the High Representative for the Least Developed Countries, Landlocked Developing Countries and Small Island Developing States). 2004. "The Criteria for the Identification of the LDCs." [www.un.org/special-rep/ohrlls/ohrlls/default.htm]

UN Permanent Forum on Indigenous Issues. 2004. Background note for UN Millennium Project. UN Millennium Project, New York.

Upadhyay, U. D. 2001. "Informed Choice in Family Planning: Helping People Decide." Population Report J–50. Johns Hopkins University, Bloomberg School of Public Health, Population Information Program, Baltimore, Md.

USAID (United States Agency for International Development). 2002. "What Happened in Uganda?" Washington, D.C. [www.usaid.gov/our_work/global_health/aids/Countries/africa/uganda_report.pdf].

USAID (United States Agency for International Development), UNAIDS (Joint United Nations Programme on HIV/AIDS), WHO (World Health Organization), UNICEF (United Nations Children's Fund), and the POLICY Project. 2004. "Coverage of Selected Services for HIV/AIDS Prevention, Care and Support in Low and Middle Income Countries in 2003." Washington, D.C.

U.S. Department of Energy. 2002. *International Energy Annual 2002.* Washington, D.C. [Retrieved on December 1, 2004,x from http://www.cia.doe.gov/iea].

U.S. PTO (United States Patent and Trademark Office). 2001. "Patent Counts by Country/State and Year: Utility Patents, January 1, 1963–December 31, 2000." Washington, D.C.

Uvin, P., P. S. Jain, and L. D. Brown. 2000. "Think Large and Act Small: Toward a New Paradigm for NGO Scaling Up." *World Development* 28 (8): 1409–19.

Van Gowder, L. 1996. "Assessment of Pre-Science and In-Science Extension Education." SD Dimensions. Food and Agriculture Organization, Rome. [www.fao.org/sd/exdirect/exan0001.htm].

Vanneman, R., and D. Barnes. 2000. *Indian District Data, 1961–1981: Machine-Readable Data File and Codebook (Release 3).* College Park, Md.: Center on Population, Gender, and Social Inequality. [ftp://cwmills.umd.edu/pub/india/].

Varshney, A. 2002. *Ethnic Conflict and Civic Life: Hindus and Muslims in India.* New Haven, Conn.: Yale University Press.

Wagstaff, A., and M. Claeson. 2004. *The Millennium Development Goals for Health—Rising to the Challenges.* Washington, D.C.: World Bank.

Walubengo, D. 1995. "Commercialization of Improved Stoves: The Case of the Kenya Ceramic Jiko (KCI)." In B. Westhoff and D. Germann, eds., *Stove Images: A Documentation of Improved and Traditional Stoves in Africa.* Brussels: Commission of the European Communities.

Water Supply and Sanitation Collaborative Council. 2000. "Vision 21: Water for People, a Shared Vision for Hygiene, Sanitation and Water Supply and a Framework for Action." Geneva. [www.worldwatercouncil.org/Vision/Documents/VISION21FinalDraft.PDF].

WBCSD (World Business Council for Sustainable Development). 2004a. "Corporate Social Responsibility." [Retrieved on November 20, 2004, from www.wbcsd.org/templates/TemplateWBCSD1/layout.asp?type=p&MenuId=MzI3&doOpen=1&ClickMenu=LeftMenu].

———. 2004b. "Rio Tinto: The Biodiversity Partnership Program." Case Study 2004. Geneva. [www.wbcsd.org/web/publications/case/rio_tinto_biodiversity_partnerships_full_case_web.pdf].

WHO (World Health Organization). 2000. *World Health Report Health Systems: Improving Performance.* Geneva.

———. 2001. *Macroeconomics and Health: Investing in Health for Economic Development.* Report of the Commission on Macroeconomics and Health. Geneva.

———. 2002. "Scaling Up the Response to Infectious Diseases." [Retrieved on December 10, 2004, from www.who.int/infectious-disease-report/2002/interventions.html].

———. 2004. *Global TB Report 2004.* [www.who.int/tb/publications/global_report/2004/en/India.pdf].

WHO (World Health Organization) and UNICEF (United Nations Children's Fund). 1996. "Revised 1990 Estimates of Maternal Mortality." Geneva.

———. 2003. *The Africa Malaria Report.* Geneva. [http://mosquito.who.int/amd2003/amr2003/pdf/amr2003.pdf].

WHO (World Health Organization) and UNICEF (United Nations Children's Fund) Joint Monitoring Programme on Water Supply and Sanitation (JMP). 2004. "Meeting the MDG Drinking-Water and Sanitation Target: A Mid-term Assessment of Progress." Geneva.

WHO (World Health Organization), UNICEF (United Nations Children's Fund), and UNFPA (United Nations Population Fund). 2003. "Maternal Mortality in 2000: Estimates Developed by WHO, UNICEF and UNFPA." Geneva. [www.reliefweb.int/library/documents/2003/who-saf-22oct.pdf].

Woo Wing, Thye, S. Li , Y. Ximing, H. Wu, and X. Xinpeng. "The Poverty Challenge for China in the New Millennium." 2004. Background paper for the UN Millennium Project Task Force on Poverty and Economic Development. UN Millennium Project, New York.

World Bank. 2001. "Market Access for Developing Countries' Exports." Research note. [www.worldbank.org/economics/marketaccess.pdf].

———. 2002. *Global Economic Prospects 2002*. Washington, D.C.

———. 2003a. *World Development Indicators 2003*. Washington, D.C.

———. 2003b. *Breaking the Conflict Trap: Civil War and Development Policy*. Policy Report. Washington, D.C.

———. 2003c. "Supporting Sound Policies with Adequate and Appropriate Financing." Report DC2003-0016 prepared for the Development Committee. [http://siteresources.worldbank.org/DEVCOMMINT/Documentation/20127712/DC2003-0016(E)-Financing.pdf].

———. 2003d. *World Development Report 2004: Making Services Work for Poor People*. Washington, D.C.

———. 2003e. Personal communication.

———. 2004a. *Doing Business in 2004: Understanding Regulation*. [http://rru.worldbank.org/DoingBusiness/Main/DoingBusiness2004.aspx].

———. 2004b. *Global Development Finance 2004*. Washington, D.C.

———. 2004c. *World Development Indicators 2004*. Washington, D.C.

———. 2004d. *World Development Report 2005: A Better Investment Climate for Everyone*. New York: Oxford University Press.

———. 2005. *Global Economic Prospects 2005: Overview and Global Outlook*. Washington, D.C.

World Bank Development Data Group. 2004. "The Marrakech Action Plan for Statistics: Better Data for Better Results: An Action Plan for Improving Development Statistics." Presented at the Second International Roundtable on Managing for Development Results, February 4–5, Marrakech, Morocco. [www.mfdr.org/documents/MarrakechActionPlanforStatistics.pdf].

World Energy Council, UN DESA (Department of Economic and Social Affairs), and UNDP (United Nations Development Programme). 2000. "World Energy Assessment: Energy and the Challenge of Sustainability." [www.undp.org/seed/eap/activities/wea/drafts-frame.html].

World Summit on Sustainable Development. 2002. "Johannesburg Plan of Implementation." [www.un.org/esa/sustdev/documents/WSSD_POI_PD/English/POIToc.htm].

Yousef, T. 2004. "Development, Growth and Policy Reform in the Middle East and North Africa since 1950." *Journal of Economic Perspectives* 18 (3): 91–115.

Zhou, L. 2003. "An Estimation of Global Biodiversity Conservation Costs." World Bank, Washington, D.C.

Zuckerman, E. 2001. "Why Engendering PRSPs Reduces Poverty, and the Case of Rwanda." United Nations University, World Institute for Development Economics and Research, Helsinki.

Zuckerman, E. and A. Garrett. 2003. "Do Poverty Reduction Strategy Papers (PRSPs) Address Gender? A Gender Audit of 2002 PRSPs." Gender Action. [www.genderaction.org].

Acknowledgments

Several people made particularly noteworthy contributions in the drafting process for this report. Macartan Humphreys played a central role in the drafting of chapter 12 on strategies for countries affected by conflict. Nirupam Bajpai, Shuming Bao, and Wing Thye Woo provided important analysis to inform chapter 11 on investment priorities in other regions. Deborah Balk, Bob Chen, Marc Levy, Alex de Sherbinin, Adam Storeygard, and their colleagues at the Center for International Earth Science Information Networks (CIESIN) at Columbia University conducted much of the report's geospatial data analysis. Brian Hammond and his colleagues at the Organisation for Economic Co-operation and Development (OECD) Development Assistance Committee (DAC) gave tremendously of their time and expertise to inform the analysis in chapter 13 on fixing the donor system. Dani Kaufmann provided very helpful inputs to chapter 7 on governance. For all these chapters any remaining errors are the authors' own.

The UN Millennium Project benefited from active intellectual collaboration with François Bourguignon, Chief Economist of the World Bank; Nicholas Stern, his distinguished predecessor in that position and now at the U.K. Treasury; and Raghuram Rajan, Chief Economist of the International Monetary Fund (IMF). All were marvelously collaborative and thoughtful in sharing insights and comments on interim products and jointly pursuing the best possible analytical frameworks for achieving the MDGs. Many co-organized seminars and meetings benefited from the contributions of several of their colleagues, including Jim Adams, Shaida Badiee, Barbara Bruns, Mariam Claeson, Shanta Devarajan, Shahrokh Fardoust, Pablo Gottret, Rudolf Knippenberg (UNICEF), Ibrahim Levent, Hans Lofgren, Gobind Nankani, John Page, Ramahatra Rakotomalala, Peter Roberts, Agnes Soucat, Eric Swanson, Jee-Peng Tan, Hans Timmer, and Dominique van der Mensbrugghe at the

World Bank; and Sanjeev Gupta, Peter Heller, and Arvind Subramanian from the IMF.

The UN Millennium Project also thanks its partners in developing a country-level MDG needs assessment methodology, the summary results of which are presented in chapter 17 on expanding the financial envelope to achieve the MDGs: Anwara Begum and M. Salimullah from the Bangladesh Institute for Development Studies; Kao Kim Hourn and Ray Zepp from the University of Cambodia; Ernest Aryeetey and Michael Nimo from the Institute of Statistical, Social and Economic Research in Ghana; Samar Datta, Ravindra Dholakia, and Akhilesh Kumar from the Indian Institute of Management, Ahmedabad; Haidari K. R. Amani, Flora Lucas Kessy, and Deogratias Macha from the Economic and Social Research Foundation in Tanzania; and Godfrey Bahiigwa, Lawrence Bategeka, and Nathan Okarut from the Economic Policy Research Center in Uganda. The UN Millennium Project is indebted to McKinsey & Company, which developed the analysis underlying the energy section of the needs assessment; to Eva Weissman for invaluable assistance in the child health and maternal health analyses; and to David Simon for analysis of official development assistance flows and their sectoral allocation. Shan Cao, Andrew Charlton, Stacy Fehlenberg, Joseph Kennedy, and Pierre Yared performed excellent research assistance in the needs assessment effort.

Countless UNDP staff supported the Project's efforts. Elli Kaplan was invaluable in supporting the Project's early stages. Jeffrey Avina, Ade Lekoetje, Elizabeth Lwanga, Jacques Loup, Lamin Manneh, and Comfort Tetteh were also tremendous contributors to the UN Millennium Project's day-to-day work. Sally Fegan-Wyles and her team at the UN Development Group were ever helpful and supportive. Major substantive contributions were made from throughout the Bureau for Development Policy, including: Susan McDade and the energy group; Antoine Heuty, Terry McKinley, Rathin Roy, and colleagues in the poverty group; Gita Welch and the institutional development group; and Terence Jones and his team from the capacity development group. Djibril Diallo, David Morrison, Bill Orme, and their COA colleagues provided invaluable communications support. The Project also thanks the MDG focal points from the other regional bureaux: Ghaith Fariz, Enrique Ganuza, Balasubramanium Murali, and Norimasa Shimomura. Pedro Conceição, Moez Doraid, Gulden Turkoz-Cosslett, Mattias Johansson, Inge Kaul, Ronald Mendoza, Omar Noman, Alejandra Pero, Bharati Sadasivam, Amina Tirana, and Caitlin Wiesen were also extremely generous with their contributions. Nissim Ezekiel and the secretariat of the Commission on Private Sector and Development offered many helpful suggestions, including Jan Krutzinna, Naheed Nenshi, Yann Risz, and Sahba Sobhani.

The UN Millennium Project collaborated closely with the staff of the Human Development Report Office in the production of *Human Development Report 2003*. Several parts of this report build on work conducted during the production of that Report. We thank Sakiko Fukuda-Parr and her team, including: Silva Bonacito, Emmanuel Boudard, Carla De Gregorio, Haishan Fu, Claes Johansson, Christopher Kuonqui, Santosh Mehrotra, Tanni Mukhopadhyay, Stefano Pettinato, David Stewart, Aisha Talib, Nena Terrell, and Emily White.

This report was edited and produced by the stellar team of Bruce Ross-Larson, Meta de Coquereaumont, Mary Goundrey, Thomas Roncoli, Christopher Trott, Timothy Walker, and Elaine Wilson at Communications Development Incorporated in Washington, D.C.

Advisory inputs

Many colleagues shared important comments on earlier drafts of this report. Members of the UN Millennium Project's United Nations Experts Group have provided invaluable advice since the Project's inception, always under the skilled chairmanship of Jan Vandemoortele. We thank all Experts Group members for their generous and consistent contributions: Ifzal Ali, Adnan Z. Amin, Patrick Asea, Daniel Biau, François Bourguignon, James P. Callahan, Andrew Cassels, Jan Cedergren, Hans D'Orville, David T. Edwards, Marika Fahlen, Orobola Fasehun, Luiz L. Fernandes Pinheiro, Charles Gore, Edward Heinemann, Raj Jumar, Ian Kinniburgh, Eddy Lee, Patrick Low, Richard Morgan, Harish Parvathaneni, Prabhu Pingali, Raghuram Rajan, Joanne Sandler, Francisco Sercovich, Mari Simonen, Joseph Smolik, Dianne Spearman, Carlos Eduardo Velez, and Gustavo R. Zlauvinen.

Special thanks are also due to colleagues at the UN Secretariat: in particular Henk-Jan Brinkman, Marta Mauras, Robert Orr, and Abiodun Williams in the Executive Office of the Secretary-General. Ibrahim Gambari, Eloho Otobo, and Yvette Stephens provided tremendous support from the office of the Secretary-General's Special Advisor on Africa. In the Department of Public Information, Sue Markham, Pragati Pascale, and Tim Wall offered tremendous assistance. Numerous colleagues in the Department of Economic and Social Affairs provided tremendous support throughout the Project, including Joseph Chamie and his team at the Population Division; Robert Johnston, Francesca Perucci, and the team in the Statistics Division; and Johan Schölvinck.

The UN Millennium Project also benefited from collaboration with colleagues at the Millennium Campaign: including Eveline Herfkens, Salil Shetty, Fernando Casado, Nisha Chatani-Rizvi, Patricia Garce, Lucille Merks, Marina Ponti, Ingrid Sanders, Marisol Sanjines, Hellen Wangusa, Carol Welch, and Erna Witoelar.

The UN Millennium Project is grateful for its collaboration with and comments from the Secretariat of the New Partnership for African Development (NEPAD), including its distinguished chair Wiseman Nkuhlu, its agricultural advisor Richard Mkandawire, and its indefatigable MDG specialist, Khadija Bah.

The UN Millennium Project benefited from several high-level consultations with members of the OECD/DAC. The Project gratefully acknowledges the Government of Sweden for hosting a high-level discussion in Stockholm in February 2004. The Project also thanks Richard Manning, Brian Hammond, and colleagues for convening a meaning of DAC leaders in Paris in July 2004. Masood Ahmed, Richard Martini, Sharon White, and colleagues in the UK Department for International Development also hosted an extremely constructive final consultation with DAC members in London during October 2004. The Project also thanks Jean-Pierre Landau for numerous helpful conversations.

Pilot country inputs

The UN Millennium Project's pilot country work provided an invaluable forum for testing many of the ideas highlighted in this report and for gathering new information. In Cambodia, Dominican Republic, Ethiopia, Kenya, Ghana, Senegal, Tajikistan, and Yemen, the UN Resident Coordinator, UN Country Team members, the World Bank, and the International Monetary Fund all provided significant support to the collaboration. The Project is especially grateful to the heads of government in several countries for their direct and active support, including President Leonel Fernández of the Dominican Republic, Prime Minister Meles Zenawi of Ethiopia, President John Kufuor of Ghana, President Mwai Kibaki of Kenya, President Abdoulaye Wade of Senegal, and President Emomali Rakhmonov of Tajikistan. Government focal points in the pilot countries include John Gagain in the Dominican Republic; Ato. Mekonnen Manyazewal and Ato. Getachew Adem in Ethiopia; S. Nii-Noi Ashong and George Gyan-Baffour in Ghana; David Nalo and George Anyango in Kenya; Abou Lom in Senegal; Nozigul Khushvakhtova in Tajikistan; and Ahmed Mohammed Sofan and Mutahar Al-Abassi in Yemen.

The UN Resident Coordinator and the UN Country Teams have been instrumental in leading the pilot country work. We particularly thank Douglas Gardner and Barbara Orlandini in Cambodia; Niky Fabiancic in the Dominican Republic; Samuel Nyambi, Modibo Toure, Bjorn Ljungqvist, and Vinetta Robinson in Ethiopia; Alfred Fawundu and Kamil Kamaluddeen in Ghana; Paul André de la Porte and Ojijo Odhiambo in Kenya; Ahmed Razhaoui, Albéric Kacou, Luc Grégoire, and Diene Keita in Senegal; William Paton, Tuya Altangerel, Oliver Babson, Temur Basilia, and Johannes Chudoba in Tajikistan; and Flavia Pansieri, James Rawley, Samuel Choritz, Sammy Khan, and Abdo Seif in Yemen.

The UN Millennium Project's MDG Technical Support Center in Nairobi has helped to guide the pilot country work in Africa. We thank Glenn Denning, Mi Hua, Patrick Milimo, Eileen Petit-Mshana, and Salina Sanou for their contributions. The Project thanks Lenora Suki and Tarik Yousef for their efforts in helping to lead the Project's collaborations in the Dominican Republic and Yemen respectively. Several WHO staff, especially Rebecca

Dodd, Jeanette de Putter, and Sergio Spinaci, have collaborated closely on the health-related activities in Yemen.

Report comments

We gratefully acknowledge the thoughtful comments and suggestions received from many governments; UN agencies, funds, and programs; civil society organizations; and individuals. We thank the governments of Australia, Belgium, Canada, CARICOM member states, China, Denmark, Finland, France, Germany, Ireland, Italy, Japan, the Netherlands, New Zealand, Norway, South Africa, Spain, Switzerland, Tunisia, the United Kingdom, the United States, and the members of the European Commission for their comments.

The UN Millennium Project benefited tremendously from many group consultations at the United Nations and thanks all who participated in and organized these events. This includes meetings with the ambassadors of the African Union, convened by Michel Kafando and Crispin Grey-Johnson; the ambassadors of the Caribbean Community, convened by Christopher F. Hackett; the ambassadors of the Pacific Islands Forum Group, convened by Ali'ioaiga Feturi Elisaia; ambassadors of the UN Economic and Social Council, convened by Marjatta Rasi; members of the Second Committee of the General Assembly, convened by Marco Balarezo; and members of the Group of 77 and China, convened by Abdulaziz Al-Nasser. We also thank UN delegates representing member states of the European Union, as convened by Koen Davidse.

We would also like to thank Kanta Adhin, Javed Ahmad, Benjamin Allen, Ifzal Ali, Aasmund Andersen, William Andrianasolo, James Banda, Tony Banks, Pierre Belanger, Clements Bidonge, Kate Bird, Bineswaree (Aruna) Bolaky, Catherine Budgett-Meakin, Barbara Burungi, M. Bukuru, Eva Busza, Bernardo Cachaca, Wendy Caird, Joana Chamusca, Erin Chapman, Bill Christeson, Anthony Costello, Jacek Cukrowski, Susanne Dam-Hansen, Denis Daumerie, Rossana Dudziak, Zamira Eshmambetova, Marcos A Espinal, Udo Etukudo, Richard Feachem, Virginia Floyd, Luc Franzoni, Dennis Garrity, Axumite Gebre-Egziabher, Adrienne Germain, Linda Ghanimé, Stefan Giljum, Genevieve Grabman, Peter Gustafsson, Toni Haapane, Lawrence Haddad, Ronnie Hall, Afaf Abu-Hasabo, Cecil Haverkamp, Ron Heller, Karen Judd, Inge Kaul, Jeff Keenan, Augusta Khew, Shannon Kowalski-Morton, Hannu Kyröläinen, Robert Leigh, Jostein Leiro, Lim Li Lin, Jon Linden, Dermot Maher, James Manor, Mariam Mayet, Christine McNab, Lenni Montiel, Tadayuki Miyashita, Ronan Murphy, Adib Nehmeh, Norm Nicholson, Samantha Page, Erik Parsons, Joanna Patrick, Bob Perciasepe, Peter Piot, Rathi Ramanathan, Mary Robinson, Rick Rowden, Rabbi Royan, Domenico Siniscalco, Charlotte Hord Smith, William Smith, Jamil Sofi, Elsa Stamatopoulou, Carsten Staur, Thomas Theisohn, Adama Toe, John Tucker, Happy James Tumwebaze, Therese Turner-Jones, Andras Uthoff, Louisa Vinton, Rob Ward, Robert Watson, Patrick Webb, Diana Weil, Pera Wells, Caron Whitaker, and David Woollcombe.

We also thank faculty at the Institute for Development Studies, Sussex, and the faculty and students of the Graduate School of International Studies at the University of Denver for their detailed analysis and comments on previous drafts.

For the needs assessment work, in addition to task force members, we also gratefully acknowledge comments and suggestions by Tahgreed Adam, Walid Badawi, Christopher Banes, Stefano Bertozzi, Razina Bilgrami, Jonathan Campaign, Tamo Chattopadhay, Mark Connolly, Ingrid Cyimana, Billy Cobbett, Joel Cohen, Chris Curtis, Ernest Darkoh, Don de Savigny, Richard Deckelbaum, Simon Ellis, Patrice Engle, David Evans, Katherine Floyd, Joe Flood, Tamara Fox, Linda Ghanime, Rainer Gross, Juan Pablo Gutierrez, Charlie Heaps, John Hendra, Mark Henderson, Andrew Hudson, Jose Hueb, Todd Johnson, Eileen Kennedy, Will Keogh, Zahia Khan, Chistoph Kurowski, Lilani Kumuranyake, Valerie Leach, Rolf Luyendijk, Pim van der Male, William McGreevey, Metsi Mekheta, Takaaki Miyaguchi, Cielo Morales, Maryam Niamir-Fuller, Elizabeth Anne Paxton, Vinod Paul, Kyoko Postill, David Redhouse, Sanjay Reddy, Harri Seppanen, Kavita Sethuraman, Manohar Sharma, Susmita Shekhar, Henri Smets, Lara Stabinski, John Stover, Daouda Toure, Juha Uitto, Meike van Ginneken, Netsanet Walelign, Jake Werksman, Edward Wilson, Meg Wirth, and Aster Zaoude.

Administrative support

This report could not have been produced without the tireless administrative support of Jennifer Copeland, Rosemary Estevez-Vidal, Hnin Hla Phyu, and Ferima Traore in the UN Millennium Project Secretariat; Lauren Canning-Luckenbach and Alan Lee in the Office of the UNDP Administrator; Patricia Maw and Alex Nitorreda in the MDGs Unit of UNDP; and Ji Mi Choi, Deborah Creque, Heidi Kleedtke, and Martha Synnott at the Earth Institute at Columbia University. Dan Nienhauser of the Earth Institute was a staunch supporter of the Project, helping to streamline many of its administrative needs.

Task Force on Hunger
Halving hunger: it can be done

Task Force on Education and Gender Equality
Toward universal primary education: investments, incentives, and institutions

Task Force on Education and Gender Equality
Taking action: achieving gender equality and empowering women

Task Force on Child Health and Maternal Health
Who's got the power? Transforming health systems for women and children

Task Force on HIV/AIDS, Malaria, TB, and Access to Essential Medicines
Working Group on HIV/AIDS
Combating AIDS in the developing world

Task Force on HIV/AIDS, Malaria, TB, and Access to Essential Medicines
Working Group on Malaria
Coming to grips with malaria in the new millennium

Task Force on HIV/AIDS, Malaria, TB, and Access to Essential Medicines
Working Group on TB
Investing in strategies to reverse the global incidence of TB

Task Force on HIV/AIDS, Malaria, TB, and Access to Essential Medicines
Working Group on Access to Essential Medicines
Prescription for healthy development: increasing access to medicines

Task Force on Environmental Sustainability
Environment and human well-being: a practical strategy

Task Force on Water and Sanitation
Health, dignity, and development: what will it take?

Task Force on Improving the Lives of Slum Dwellers
A home in the city

Task Force on Trade
Trade for development

Task Force on Science, Technology, and Innovation
Innovation: applying knowledge in development

The UN Millennium Project is an independent advisory body commissioned by the UN Secretary-General to propose the best strategies for meeting the Millennium Development Goals (MDGs). The MDGs are the world's targets for dramatically reducing extreme poverty in its many dimensions by 2015—income poverty, hunger, disease, exclusion, lack of infrastructure and shelter—while promoting gender equality, education, health, and environmental sustainability.

The UN Millennium Project is directed by Professor Jeffrey D. Sachs, Special Advisor to the Secretary-General on the Millennium Development Goals. The bulk of its analytical work has been carried out by 10 thematic task forces comprising more than 250 experts from around the world, including scientists, development practitioners, parliamentarians, policymakers, and representatives from civil society, UN agencies, the World Bank, the International Monetary Fund, and the private sector. The UN Millennium Project reports directly to UN Secretary-General Kofi Annan and United Nations Development Programme Administrator Mark Malloch Brown, in his capacity as Chair of the UN Development Group.

For Product Safety Concerns and Information please contact our EU representative GPSR@taylorandfrancis.com Taylor & Francis Verlag GmbH, Kaufingerstraße 24, 80331 München, Germany

Batch number: 10732772

Printed by Printforce, the Netherlands